$6.022 \times 10^{23}.$

DEVICE ELECTRONICS FOR INTEGRATED CIRCUITS

RICHARD S. MULLER , writer
University of California
Berkeley, California

THEODORE I. KAMINS , writer
Hewlett-Packard Laboratories
Palo Alto, California

Nguyễn văn Thắng , editor and high-lighter.
University of California.
Berkeley, California
Fall, 1982. Super Senior.

JOHN WILEY & SONS
New York • Chichester • Brisbane • Toronto

To our TEACHERS . . .
who opened new vistas,
and to our STUDENTS . . .
who will see further than we.

Library of Congress Cataloging in Publication Data

Muller, Richard S
  Device electronics for integrated circuits.

  Includes index.
  1. Semiconductors. 2. Integrated circuits.
I. Kamins, Theodore I., 1941–   joint author.
II. Title.
TK7871.85.M86   621.381′71   77-1332
ISBN 0-471-62364-4

Printed in the United States of America

10 9 8 7

PREFACE

If, as we believe, the past is guide to the future, then major advances in integrated circuits will be made by individuals who have a firm grasp of device electronics. For this reason, a course emphasizing the concepts that underlie the operation and use of integrated-circuit devices has been offered to students at the University of California, Berkeley for several years. This book is the result of teaching students whose diverse interests range from basic device electronics to circuit design. About ten years' experience in teaching the course has given us the opportunity to experiment with a number of approaches to making manageable and, more important, making *digestible* the many facts of solid-state physics that are fundamental to device electronics.

After trying several alternatives, the course (and this book) evolved in the following way. First, we narrowed our focus to the most important devices used in silicon-integrated circuits. This selection left us with material that could be adequately introduced in roughly two quarters of the university calendar. Furthermore, this topical coverage has strong student appeal, a fact consistent with a precept that was well-phrased by Anatole France: "The whole art of teaching is only the art of awakening the natural curiosity of young minds."

Our next step in organizing the material was to prepare a list of the devices of interest and to order the list in a sequence determined by the number of physical concepts necessary for the reader to understand the underlying basic ideas. The book was then put together by alternating a discussion of physical concepts with a description of the application of these concepts to specific devices. Device applications are dealt with in the concluding sections of each chapter. This approach has the following major advantages:

1. It exposes the reader to material that has more permanence than does detailed description of a particular device.
2. It emphasizes the association between the facts of science and the techniques of engineering.
3. It presents the material in a *fact-application* sequence that stimulates understanding and retention.

However, in using this technique, it is difficult to maintain an even level of discussion. The problem arises because the operation of some devices can be explained in terms of only a few concepts, even though from many aspects the device is still quite complex. For example, only a few physical principles are needed to explain the Schottky-barrier diodes covered in Chapter 2, but the complexity of analysis required for rigor in the discussion can make the topic seem unduly confusing. To avoid this problem and to provide a more uniform level of discussion, we have marked certain sections of the text with daggers (†). These sections contain material of generally higher sophistication than that in the unmarked sections and can be omitted by the reader who is interested mainly in first principles. Whenever results from the more sophisticated topics are necessary for the continuity of the presentation, we have included a more easily grasped explanation outside of the sections marked with a dagger.

A summary is given at the end of each chapter. It is intended to help the student review the material and to focus his attention on the essential concepts developed in the chapter.

Most of the problems at the end of each chapter give the reader an opportunity to apply the concepts that have been developed. There are only a few straight-forward "plug-in" problems that follow directly from formulas developed in the text. These have usually been included to give the student a feeling for the magnitudes of various quantities. Some problems ask him to carry out portions of derivations only partially developed in the text. These can be used gainfully to emphasize certain treatments that may hold special interest. Challenging problems are occasionally included to extend the reader's comprehension and to deduce new principles. We have marked with a dagger both these problems and those that emphasize material from a section marked with a dagger. An instructor's manual containing solutions to all of the problems is available from the publisher. The manual contains suggestions on how to use some of the problems for lecture discussions.

The students using this book at the University of California, Berkeley, are mostly seniors, but some juniors and a few graduate students are also enrolled each quarter. All students are required to have a standard sophomore background in the sciences that encompasses basic chemistry and physics including introductory quantum mechanics, courses in differential and integral calculus, a first course in electric circuit principles, and a course in the properties of materials that contains a discussion of basic crystal structure. We have found that the most important background needed for the course is an understanding of electromagnetic theory and of the elements of modern physics, particularly the notion of allowed energy values in systems of electrons.

Below is an overview of the topics covered in the book presented according to chapter.

Chapter 1 ELECTRONICS AND TECHNOLOGY. This chapter deals with the physical electronics of solids that comprise the necessary background material for the book. Included are sections on material and processing technology that are optional to further discussion. The device described in this chapter is an integrated-circuit resistor.

Chapter 2 METAL-SEMICONDUCTOR CONTACTS. In this chapter the student is introduced to the concept of thermal equilibrium between dissimilar solids. The electronics of Schottky-barrier contacts are described in detail. Optional sections present a rigorous derivation of the current-voltage relationship for the junction, the nature of Schottky ohmic contacts, and surface effects at metal-semiconductor contacts. Practical Schottky diodes are considered in the device section.

Chapter 3 *pn* JUNCTIONS. The topic of this chapter is the electronics associated with inhomogeneously doped semiconductors. The behavior of *pn* junctions under reverse bias is described in detail. A section on junction breakdown is included, but may be omitted without loss of continuity. In the concluding section the junction field-effect transistor is discussed.

Chapter 4 CURRENTS IN *pn* JUNCTIONS. In this chapter, the continuity equations for free carriers are derived. The concepts of generation and recombination are introduced and then applied to describe the electronics of *pn* junctions under forward (injecting) bias. A discussion of minority-carrier storage and circuit modeling of diodes provides a foundation for the later development of transistor equivalent circuits. Optional sections include a full description of Shockley-Hall-Read recombination theory and space-charge-region generation and recombination. The device discussion focuses on the integrated-circuit uses of junction diodes, particularly for the junction isolation of devices.

Chapter 5 BIPOLAR TRANSISTORS I: BASIC OPERATION. The framework for the discussion of junction transistors in this chapter is the integrated-charge model. The familiar homogeneously doped transistor is treated as a special case of the general theory. This approach to bipolar transistors creates a smooth transition into topics such as bias ranges, current gain, and the Ebers-Moll model. The device discussion focuses on properties of planar integrated-circuit transistors for amplifying and switching purposes. Only one section of this chapter, that dealing with reciprocity in the transistor, is properly optional to further discussion.

Chapter 6 BIPOLAR TRANSISTORS II: LIMITATIONS AND MODELS. Most of this chapter could be omitted from a first course. The material discussed is, however, very much concerned with the practical aspects of bipolar transistors for integrated circuits. The topical coverage is oriented toward computer modeling of the transistor. For example, the Early effect is introduced in terms of the Early voltage, and effects at low and high emitter-base bias are considered with the use of the integrated-charge model. The same approach is taken to describe the electronics of the base transit time. This topic leads naturally to the charge-control model for the transistor and to examples of its use. The small-signal transistor model (hybrid-pi) is deduced from the charge-control model, and equivalences between the various representations for the transistor are discussed. The final model discussed is for computer simulation. The device discussion focuses on *pnp* transistors for integrated circuits, and introduces circuit applications such as integrated-injection logic.

Chapter 7 PROPERTIES OF THE OXIDE-SILICON SYSTEM. This chapter provides the necessary additional background in physical electronics that is needed for a discussion of MOS devices. The control exercised by the gate to accumulate, deplete or invert the semiconductor surface is described, and the concepts underlying the flat-band and threshold voltages in an MOS system are presented. An optional section describes surface effects on *pn* junctions. The chapter concludes with a discussion of MOS capacitors for integrated circuits and charge-coupled devices (CCDs).

Chapter 8 THE INSULATED-GATE FIELD-EFFECT TRANSISTOR. The properties of IGFETs are deduced initially from an approximate charge-control model, and then more rigorously from a distributed model. The parameters of the device and a model for circuit design are described. A discussion of technologies and an optional section concerned with circuit applications are included. Also optional is a discussion of second-order effects in the device. A final section describes the design and properties of ion-implanted, *n*-channel IGFETs.

The development of this book was greatly aided by the supportive comments of many colleagues at the University of California. Earlier versions of the manuscript were used in teaching by Professors T. Van Duzer, R. W. Brodersen, T. E. Everhart, and C. Hu. Each made useful suggestions for improvement. Professors A. R. Neureuther, R. G. Meyer, and D. A. Hodges provided helpful criticism of the later chapters in the book. The manuscript was also used at Stanford University by one of the authors (T. I. Kamins) and by Professors J. B. Angell, R. W. Dutton, and J. G. Linvill. Several helpful suggestions came from colleagues in industrial laboratories. These include Dr. J. Kerr at Signetics Corporation, Dr. J. Graul at Siemens Research Laboratory, Munich, Germany, Mr. E. H. Nicollian at the Bell Telephone Laboratories, Murray Hill, New Jersey, Mr. H. Jakobsen at the Central Institute for Industrial Research, Oslo, Norway, and Mr. T. Masuhara at the Hitachi Research Laboratory, Tokyo, Japan. Many comments from students were influential in improving the manuscript, and special appreciation is expressed to several graduate students at the University of California, Berkeley, who provided useful, detailed criticism. These include R. Amantea, S. H. Kwan, W. Loesch, K. W. Yeh, S. P. Fan, and especially R. Coen who very thoroughly checked several of the chapters and who diligently read and solved many of the problems. The problems were checked further and the solutions manual was written with meticulous care by K. C. Hsieh. F. Kashkooli of Signetics Corporation and J. Solomon of National Semiconductor responded generously to requests for illustrations.

As any author knows, textbooks are only harvested after months of nurture. It seems appropriate to acknowledge here the contribution of the University sabbatical system which afforded the opportunity to carry a rather large effort to completion. The staff at Wiley, our editor G. Davenport and his assistant Mrs. E. King, our copy editor E. Patti, and our production manager P. Klein have all been extremely helpful in guiding this project through its publication phase. Our appreciation is also extended to Bettye Fuller who ably typed several sections. Mrs. Joyce R. Muller gave devoted effort to shaping the manuscript and typed most of it several times. Her careful reading, copy editing, typing, and suggestions for revision and rewording have been responsible for improvements throughout the entire book.

Richard S. Muller
Theodore I. Kamins

CONTENTS

SELECTED LIST OF SYMBOLS

Symbol	Definition	Page
a	gradient of dopant density	121
BV	breakdown voltage	133
C_s	surface concentration of dopant	46
$C_{jc}, (C_{je})$	small-signal junction capacitance at the collector (emitter)	244, 281
C_{ox}	oxide capacitance	314
C_D	diffusion capacitance	276
$D_n, (D_p)$	diffusion constant for electrons (holes)	30, 31
D_s'	density of surface states	94
$E_a, (E_d)$	acceptor (donor) energy	13, 12
$E_c, (E_v)$	energy at the conduction-band (valence-band) edge	12, 13
E_f	Fermi energy	17
E_i	intrinsic Fermi energy	20
E_0	vacuum reference energy	69
$\mathscr{E}_l$	electric field at which drift velocity reaches a limiting value	251
$\mathscr{E}_s, (\mathscr{E}_{ox})$	electric field in a semiconductor (an oxide)	322, 327
$f_D(E)$	Fermi-Dirac distribution function	17
f_T	frequency at which β_F approaches unity	282
$g(E)$	density of allowed states	17
$g_m, (g_{m\,sat})$	transconductance (in the current saturation region)	143, 144
$G, (R)$	generation (recombination) rate	15
$i_B, (i_C), (i_E)$	total base (collector) (emitter) current	265
I_n	electron current traversing base	204
I_{pE}	emitter hole current (npn transistor)	213
I_{rB}	base recombination current	212
$I_D, (I_{D\,sat})$	drain current (in saturation)	139, 142
$I_{CS}, (I_{ES})$	collector (emitter) saturation current	223
$J_n, (J_p)$	electron (hole) current density	23
k'	constant for simplified MOS equations	359
l	carrier mean-free path	29
$L, (L')$	channel length (channel length to pinch-off point)	352, 377
$L_n, (L_p)$	diffusion length for minority-carrier electrons (holes)	168
L_D	Debye length	85
$m_n^*, (m_p^*)$	effective mass of an electron (a hole)	12, 19
m_0	free electron mass	3
$n, (p)$	electron (hole) density	11
$n', (p')$	excess electron (hole) density	161
$n_i, (p_i)$	intrinsic electron (hole) density	11
$n_o, (p_o)$	electron (hole) density at thermal equilibrium	161
$n_s, (p_s)$	surface electron (hole) density	162
$N_a, (N_d)$	acceptor (donor) density	13, 12
$N_c, (N_v)$	effective density of states in the conduction (valence) band	16
N_t	density of trapping states	156
q	electronic charge	3

CHAPTER 1

ELECTRONICS AND TECHNOLOGY

From everyday experience we know that the electrical properties of materials vary widely. If we measure the current I flowing through a bar of homogeneous material with uniform cross section when a voltage V is applied across it, we can find its resistance $R = V/I$. The resistivity ρ—a basic electrical property of the material comprising the bar—is related to the resistance of the bar by a geometric ratio

$$\rho = R \frac{A}{L} \qquad (1.1)$$

where L and A are the length and cross-sectional area of the sample.

The resistivities of common materials used in solid-state devices cover a very wide range of values. For example, there is the range of resistivities encountered at room temperature for the materials used to fabricate typical silicon integrated

1

circuits. Deposited metal stripes, made from very low-resistivity materials, connect elements of the integrated circuit; aluminum, which is most frequently used, has a resistivity at room temperature of about 10^{-6} Ω-cm. On the other end of the resistivity scale are the insulating materials, such as silicon dioxide, which serve to isolate portions of the integrated circuit. Silicon dioxide has a resistivity 22 orders of magnitude higher than does aluminum—about 10^{16} Ω-cm. The resistivity of the plastics often used to encapsulate integrated circuits may be as high as 10^{18} Ω-cm. Thus, a typical integrated circuit may contain materials with resistivities varying over 24 orders of magnitude—an extraordinarily wide range for a common physical property.

For electrical applications, materials are generally classified according to their resistivities. Those with resistivity values less than about 10^{-2} Ω-cm are called conductors, while those having resistivities greater than about 10^5 Ω-cm are called insulators. In the intermediate region is a class of materials called semiconductors that have profound electrical importance because their resistivities can be varied by design under very precise control. Equally important, they can be made to conduct by one of two types of current carriers. The utilization of these unique properties of semiconductors is the primary topic of this book.

In the first portion of this chapter we consider what gives rise to the enormous range of resistivities in solids. We then briefly review some important concepts from semiconductor physics. To focus on devices, however, our discussion of materials is brief, and we assume that the reader has had some exposure to basic semiconductor and solid-state physics such as that contained in reference 1. The treatment here is merely a review of some of the more important concepts in a form useful for later application.

After this brief review of semiconductor physics we look in detail at silicon planar technology. Perfected by more than a decade of refinement, this technology, in which all operations are performed on one surface of a single crystal of silicon (hence the name *planar*), is still being improved. The capabilities of the planar process and the accumulated experience in employing these capabilities assure a continuing importance to an understanding of the electronics of devices made by this process.

1.1 PHYSICS OF SEMICONDUCTOR MATERIALS

An understanding of the physics of electrons in solids can be attained by first considering electrons in an isolated atom. By perturbing the electrons in this system in a logical manner we can obtain many of the important electronic properties of solids.

We begin therefore by considering the allowed energies of an electron influenced only by an isolated atomic core. Then we look at the effect of bringing other atoms near the first atom so that the core of one atom influences the electrons associated with the other atoms, as is true in a solid material. In this manner we can study the effect of the atomic cores in a crystal on the behavior of the associated electrons. We then investigate the effect on the electrons of applying an electric field to the solid material.

Two complementary models are used in the discussion of semiconductor physics: the *energy-band model* and the *crystal-bonding model*. Let us first consider the energy-band description and bring in the bonding model later to introduce other concepts.

Band Model of Solids

We know that an electron acted on by the Coulomb potential of an atomic core may have only certain allowed energies (Fig. 1.1). In particular, the electron can occupy one of a series of energy levels

$$E_n = \frac{-Z^2 m_0 q^4}{8\epsilon_0{}^2 h^2 n^2} \tag{1.1.1}$$

below a reference energy taken as zero. In Eq. (1.1.1) Z is the net charge on the core, m_0 the free electron mass, q the magnitude of its charge, ϵ_0 the permittivity of free space, h Planck's constant, and n a positive integer. For the hydrogen atom with $Z = 1$, the allowed energies are $-2.19 \times 10^{-18}/n^2$ joules or $-13.6/n^2$

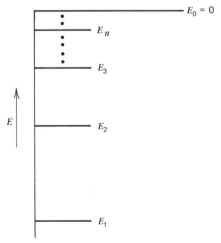

Figure 1.1   Allowed energy levels of an electron acted on by the Coulomb potential of an atomic core.

electron volts (eV) below the zero reference level.* At very low temperatures, when more than one electron are associated with the atom, the electrons fill the allowed levels starting with the lowest energies. By the Pauli exclusion principle, at most two electrons (of opposite spins) may occupy any energy level.

Let us now consider the electron in the highest occupied energy level of an atom and neglect the lower filled levels. When two isolated atoms are separated by a very large distance, the electron associated with each atom has the energy E_n given by Eq. (1.1.1). If two atoms approach one another, however, there will be a force on the first electron resulting from the second atomic core. The potential that determines the energy levels of the electron is therefore changed. All allowed energy levels for the electron are consequently modified because of this change in potential.

We must also consider the Pauli exclusion principle when describing a two-atom system. An energy level E_n, which can contain at most two electrons of opposite spin, is associated with each isolated atom so that the total system can contain at most four electrons. When the two atoms are brought together to form one system, however, only two electrons can be associated with the allowed energy level E_n. Therefore, the allowed energy level E_n of the isolated atoms must split into two levels with slightly different energies in order to retain space for a total of four electrons. Therefore, bringing two atoms close together not only slightly perturbs each energy level of the isolated atom but also splits each of the energy levels of the isolated atoms into two slightly separated energy levels. As the two atoms are brought closer together, stronger interactions are expected and the splitting becomes greater.

As more atoms are added to form a crystalline structure, the forces encountered by each electron are altered further and additional changes in the energy levels occur. Again, the Pauli exclusion principle demands that each allowed electron energy level have a slightly different energy so that many distinct, closely spaced energy levels characterize the crystal. Each of the original quantized levels of the isolated atom is split many times, and each contains one level for each atom in the system. When N atoms are included in the system, the original *energy level E_n* will split into N different allowed levels, forming an *energy band*, which may contain at most $2N$ electrons (because of spin degeneracy). Since the number of atoms in a crystal is generally large—of the order of 10^{22} cm^{-3}—and the total extent of the energy band is of the order of a few electron volts, the separation between the N different energy levels within the bands is very much smaller than the thermal energy possessed by an electron at room temperature, and the electron

* The use of the unit electron volt (1 eV $= 1.6 \times 10^{-19}$ joules; see the conversion factors on the inside front cover) is very convenient in semiconductor physics since it often avoids cumbersome exponents. Although not a pure rationalized unit, the electron volt is widely employed and is used extensively in our discussion.

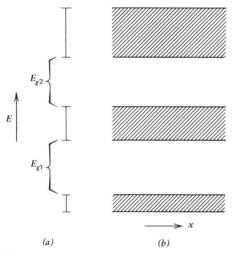

E_{g2}

E

E_{g1}

x

(a)   (b)

Figure 1.2 Broadening of allowed energy levels into allowed energy bands separated by forbidden energy gaps as more atoms influence each electron in a solid: (a) one-dimensional representation; (b) two-dimensional diagram in which energy is plotted versus distance.

may easily move between levels. Thus, we may speak of a continuous band of allowed energies containing space for 2N electrons. This allowed band is bounded by maximum and minimum energies, and it may be separated from adjacent allowed bands by *forbidden-energy gaps*, as shown in Fig. 1.2a, or it may overlap other bands. The detailed behavior of the bands, whether they overlap or form gaps and, if so, the size of the gaps, fundamentally determines the electronic properties of a given material. It is the essential distinguishing feature between conductors, insulators, and semiconductors.

Although each energy level of the original isolated atom splits into a band composed of 2N levels, the range of allowed energies of each band may be different. The higher energy bands generally span a wider energy range than do those at lower energies. The cause for this difference can be seen by considering the Bohr radius r_n associated with the nth energy level:

$$r_n = \frac{n^2 \epsilon_0 h^2}{Z \pi m_0 q^2} = \frac{n^2}{Z} \times 0.529 \text{ Å} \tag{1.1.2}$$

For higher energy levels (larger n) the electron is less tightly bound and may wander farther from the atomic core. If the electron is less tightly confined, it comes closer to the adjacent atoms and is more strongly influenced by them. This greater interaction, of course, causes a larger change in the energy levels so that the wider energy bands correspond to the higher energy electrons of the isolated atoms.

We represent the energy bands at the equilibrium spacing of the atoms by the one-dimensional picture in Fig. 1.2a. The highest allowed level in each band is separated by an energy gap E_g from the lowest allowed level in the next band. It is convenient, however, to extend the band diagram into a two-dimensional picture (Fig. 1.2b) where the vertical axis still represents the electron energy while the horizontal axis now represents position in the semiconductor crystal. This version of the diagram serves to emphasize that electrons in the bands are not associated with any of the nuclei, but are confined only by the crystal boundaries. This type of diagram is especially useful when we consider combining materials with different energy-band structures into semiconductor devices. In this heuristic derivation of the basis for energy bands in elemental solids, we have implicitly assumed that each atom is like its neighbor in all respects including orientation; that is, we are considering *perfect crystals*. In practice, a very high level of crystalline order in which defects are measured in parts per billion or less is normal for device-grade semiconductor materials.

The formation of energy bands from discrete levels occurs whenever the atoms of any element are brought together to form a solid. However, differing numbers of electrons within the energy bands of different solids strongly influence their electrical properties. Consider first, for example, an alkali metal composed of N atoms each with one valence electron in the outer shell. When the atoms are brought close together, an energy band forms from this energy level. In the simplest case this band has space for $2N$ electrons. The N available electrons then fill the lower half of the energy band (Fig. 1.3a), and there are empty states very near some of the filled states. The electrons near the top of the filled portion of the band can easily gain small amounts of energy from an applied electric field and move into these empty states. They behave, therefore, almost as free electrons and can be transported throughout the crystal under the influence of an externally applied electric field. In general, metals are characterized by partially filled energy bands similar to the case just described. They are, therefore, highly conductive.

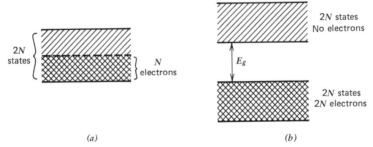

(a)   (b)

Figure 1.3   Energy-band diagrams: (a) N electrons filling half of the $2N$ allowed states, as might occur in a metal. (b) A completely empty band separated by an energy gap E_g from a band whose $2N$ states are completely filled by $2N$ electrons, representative of an insulator.

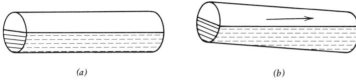

(a)   (b)

Figure 1.4   Electron motion in an allowed band is analogous to fluid motion in a glass tube with sealed ends; the fluid can move in a half-filled tube just as electrons can move in a metal.

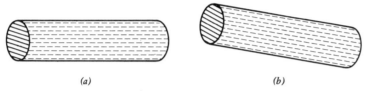

(a)   (b)

Figure 1.5   No fluid motion can occur in a completely filled tube with sealed ends.

Markedly different electrical behavior occurs in materials in which the valence (or outermost shell) electrons completely fill an allowed energy band and there is an energy gap to the next higher band. In this case, characteristic of insulators, the closest allowed band above the filled band is completely empty at low temperatures as shown in Fig. 1.3*b*. Thus, the lowest-energy empty states are separated from the highest filled states by the energy gap E_g. In insulating materials E_g is generally greater than 5 eV ($\sim$9 eV for SiO_2),* much larger than typical thermal or field-imparted energies (tenths of an eV or less). In the idealization that we are considering there are no electrons close to empty allowed states and, therefore, none available to gain small energies from an externally applied field. Therefore, no electrons may carry an electric current and the material is an insulator.

An analogy may be helpful. Consider a horizontal glass tube with sealed ends representing the allowed energy states and fluid in the tube representing the electrons in a solid. In the case analogous to a metal, the tube is partially filled (Fig. 1.4). When a force (gravity in this case) is applied by tipping the tube, the fluid can easily move along the tube. In the situation analogous to an insulator, the tube is completely filled with fluid (Fig. 1.5). When the filled tube is tipped, the fluid cannot flow since there is no empty volume into which it can move; that is, there are no empty allowed states.

Semiconductors have band structures similar to insulators. The difference between these two classifications arises from the size of the forbidden energy gap and the ability to populate a nearly empty band by adding conductivity-enhancing impurities to a semiconductor. In a semiconductor the energy gap separating the highest band that is filled at absolute zero temperature from the lowest empty

---

* Specifying band gaps in amorphous materials may cause anxiety to theorists, but there is useful engineering purpose to doing so.

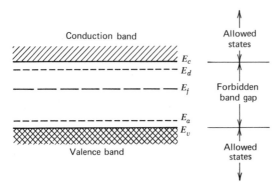

Figure 1.6   Energy-band diagram for a semiconductor showing the lower edge of the conduction band (E_c), a donor level (E_d) within the forbidden gap, the Fermi level (E_f), an acceptor level (E_a), and the top edge of the valence band (E_v).

band is typically of the order of one electron volt (silicon: 1.1 eV; germanium: 0.7 eV). In an impurity-free semiconductor the uppermost filled band is populated by electrons that were the valence electrons of the isolated atoms; this is known as the *valence band.*

The band structure for a semiconductor is shown in Fig. 1.6. At any temperature above absolute zero, the valence band is not entirely filled because a small number of electrons possess enough thermal energy to be excited across the forbidden gap into the next allowed band. The smaller the energy gap and the higher the temperature, the greater the number of electrons that can jump between bands. The electrons in the upper band can easily gain small amounts of energy and can respond to an applied electric field to constitute a current. This band is therefore called the *conduction band* because the electrons that populate it are conductors of electricity. The current density J (current per unit area) flowing in the conduction band can be found by summing the charge ($-q$) times the net velocity (v_i) of each electron populating the band. The summation is then taken over all electrons per unit volume in the conduction band.

$$J_{cb} = \frac{I}{A} = \sum_{cb} (-q)v_i \qquad (1.1.3)$$

Since there is only a small number of electrons in this band, however, the current for a given field is considerably less than that in a metal.

Holes

When electrons are excited into the conduction band, empty states are left in the valence band. When an electric field is applied, nearby electrons can move into these empty states and constitute a current. This current can be formulated by summing the motion of all electrons per unit volume in the valence band:

$$J_{vb} = \sum_{vb} (-q)v_i \tag{1.1.4}$$

Since there are many electrons in the valence band but only a few empty states, it is easier to describe the conduction resulting from electrons interacting with these empty states than to describe the motion of all the electrons in the valence band. Mathematically, we may describe the current in the valence band as the current that would flow if the band were completely filled minus that associated with the missing electrons. Again, summing over the populations per unit volume, we find

$$J_{vb} = \sum_{vb} (-q)v_i = \sum_{Filled\ band} (-q)v_i - \sum_{Empty\ states} (-q)v_i \tag{1.1.5}$$

Since no current can flow in a completely filled band (because no net energy can be imparted to the electrons populating it), the current in the valence band can be written as

$$J_{vb} = 0 - \sum_{Empty\ states} (-q)v_i = \sum_{Empty\ states} qv_i \tag{1.1.6}$$

where now the summation is over the empty states per unit volume. We have found in Eq. (1.1.6) that we can express the motion of charge in the valence band in terms of the vacant states by treating the states as if they were particles with positive charge. These "particles" are called *holes*; they can only be discussed in connection with the energy bands of a solid and cannot exist in free space. Note that energy-band diagrams, such as those shown in Figs. 1.3 and 1.6., are drawn for electrons so that the energy of an electron increases as we move toward the top of the diagram. However, because of its opposite charge, the energy of a hole increases as we move downward on this same diagram.*

The concept of holes can be illustrated by continuing our analogy of fluids in glass tubes. We start with two sealed tubes—one completely filled, and the other completely empty (Fig. 1.7a). When we apply a force by tipping the tubes, no motion can occur (Fig. 1.7b). If we transfer a small amount of fluid from the lower tube to the upper tube (Fig. 1.7c), however, the fluid in the upper tube can now move when the tube is tilted (Fig. 1.7d). This flow corresponds to electron conduction in the conduction band of the solid. In the lower tube, a bubble is left because of the fluid we removed. This bubble is analogous to holes in the valence band. It cannot exist outside of the nearly filled tube, just as it is only useful to discuss holes in connection with a nearly filled valence band. When the tube is tilted, the fluid in the tube moves downward, but the bubble moves in the opposite direction,

---

* As in Fig. 1.6 we often simplify the energy-band diagram by showing only the upper bound of the valence band (denoted by E_v) and the lower bound of the conduction band (denoted by E_c) since we are primarily interested in states near these two levels and in the energy gap E_g separating them.

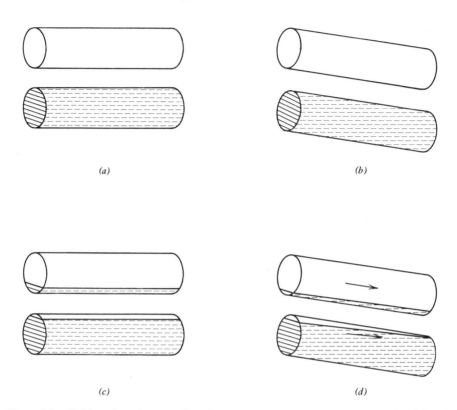

Figure 1.7 Fluid analogy for a semiconductor. (*a*) and (*b*) No flow can occur in either the completely filled or completely empty tube. (*c*) and (*d*) Fluid can move in both tubes if some of it is transferred from the filled tube to the empty one, leaving unfilled volume in the lower tube.

as if it had a mass of opposite sign to that of the fluid. Similarly the holes in the valence band move in a direction opposite to that of the electrons, as if they had a charge of the opposite sign. Just as it is easier to describe the motion of the small bubble than that of the large amount of fluid, it is easier to discuss the motion of the few holes rather than the motion of the electrons, which nearly fill the valence band.

Bond Model

The discussion of free holes and electrons in semiconductors can also be phrased in terms of the behavior of completed and broken electronic bonds in a semiconductor crystal. This viewpoint, which is often called the *bond model*, fails to account for important quantum-mechanical constraints on the behavior of electrons in crystals, but it does illustrate several useful qualitative concepts.

To discuss the bond model, we consider the diamond-type crystal structure that is common to silicon and germanium. In the diamond structure, each atom has four covalent bonds with its nearest neighbors. There are two tightly bound electrons associated with each bond—one from each atom. At absolute zero temperature, all electrons are held in these bonds, and none are therefore free to move about the crystal in response to an applied electric field. At higher temperatures, thermal energy breaks some of the bonds and creates nearly free electrons, which can then contribute to the current under the influence of an applied electric field. This current corresponds to the current associated with the conduction band in the energy-band model.

After a bond is broken by thermal energy and the freed electron moves away, an empty bond is left behind. An electron from an adjacent bond can then jump into the vacant bond, leaving a vacant bond behind it. The vacant bond, therefore, moves in the opposite direction to the electrons. If a net motion is imparted to the electrons by an applied field, the vacant bond can continue moving in the direction opposite to the electrons as if it had a positive charge. This vacant bond corresponds to the hole that we have been discussing in terms of the valence band of the energy-band picture.

Donors and Acceptors

Thus far we have discussed a pure semiconductor material in which each electron excited into the conduction band leaves a vacant state in the valence band. Consequently, the number of electrons n in the conduction band equals the number of holes p in the valence band. Such a material is called an *intrinsic* semiconductor, and the electron and hole populations in it (carriers cm^{-3}) are usually subscripted i, as n_i and p_i. However, the most important uses of semiconductors arise from the interaction of adjacent semiconductor materials having differing densities of the two types of charge carriers. We can achieve such a structure either by physically joining two materials with different band gaps or by varying the number of carriers in one semiconductor material.

The most useful means for controlling the number of carriers in a semiconductor is by incorporating substitutional impurities, that is, impurities that occupy lattice sites in place of the atoms of the pure semiconductor. For example, if we replace one silicon atom (four valence electrons) with an impurity atom from group V in the periodic table, such as phosphorus (five valence electrons), then four of the valence electrons from the impurity atom fill bonds between the impurity atom and the adjacent silicon atoms. The fifth electron, however, is not covalently bonded to near neighbors; it is only weakly bound to the impurity atom by the excess positive charge on the nucleus. Only a small amount of energy is required to break this weak bond so that the fifth electron can wander about the crystal and contribute to electrical conduction. Since the substitutional group V impurities donate electrons to the silicon, they are known as *donors*.

In order to estimate the amount of energy needed to break the bond to a donor atom, we consider the net Coulomb potential that the electron experiences because of the core of its parent atom. We assume that the electron is attracted by the single net positive charge of the impurity atom core weakened by the polarization effects of the background of silicon atoms. The energy binding the electron to the core is then

$$E = \frac{m_n^* q^4}{8h^2 \epsilon_0^2 \epsilon_{rs}^2} = \frac{13.6}{\epsilon_{rs}^2} \frac{m_n^*}{m_0} \text{ eV} \tag{1.1.7}$$

where ϵ_{rs} is the relative permittivity of the semiconductor and m_n^* is the effective mass of the electron in the semiconductor conduction band. The use of an effective mass accounts for the influence of the crystal lattice on the motion of an electron. For silicon with $\epsilon_{rs} = 11.7$ and $m_n^* = .26\ m_0$, $E \approx .03$ eV, which is only about 3% of the silicon gap energy (1.1 eV). (More detailed calculations and measurements indicate that the binding energy for typical donors is somewhat more: 0.044 eV for phosphorus, 0.049 eV for arsenic, and 0.039 eV for antimony.) The small binding energies make it much more likely that the weak bond connecting the fifth electron to the donor will be broken than will the silicon-silicon bonds.

According to the energy-band model, it requires only a small amount of energy to excite the electron from the donor atom into the conduction band, while a much greater amount of energy is required to excite an electron from the valence band to the conduction band. Therefore, we may represent the state corresponding to the electron when bound to the donor atom by a level E_d about 0.05 eV below the bottom of the conduction band E_c (Fig. 1.6). The density of donors (atoms cm^{-3}) is generally designated by N_d. Thermal energy at temperatures greater than about 150°K is generally sufficient to excite electrons from the donor atoms into the conduction band. Once the electron is excited into the conduction band, a fixed, positively charged atom core is left behind in the crystal lattice. The allowed energy state provided by a donor (*donor level*) is, therefore, neutral when occupied by an electron and positively charged when empty.

If most impurities are of the donor type, the number of electrons in the conduction band is much greater than the number of holes in the valence band. Electrons are then called the *majority carriers*, and holes are called the *minority carriers*. The material is said to be an *n-type* semiconductor since most of the current is carried by the *negatively charged* electrons. A graph showing the conduction electron concentration versus temperature for silicon and germanium is sketched in Fig. 1.8. Since the hole density is at most equal to n_i, this figure shows clearly that electrons are far more numerous than holes when the temperature ranges from a value sufficient to ionize the donor atoms (about 100°K) up to values that free many electrons from silicon-silicon bonds (about 600°K).

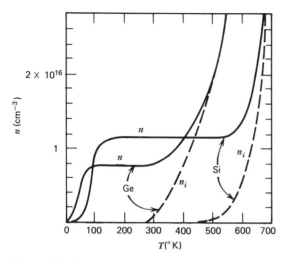

Figure 1.8   Electron concentration versus temperature for two doped semiconductors: (a) Silicon doped with 1.15×10^{16} arsenic atoms cm^{-3}.[2] (b) Germanium doped with 7.5×10^{15} arsenic atoms cm^{-3}.[3]

In an analogous manner, an impurity atom with three valence electrons, such as boron, can replace a silicon atom in the lattice. The three electrons fill three of the four covalent silicon bonds, leaving one bond vacant. If another electron moves to fill this vacant bond from a nearby bond, the vacant bond has moved, carrying with it positive charge and contributing to hole conduction. Just as a small amount of energy was necessary to initiate the conduction process in the case of a donor atom, a small amount of energy is necessary to excite an electron from the valence band into the vacant bond caused by the trivalent impurity. This energy is represented by an energy level E_a slightly above the top of the valence band E_v (Fig. 1.6). An impurity that contributes to hole conduction is called an *acceptor impurity* since it leads to vacant bonds, which easily accept electrons. The acceptor concentration (atoms cm^{-3}) is denoted as N_a. If most of the impurities in the solid are of this type, the material is called a *p-type* semiconductor since most of the conduction is carried by *positively charged* holes. An acceptor level is neutral when empty and negatively charged when occupied by an electron.

Semiconductors in which conduction results primarily from carriers contributed by impurity atoms are said to be *extrinsic*. The donor and acceptor impurity atoms, which are intentionally introduced to change the charge-carrier concentration, are called *dopant* atoms.

Other impurity atoms or crystalline defects may provide charge-neutral energy states in which electrons are tightly bound; this means that it takes appreciable energy to excite an electron from a bound state to the conduction band. Such *deep donors* may be represented by energy levels well below the conduction-band edge in contrast to the *shallow donors* previously discussed, which had energy levels only a few times the thermal energy kT below the conduction-band edge. Similarly, *deep acceptors* are located well above the valence-band edge. Since deep levels are not always related to impurity atoms in the same straightforward manner as shallow donors and acceptors, the distinction between the terms donor and acceptor is made on the basis of the possible charge states the level may take. A deep level is called a donor if it is neutral when occupied by an electron and positively charged when empty, while a deep acceptor is neutral when empty and negative when occupied by an electron.

Thermal-Equilibrium Statistics

Before proceeding to a more detailed discussion of electrical conduction in a semiconductor, let us consider three additional concepts: first, the concept of thermal equilibrium; second, the relationship at thermal equilibrium between the majority- and minority-carrier concentrations in a semiconductor; and third, the use of Fermi statistics and the Fermi level to specify the carrier concentrations.

Thermal Equilibrium. We have seen that free-carrier densities in semiconductors correspond to the populations of allowed states in the conduction and valence bands. The densities are therefore dependent upon the net energy in the semiconductor. This energy is stored in crystal-lattice vibrations (phonons) as well as by the electrons. Although a semiconductor crystal may be excited by external sources of energy such as incident photoelectric radiation, there are many situations in which the total energy is a function only of the crystal temperature. In this case the semiconductor will spontaneously (but not instantaneously) assume a state known as *thermal equilibrium*. Thermal equilibrium is a dynamic situation in which every process is balanced by its inverse process. For example, at thermal equilibrium if electrons are being excited from a lower energy E_1 to a higher energy E_2, then there must be equal traffic of electrons from the states at E_2 to those at E_1. Likewise, if energy is being transferred into the electron population from the crystal vibrations (phonon population), then at thermal equilibrium an equal flow of energy is occurring in the opposite direction. A useful thought picture for thermal equilibrium is that a moving picture taken of any event can be run either backwards or forwards without the viewer being able to detect it; thermal equilibrium means that time can run toward the past as well as into the future! In the following we consider some properties of hole and electron populations in semiconductors at thermal equilibrium.

Mass-Action Law. At most temperatures of interest to us, there is sufficient thermal energy to excite some electrons from the valence band to the conduction band. A dynamic equilibrium exists in which some electrons are constantly being excited into the conduction band while others are losing energy and falling back across the energy gap to the valence band. The excitation of an electron from the valence band to the conduction band corresponds to the generation of a hole and an electron, while an electron falling back across the gap corresponds to electron-hole recombination, since it annihilates both carriers. The rate of generation of electron-hole pairs G depends on the temperature T but is to first order independent of the number of carriers already present. We therefore write

$$G = f_1(T) \tag{1.1.8}$$

where $f_1(T)$ is a function that is determined by crystal physics and temperature. The rate of recombination R, on the other hand, depends on the concentration of electrons n in the conduction band and also on the concentration of holes p in the valence band, since both species must interact for recombination to occur. We therefore represent the recombination rate as a product of these concentrations as well as other factors that are included in $f_2(T)$:

$$R = npf_2(T) \tag{1.1.9}$$

At equilibrium the generation rate must equal the recombination rate. Equating G and R in Eqs. (1.1.8) and (1.1.9), we have

$$npf_2(T) = f_1(T)$$

or

$$np = \frac{f_1(T)}{f_2(T)} = f_3(T) \tag{1.1.10}$$

Equation (1.1.10) expresses the important result that the product of the hole and electron densities in a given semiconductor is a function only of temperature at thermal equilibrium.

In an intrinsic (i.e., undoped) semiconductor all carriers result from excitation across the forbidden gap. Consequently, $n = p = n_i$, where the subscript i reminds us that we are dealing with intrinsic material. Applying Eq. (1.1.10) to intrinsic material, we have

$$n_i p_i = n_i^2 = f_3(T) \tag{1.1.11}$$

The intrinsic carrier concentration depends on temperature because thermal energy is the source of carrier excitation across the forbidden energy gap. The intrinsic concentration is also a function of the size of the energy gap since fewer electrons can be excited across a larger gap. Under most conditions we shall soon be able to show that n_i^2 is given by the expression

$$n_i^2 = N_c N_v \exp\left(\frac{-E_g}{kT}\right) \tag{1.1.12}$$

where N_c and N_v are related to the density of allowed states near the edges of the conduction band and valence band, respectively. Although N_c and N_v vary somewhat with temperature, n_i is much more temperature dependent because of the exponential term in Eq. (1.1.12). For silicon with $E_g = 1.1$ eV, n_i doubles for every 8°C increase in temperature near room temperature. Since the intrinsic carrier concentration n_i is constant for a given semiconductor at a fixed temperature, it is useful to replace $f_3(T)$ by n_i^2 in Eq. (1.1.10). Therefore the relation

$$np = n_i^2 \qquad (1.1.13)$$

holds for both extrinsic and intrinsic semiconductors; it shows that increasing the number of electrons in a sample by adding donors causes the hole concentration to decrease so that the product np remains constant. This result, often called the *mass-action law*, has its counterpart in the behavior of interacting chemical species, such as the concentrations of hydrogen and hydroxyl ions (H^+ and OH^-) in acidic or basic solutions. As we see from our derivation, the law of mass action is a straightforward consequence of equating generation and recombination, that is, of thermal equilibrium.

In the neutral regions of a semiconductor (i.e., regions free of field gradients), the number of positive charges must be exactly balanced by the number of negative charges. Positive charges exist on ionized donor atoms and on holes, while negative charges are associated with ionized acceptors and electrons.* If there is charge neutrality in a region where all dopant atoms are ionized,

$$N_d + p = N_a + n \qquad (1.1.14)$$

Rewriting Eq. (1.1.14) and using the mass-action law, Eq. (1.1.13), we obtain the expression

$$n - \frac{n_i^2}{n} = N_d - N_a \qquad (1.1.15)$$

which may be solved for the electron concentration n:

$$n = \frac{N_d - N_a}{2} + \left[\left(\frac{N_d - N_a}{2} \right)^2 + n_i^2 \right]^{1/2} \qquad (1.1.16)$$

In an n-type semiconductor $N_d > N_a$. From Eq. (1.1.16) we see that the electron density depends on the net excess of ionized donors over acceptors. Thus, a piece of p-type material containing N_a acceptors may be converted into n-type material by adding an excess of donors so that $N_d > N_a$. In Section 1.3 we shall see how this conversion is carried out in fabricating silicon integrated circuits.

---

* When we speak of *electrons* in our discussion of devices in subsequent chapters, we refer generally to electrons in the conduction band; exceptions are explicitly noted. The term *holes* always denotes vacancies in the valence band.

For silicon at room temperature, n_i is 1.45×10^{10} cm^{-3} while the net donor density in n-type silicon is typically about 10^{15} cm^{-3} or greater; this means that $(N_d - N_a) \gg n_i$; Eq. (1.1.16) reduces to $n \approx (N_d - N_a)$. Consequently, from Eq. (1.1.13),

$$p = \frac{n_i^2}{n} \approx \frac{n_i^2}{N_d - N_a} \qquad (1.1.17)$$

so that $p = 2 \times 10^5$ cm^{-3} when $N_d - N_a = 10^{15}$ cm^{-3}, nearly 10 orders of magnitude below the majority-carrier population. In general, the concentration of one type of carrier is many orders of magnitude greater than that of the other in extrinsic semiconductors.

Fermi Level. The numbers of free carriers (electrons and holes) in any macroscopic piece of semiconductor are relatively large—large enough to make use of the laws of statistical mechanics to determine physical properties. One important property of electrons in crystals is their distribution at thermal equilibrium among the allowed energy states. Basic considerations of ways to populate allowed energy states with particles subject to the Pauli exclusion principle leads to a distribution function for electrons in energy that is called the Fermi-Dirac distribution function. It is denoted by $f_D(E)$ and has the form

$$f_D(E) = \frac{1}{1 + \exp\left[(E - E_f)/kT\right]} \qquad (1.1.18)$$

where E_f is a reference energy called the Fermi energy or Fermi level. From Eq. (1.1.18), we see that $f_D(E_f)$ always equals $\frac{1}{2}$. The Fermi-Dirac distribution function, often referred to simply as the Fermi function, describes the probability that a state at energy E is filled by an electron. As shown in Fig. 1.9a, the Fermi function approaches unity at energies much lower than E_f, indicating that the lower-energy states are mostly filled. It is very small at higher energies, indicating that few electrons are found in high-energy states at thermal equilibrium—in agreement with physical intuition. At absolute zero temperature all allowed states below E_f are filled and all states above it are empty. At finite temperatures, the Fermi function does not change so abruptly; there is a small probability that some states above the Fermi level are occupied and some states below it are empty.

The Fermi function represents only a probability of occupancy. It does not contain any information about the states available for occupancy and, therefore, cannot of itself specify the electron population at a given energy. An application of quantum physics to a given system will reveal information about the density of available states as a function of energy. We denote this function by $g(E)$. A sketch of $g(E)$ for an intrinsic semiconductor is shown in Fig. 1.9b. Note that it is zero in the forbidden gap ($E_c > E > E_v$), but it rises sharply within both the valence band ($E < E_v$) and the conduction band ($E > E_c$). The actual distribution of electrons as a function of energy can then be found from the product of the density of allowed

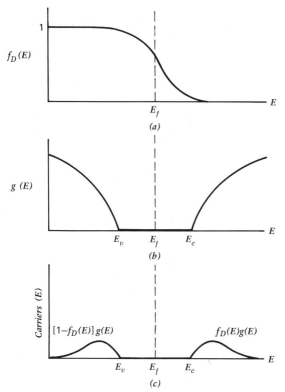

Figure 1.9 (*a*) Fermi-Dirac distribution function describing probability that an allowed state at energy E is occupied by an electron. (*b*) The density of allowed states for a semiconductor as a function of energy; note that $g(E)$ is zero in the forbidden gap between E_v and E_c. (*c*) The product of the distribution function and the density-of-states function.

states $g(E)$ within a small energy interval dE and the probability that these states are filled $f_D(E)$. The total density of electrons in the conduction band may be obtained by multiplying the density-of-states function in the conduction band $g(E)$ by the Fermi function and integrating over the conduction band:

$$n = \int_{cb} f_D(E)g(E)\,dE \qquad (1.1.19)$$

Similarly, the density of holes in the valence band is found by multiplying the density-of-states function in the valence band by the probability that these states are empty $[1 - f_D(E)]$ and integrating over the valence band.

In *n*-type material that is not too highly doped, few of the allowed states in the conduction band are filled. The Fermi function in the conduction band is very small, and the Fermi level is well below the bottom of the conduction band. Then $(E_c - E_f) \gg kT$, and the Fermi function given by Eq. (1.1.18) reduces to the mathematically simpler Maxwell-Boltzmann distribution function:

$$f_M(E) = \exp\left[\frac{-(E - E_f)}{kT}\right] \qquad (1.1.20)$$

This thermal-equilibrium distribution function can also be derived independently by omitting the limitations imposed by the Pauli exclusion principle; that is, the Boltzmann function applies to the case that any number of electrons may exist in an allowed state.

Using Eq. (1.1.20) in the integration described in Eq. (1.1.19) and making several approximations, the carrier concentration in the conduction band can be expressed in terms of the Fermi level by

$$n = N_c \exp\left[-\frac{(E_c - E_f)}{kT}\right] \qquad (1.1.21)$$

Similarly, in moderately doped p-type material, the Fermi level is significantly above the top of the valence band, and

$$p = N_v \exp\left[-\frac{(E_f - E_v)}{kT}\right] \qquad (1.1.22)$$

where $(E_c - E_f)$ expresses the energy between the bottom edge of the conduction band and the Fermi level, and $(E_f - E_v)$ is the energy separation from the Fermi level to the top of the valence band. The quantities N_c and N_v, called the effective densities of states at the conduction- and valence-band edges, respectively, are given by the expressions

$$N_c = 2\left(\frac{2\pi m_n^* kT}{h^2}\right)^{3/2} \qquad (1.1.23)$$

and

$$N_v = 2\left(\frac{2\pi m_p^* kT}{h^2}\right)^{3/2} \qquad (1.1.24)$$

where m_n^* and m_p^* denote the effective masses of electrons and holes. These effective masses are related to m^* as introduced in Eq. (1.1.7) but differ slightly from it because of factors we shall not consider here. As can be seen from Eqs. (1.1.21) and (1.1.22), the quantities N_c and N_v effectively concentrate all of the distributed conduction- and valence-band states at E_c and E_v. They may be used to calculate thermal-equilibrium densities whenever the Fermi level is a few kT or more removed from a band edge.

Except for slight differences in the m^* values, all terms in Eqs. (1.1.23) and (1.1.24) are equal so that $N_c \simeq N_v$. Hence, in an n-doped material for which

$n \gg p$, $(E_c - E_f) \ll (E_f - E_v)$; this means that the Fermi level is much closer to the conduction band than it is to the valence band. Similarly, the Fermi level is nearer to the valence band than to the conduction band in a p-type semiconductor.

In an intrinsic semiconductor $n = p$. Therefore, $(E_c - E_f) \approx (E_f - E_v)$ and the Fermi level is nearly at the middle of the forbidden gap $[E_f = (E_c + E_v)/2]$. We denote this *intrinsic Fermi level* by the symbol E_i. Just as the quantity n_i is useful in relating the carrier concentrations even in an extrinsic semiconductor [Eq. (1.1.13)], E_i is frequently used as a reference level when discussing extrinsic semiconductors. In particular, since

$$n_i = N_c \exp\left[\frac{-(E_c - E_i)}{kT}\right] = N_v \exp\left[\frac{-(E_i - E_v)}{kT}\right] \tag{1.1.25}$$

the expressions for the carrier concentrations n and p in an extrinsic semiconductor [Eqs. (1.1.21) and (1.1.22)] may be rewritten in terms of the intrinsic carrier concentration and the intrinsic Fermi level:

$$n = n_i \exp\left[\frac{(E_f - E_i)}{kT}\right] \tag{1.1.26}$$

and

$$p = n_i \exp\left[\frac{(E_i - E_f)}{kT}\right] \tag{1.1.27}$$

Thus, the energy separation from the Fermi level to the intrinsic Fermi level is a measure of the departure of the semiconductor from intrinsic material. Since E_f is above E_i in an n-type semiconductor, $n > n_i > p$, as we found before.

When the semiconductor contains a large dopant concentration ($N_d \to N_c$ or $N_a \to N_v$), we may no longer ignore the limitations imposed by the Pauli exclusion principle. That is, the Fermi-Dirac distribution function may not be approximated by the Maxwell-Boltzmann distribution function. Equations (1.1.21–22) and (1.1.26–27) are then no longer valid. More exact expressions must be used or the limited validity of the simplified expressions must be realized. Very highly doped semiconductors ($N_d \gtrsim N_c$ or $N_a \gtrsim N_v$) are referred to as *degenerate* semiconductors because the Fermi level can be shown to exist within the conduction or valence band. This condition is the same as that for metals and, consequently, many of the electronic properties of semiconductors degenerate under heavy doping into those of metals.

1.2 FREE CARRIERS IN CRYSTALS

Our first reference to the electronic properties of solids earlier in this chapter was to the familiar linear relationship that is often found between the current flowing through a sample and the voltage applied across it. This relationship is known as Ohm's law: $V = IR$. Although a thorough derivation of the physics of ohmic-current flow can be quite complex, an approximate representation of the process will provide adequate background for our purposes. To accomplish this we first develop a picture of the kinetic properties of free electrons without any external fields. We then consider the imposition first of low to moderate fields, characteristic of many device applications, and finally we discuss the high-field case.

We begin by recalling that electrons (and holes) in semiconductors are almost "free particles" in the sense that they are not associated with any particular lattice site. The influences of crystal forces are incorporated in an effective mass that differs somewhat from the free-electron mass. Using once again the laws of statistical mechanics, we can assert that electrons and holes will have the thermal energy associated with classical free particles; that is, $\frac{1}{2}kT$ units of energy per degree of freedom where k is Boltzmann's constant and T is the absolute temperature. This means that electrons in a crystal are not stationary, but rather move about with random velocities. Furthermore, the mean-square thermal velocity of the electrons will be approximately* related to the temperature by the equation

$$\frac{1}{2} m_n^* v_{th}^2 = \frac{3}{2} kT \tag{1.2.1}$$

where m_n^* is the effective mass of free (conduction-band) electrons. For silicon $m_n^* = .26\, m_0$ (where m_0 is the rest mass), and v_{th} in Eq. (1.2.1) is therefore 2.3×10^7 cm s^{-1} at $T = 300°$K. The electrons may be pictured as moving in random directions through the lattice, colliding among themselves and with the lattice. At thermal equilibrium the motion of the system of electrons is completely random so that the net current in any direction is zero. Collisions with the lattice result in energy transfer between the electrons and the atomic cores that form the lattice. The time interval between collisions averaged over the entire electron population is τ_{cn}, the mean scattering time for electrons. These considerations all apply to the field-free, thermal-equilibrium crystal.

Drift Velocity

Let us now consider a small electric field applied to the lattice. The electrons are accelerated along the field direction during the time between the collisions. Figure 1.10a is a sketch of the motion typical of a crystal electron in response to a small applied field $\mathscr{E}$. Note in the figure that the field-directed motion is a small perturbation on the random thermal velocity. Therefore τ_{cn}, the mean scattering time, is not altered appreciably by the applied field.

* The formulation of Eq. (1.2.1) is slightly in error because of improper averaging. Only the order of magnitude of the result is of consequence to us, however, and it is not worthwhile to make an exact formulation.

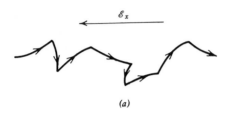

(a)

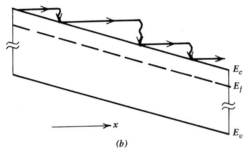

(b)

Figure 1.10   (a) The motion of an electron in a solid under the influence of an applied field. (b) Energy-band representation of the motion, indicating the loss of energy when the electron undergoes a collision.

In Fig. 1.10b electron motion in a small applied field is schematically represented on a band diagram. A constant applied field results in a linear variation in the energy levels in the crystal. Electrons (which move downward on energy level diagrams) tend to move to the right on the diagram as is appropriate for a field directed in the negative x direction. The electrons exchange energy when colliding with the lattice and drop toward their thermal-equilibrium positions. If the field is small, these energy exchanges are very small, and the lattice is not appreciably heated by the passage of the current. The slope of the energy diagrams and the energy losses associated with lattice collisions are exaggerated in Fig. 1.10b to show the process schematically. In fact, the incremental energy lost in each collision is much less than the mean thermal energy of the electrons. An electron at rest in the conduction band will be at the band edge E_c. Its kinetic energy is therefore measured by $(E - E_c)$, and the mean value at thermal equilibrium for all the electrons is just $(E - E_c) = \frac{3}{2}kT$ or about 0.04 eV at 300°K. This is less than 4% of the band-gap energy.

The net carrier velocity in an applied field is called the *drift velocity*, v_d. It can be found by equating the impulse (force × time) applied to an electron during its free flight between collisions with the momentum gained by the electron in the same period. This equality is valid because steady state will be reached when all momentum gained between collisions is lost to the lattice in the collisions. The force on an electron is $-q\mathscr{E}$ and the momentum gained is $m_n^* v_d$. Thus (cf footnote on page 21).

$$-q\mathscr{E}\tau_{cn} = m_n^* v_d \tag{1.2.2}$$

or

$$v_d = -\frac{q\mathscr{E}\tau_{cn}}{m_n^*} \tag{1.2.3}$$

Equation (1.2.3) states that the electron drift velocity v_d is proportional to the field with a proportionality factor that depends on the mean scattering time and the effective mass of the quasi-free electron. The proportionality factor is an important free-carrier property of the electron called the *mobility*. It is designated by the symbol μ_n.

$$\mu_n = \frac{q\tau_{cn}}{m_n^*} \tag{1.2.4}$$

Since $v_d = -\mu_n\mathscr{E}$, the mobility describes how strongly the motion of an electron is influenced by an applied field.

From Eq. (1.1.3) the current density flowing in the direction of the applied field can be found by summing the product of the charge on each electron times its velocity over all electrons per unit volume n.

$$J_n = \sum_{i=0}^{n} -qv_i = -nqv_d = nq\mu_n\mathscr{E} \tag{1.2.5}$$

Entirely analogous arguments apply to holes. A hole with no kinetic energy resides at the valence-band edge E_v. Its kinetic energy is therefore measured by $(E_v - E)$. If a band edge is tilted, the hole will move upward on an electron energy-band diagram. The hole mobility μ_p is defined as $\mu_p = q\tau_{cp}/m_p^*$. The total current can be written as the sum of the electron and hole components:

$$J = J_n + J_p = (nq\mu_n + pq\mu_p)\mathscr{E} \tag{1.2.6}$$

The term in parentheses in Eq. (1.2.6) is defined as the conductivity σ of the semiconductor:

$$\sigma = (q\mu_n n + q\mu_p p) \tag{1.2.7}$$

In extrinsic semiconductors, only one of the components in Eq. (1.2.7) is generally significant because of the very large ratio between the two carrier densities. The resistivity, which is the reciprocal of the conductivity, is shown as a function of the dopant concentration in Fig. 1.11 for n- and p-type silicon.

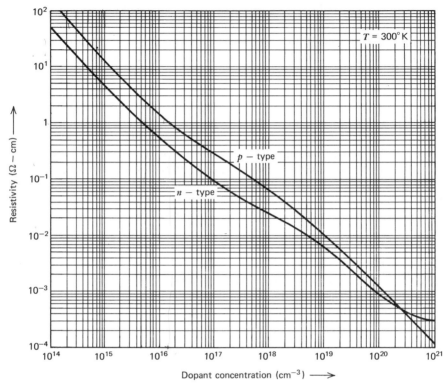

Figure 1.11 Resistivity as a function of dopant concentration at 300 °K for *n*- and *p*-type silicon.[4]

A property of a solid that is closely related to its conductivity is its *dielectric relaxation time*. The dielectric relaxation time is a measure of the time it takes for charge in a semiconductor to become neutralized by conductive processes. It is very small in metals and can be large in semiconductors and insulators. One of its uses is to obtain qualitative insight. Several device concepts may be quite readily interpreted according to the relative sizes of the charge transit time through a material and the dielectric relaxation time in the same material. Problem 1.12 provides further introduction to the dielectric relaxation time.

Mobility and Scattering

Quantum-mechanical calculations indicate that a perfectly periodic lattice would not scatter free carriers; that is, the carriers would not interchange energy with a stationary, perfect lattice. However, at any temperature above absolute zero the atoms that form the lattice vibrate. These vibrations disturb periodicity and allow energy to be transferred between the carriers and the lattice.* Since the lattice vibrations increase with increasing temperature, lattice scattering becomes more effective at higher temperatures. Theoretical analysis indicates that the mobility should decrease in proportion to T^{-n} with n = 1.5 to 2.5 when lattice scattering is dominant. Experimentally, n values ranging from 1.66 to 3 are observed, with n = 2.5 being common.

Dopant impurities, in addition to lattice vibrations, also cause local distortions in the lattice and scatter free carriers. However, unlike scattering from lattice vibrations, scattering from ionized impurities becomes less significant at higher temperatures. Since the carriers are moving faster at higher temperatures, they remain near the impurity atom for a shorter time and are therefore less effectively scattered. Consequently, when impurity scattering is dominant, the mobility increases with increasing temperature. Scattering may also be caused by collisions with unintentional impurities and by crystal defects. These defects can arise from poor control of the quality of the material, or they may be related to boundaries between grains of a polycrystalline material. An example of the latter would be the thin films of polycrystalline silicon used to form portions of many MOS integrated circuits. The grain boundaries and defects in polycrystalline material can reduce the mobility to a small fraction of its value in single-crystal material with the same dopant concentration.

Two or more of the scattering processes discussed above may be important at the same time, and their combined effect on mobility must be assessed. To do this we consider the number of particles that are scattered in a time dt. The probability that a carrier is scattered in a time interval dt by process i is dt/τ_i where τ_i is the average time between scattering events resulting from process i. The total probability dt/τ_c that a carrier is scattered in the time interval dt is then the sum of the probabilities of being scattered by each mechanism:

$$\frac{dt}{\tau_c} = \sum_i \frac{dt}{\tau_i} \qquad (1.2.8)$$

This is a reasonable way to combine the scattering probabilities since the average scattering time resulting from all of the processes acting simultaneously is less than that resulting from any one process, and is dominated by the shortest scattering time. Since the mobility μ equals $q\tau_c/m^*$, we may write

$$\frac{1}{\mu} = \sum_i \frac{1}{\mu_i} \qquad (1.2.9)$$

---

* Since this energy is supplied to the carriers by the applied field, scattering processes lead to heating of the semiconductor. The dissipation of this heat is often a limiting factor in the size of semiconductor devices. A device must be large enough to avoid heating to temperatures at which it no longer functions.

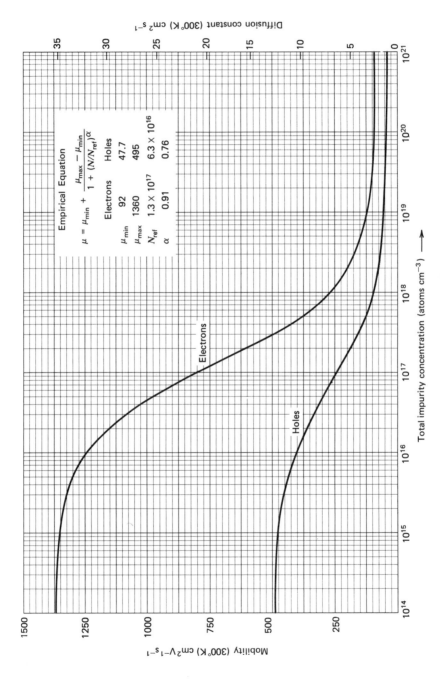

Figure 1.12 Electron and hole mobilities in silicon at 300°K as functions of the total dopant concentration. The values plotted are the results of curve fitting measurements from several sources. An empirical equation for the data is shown in the inset.

The mobility of a carrier acted on by several scattering mechanisms may thus be found by combining the reciprocal values of the mobility corresponding to each mechanism. When the mobility is limited by several scattering mechanisms, it is less than the mobility resulting from any one of the individual scattering mechanisms, as is reasonable. In addition, the mobility is dominated by the process for which τ_i is smallest.[5]

We can apply these considerations to the measured mobilities of electrons and holes in silicon at room temperature that are shown in Fig. 1.12. In lightly doped material, the mobility resulting from ionized impurity scattering is higher than that due to lattice scattering. The observed mobility for both holes and electrons therefore remains nearly constant as the dopant concentration is increased. At higher dopant concentrations, however, scattering by ionized impurities becomes comparable to that resulting from lattice vibrations, and the total mobility decreases. An important practical consequence of the dependence of mobility on dopant concentration is observed if a semiconductor is converted from one type to the other (p to n or n to p) by overcompensating the dopant impurity atoms already present. While the carrier densities depend on the difference between the concentrations of the two types of dopant impurities, $(N_d - N_a)$ [Eqs. (1.1.16) and (1.1.17)], the scattering depends on the sum of the ionized impurity concentrations, $(N_d + N_a)$. Thus, the mobilities in a compensated semiconductor may be markedly lower than those in an uncompensated material with the same net carrier densities.

Velocity Limitations. In the simplified treatment presented thus far, we have assumed (by taking τ_c to be insensitive to $\mathscr{E}$) that the velocity imparted to the free carriers by the applied field is much less than the random thermal velocity, which we found from Eq. (1.2.1) to be approximately 10^7 cm s^{-1} at room temperature. For electrons in silicon with $\mu_n = 1350$ cm^2 V^{-1} s^{-1} the drift velocity at a typical field of 100 V cm^{-1} is roughly 10^5 cm s^{-1} or about 1% of the thermal velocity, and the applied field does not appreciably change the total velocity. At high fields, however, the drift velocity can become comparable to the random thermal velocity, and thermal and drift motions can no longer be considered independently. This condition gives rise to scattering processes that cause the electrons to lose energy to the lattice more effectively. The drift velocity in this situation approaches a limiting value v_l and is no longer proportional to the electric field. Figure 1.13 shows measured values of drift velocity for electrons and holes in silicon as functions of the applied field. Noticeable deviations from a linear relationship (constant mobility) occur at about 3×10^3 V cm^{-1} for electrons and 6×10^3 V cm^{-1} for holes. Fields of this magnitude are not unusual in semiconductor devices; consequently, velocity saturation is of significant practical importance.

As a useful approximation the velocity-field relation may be divided into three regions: a low-field region where $v_d = \mu\mathscr{E}$, a high-field region where v_d approaches

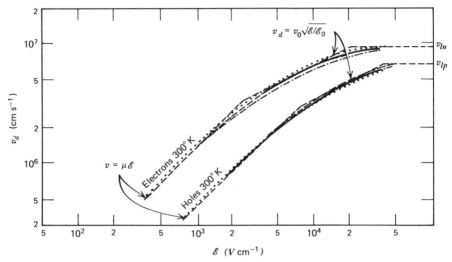

Figure 1.13   Drift velocities of electrons and holes in silicon as functions of the applied field showing the saturation of the drift velocity at high fields.[6] The dotted and dot-dash lines represent the data limits. The dashed lines show the three-part, piecewise approximations to the curves.

v_l, and an intermediate region where v_d varies roughly as the square root of field. The velocity vs field curve can be empirically fit to the expression

$$v_d = v_l(\mathscr{E} / \mathscr{E}_c)/[1 + (\mathscr{E} / \mathscr{E}_c)^\beta]^{1/\beta} \qquad (1.2.10)$$

where:	Electrons	Holes	Units
v_l	$1.53 \times 10^9 T^{-.87}$	$1.62 \times 10^8 T^{-.52}$	cm s^{-1}
$\mathscr{E}_c$	$1.01 \times T^{1.55}$	$1.24 \times T^{1.68}$	V cm^{-1}
β	$2.57 \times 10^{-2} \times T^{.66}$	$.46\ T^{.17}$	—

T is measured in degrees Kelvin

Since the kinetic energy of an electron is proportional to the square of its velocity, we can use the velocity to define a temperature as in Eq. (1.2.1). Thus, at high fields the electron system may be considered to have a higher temperature than does the lattice. Energetic electrons are, therefore, called *hot electrons*.

Diffusion Current

In the previous section we discussed drift current, which flows when an electric field is applied and which gives rise to Ohm's law behavior. Ohmic behavior is observed in metals and semiconductors and is probably familiar from direct experience. In semiconductors, however, another important component of current can exist if there is a spatial variation of carrier energies or densities within the material. This component of current is called *diffusion current*. Diffusion current is generally not an important consideration in metals because of their high conductivities. The lower conductivity and the possibility of nonuniform densities of carriers and of carrier energies, however, often makes diffusion a very important process affecting current flow in semiconductors.

To understand the origin for diffusion current, we consider the hypothetical case of an n-type semiconductor with an electron density that varies only in one dimension (Fig. 1.14). We assume that the semiconductor is at uniform temperature so that the average energy of electrons does not vary with x: only the density $n(x)$ is variable. We consider the number of electrons crossing the plane at $x = 0$ per unit collision time per unit area. Because they are at finite temperature, the electrons have random thermal motion along the single dimension, but we assume that no electric fields are applied. On the average, the electrons crossing the plane $x = 0$ from the left in Fig. 1.14 each collision time start at approximately $x = -l$ where l is the *mean-free path* of an electron, given by $l = v_{th}\tau_{cn}$. The average rate per unit area of electrons crossing the plane $x = 0$ from the left, therefore, depends upon the density of electrons that started at $x = -l$ and is

$$\tfrac{1}{2}n(-l)v_{th} \tag{1.2.11}$$

The factor $(\tfrac{1}{2})$ appears since half of the electron density travels to the left and half travels to the right after a collision at $x = -l$. Similarly, the density of electrons crossing the plane $x = 0$ from the right is given by

$$\tfrac{1}{2}n(l)v_{th} \tag{1.2.12}$$

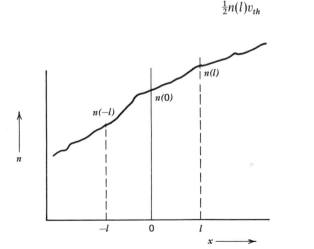

Figure 1.14 Electron concentration n versus distance x in a hypothetical one-dimensional solid. Boundaries are demarcated l units on either side of the origin where l is a collision mean-free path.

so that the net rate of particle flow from the left (denoted F) is

$$F = \tfrac{1}{2}v_{th}[n(-l) - n(l)] \tag{1.2.13}$$

Approximating the densities at $x = \pm l$ by the first two terms of a Taylor series expansion, we find

$$F = \tfrac{1}{2}v_{th}\left\{\left[n(0) - \frac{dn}{dx}l\right] - \left[n(0) + \frac{dn}{dx}l\right]\right\} = -v_{th}l\frac{dn}{dx} \tag{1.2.14}$$

Since each electron carries a charge, $-q$, the particle flow corresponds to a current

$$J_n = -qF = qlv_{th}\frac{dn}{dx} \tag{1.2.15}$$

We thus see that diffusion current is proportional to the spatial derivative of the electron density and arises because of the random thermal motion of charged particles in a concentration gradient. For an electron density that increases with x the gradient is positive, as is the current. Because we expect electrons to flow from the higher density region at the right to the lower density region at the left, and current flows in the direction opposite to that of the electrons, the direction of current indicated by Eq. (1.2.15) is physically reasonable.

We may write Eq. (1.2.15) in a more useful form by applying the theorem for the equipartition of energy to this one-dimensional case. This allows us to write

$$\tfrac{1}{2}m_n^* v_{th}^2 = \tfrac{1}{2}kT \tag{1.2.16}$$

We now use the relationship $l = v_{th}\tau_{cn}$ together with Eq. (1.2.4) to write Eq. (1.2.15) in the form

$$J_n = q\left(\frac{kT}{q}\mu_n\right)\frac{dn}{dx} \tag{1.2.17}$$

The quantity in parentheses on the right-hand side of Eq. (1.2.17) is defined as the *diffusion constant* D_n and our short derivation has established that

$$D_n = \left(\frac{kT}{q}\right)\mu_n \tag{1.2.18}$$

Equation (1.2.18) is known as the *Einstein relationship*. It relates the two important constants that characterize free-carrier transport by drift and by diffusion in a solid. Its validity can be established with rigor by considering the full-blown statistical mechanics of solids. Our derivation is aimed at intuition, not physical exactness.

If a field were present, then a drift current as well as a diffusion component would flow. In that case, the total current density at any point would be the sum of both the drift and diffusion components.

$$J_{nx} = q\mu_n n \mathscr{E}_x + qD_n \frac{dn}{dx} \qquad (1.2.19)$$

where $\mathscr{E}_x$ is the component of the electric field in the x-direction.

A similar expression may be found for the hole diffusion current so that the total hole current is written as

$$J_{px} = q\mu_p p \mathscr{E}_x - qD_p \frac{dp}{dx} \qquad (1.2.20)$$

where the negative sign arises because of the positive charge of a hole. The Einstein relation [Eq. (1.2.18)] also applies between D_p and μ_p.

Our discussion of diffusion current has been framed in terms of nonuniform carrier concentrations. Although this is the most frequent situation encountered in device analysis and Eqs. (1.2.19) and (1.2.20) are usually adequate, diffusion will also occur if carriers are equal in density in a semiconductor but are, at the same time, more energetic in one region than in another. In this case, other formulations must be used. A practical situation in which this is the case forms the basis for Problem 1.16.

1.3 SEMICONDUCTOR TECHNOLOGY

We have given several examples of the direct influence exerted by material properties on device performance. In addition, the materials used are major determinants of the technology that may be employed. In this section we consider device technology with emphasis on the silicon planar process.

The evolution of semiconductor technology was determined as much by the chemical properties of the available materials as by their electrical characteristics. The first semiconductors extensively developed were germanium and silicon, which have forbidden band gaps of 0.7 and 1.1 eV, respectively. Both of these elements crystallize in the diamond structure, each atom forming covalent bonds with its four nearest neighbors. Various properties of these semiconductors and other useful electronic materials are summarized in Table 1.1. Some additional properties of silicon are contained in Table 1.2.

Germanium was used extensively before the use of silicon became common. It is less reactive than silicon and can, therefore, be refined and processed more easily. The carrier mobilities in germanium are also higher than are those in silicon. Since the frequency response of a semiconductor device usually depends on the carrier velocity, which is proportional to the mobility and the applied electric field, the highest possible mobility is desirable. Because the mobilities are about three times as great in germanium as in silicon, the frequency response of devices fabricated in germanium can be superior to that of devices fabricated in silicon.

Table 1.1 Properties of Semiconductors and Insulators (at 300°K Unless Otherwise Noted)

Property	Symbol	Units	Si	Ge	GaAs	GaP	SiO$_2$	Si$_3$N$_4$
Crystal structure			Diamond	Diamond	Zincblende	Zincblende	Amorphous for most IC applications	Hexagonal
Atoms per unit cell			8	8	8	8		
Atomic number	Z		14	32	31/33	31/15	14/8	14/7
Atomic or molecular weight	MW	g/g-mole	28.09	72.59	144.64	100.70	60.08	140.28
Lattice constant	a_0	Å	5.4307	5.6575	5.6532	5.4505		7.75
Atomic or molecular density	N_0	cm^{-3}	5.00×10^{22}	4.42×10^{22}	2.21×10^{22}	2.47×10^{22}	2.20×10^{22}	1.48×10^{22}
Density		g cm^{-3}	2.328	5.323	5.316	4.13	2.19	3.44
Energy gap 300°K	E_g	eV	1.124	0.67	1.43	2.24	~9	4.7
0°K	E_g	eV	1.170	0.744	1.53	2.40		
Temperature dependence	$\Delta E_g/\Delta T$	eV °K^{-1}	-2.3×10^{-4}	-3.7×10^{-4}	-5.0×10^{-4}	-5.4×10^{-4}		
Relative permittivity	ϵ_r		11.7	15.8	13.1	10.2	3.9	7.5
Index of refraction	n		3.44	3.97	3.3	3.3	1.46	2.0
Melting point	T_m	°C	1412	937	1237	1467	~1700	~1900
Vapor pressure		Torr (mm Hg) (at °C)	10^{-7} (1050) 10^{-5} (1250)	10^{-9} (750) 10^{-7} (880)	1 (1050) 100 (1220)	10^{-6} (770) 10^{-4} (920)		
Specific heat	C_p	Joule (g °K)$^{-1}$	0.70	0.32	0.35		1.4	0.17
Thermal conductivity	κ	W(cm °K)$^{-1}$	1.412	0.606	0.455	0.97	0.014	0.185(?)
Thermal diffusivity	D_{th}	cm^2 s^{-1}	0.87	0.36	0.44		0.004	0.32(?)

Coefficient of linear thermal expansion	α'	$°K^{-1}$	2.5×10^{-6}	5.7×10^{-6}	5.9×10^{-6}	5.3×10^{-6}	5×10^{-7}	2.8×10^{-6}
Intrinsic carrier concentration	n_i	cm^{-3}	1.45×10^{10}	2.4×10^{13}	9.0×10^{6}			
Lattice mobility								
Electron	μ_n	$cm^2 [V\,s]^{-1}$	1350	3900	8800	300	20	
Hole	μ_p	$cm^2 [V\,s]^{-1}$	480	1900	400	100	$\sim 10^{-8}$	
Effective density of states								
Conduction band	N_c	cm^{-3}	3.22×10^{19}	1.02×10^{19}	4.7×10^{17}			
Valence band	N_v	cm^{-3}	1.83×10^{19}	5.64×10^{18}	7.0×10^{18}			
Electric field at breakdown	$\mathscr{E}_1$	$V\,cm^{-1}$	3×10^{5}	8×10^{4}	3.5×10^{5}		$6 - 9 \times 10^{6}$	
Effective mass								
Electron	m_n^*/m_0		1.08^a 0.26^b	0.55^a 0.12^b	0.068	0.5		
Hole	m_p^*/m_0		0.81^a 0.386^b	0.3	0.5	0.5		
Electron affinity	qX	eV	4.15	4.00	4.07	~ 4.3	1.0	
Average energy loss per phonon scattering		eV	0.063	0.037	0.035			
Optical phonon mean-free path								
Electron	l_{ph}	Å	62	65	35			
Hole	l_{ph}	Å	45	65	35			

Sources: A. S. Grove, *Physics and Technology of Semiconductor Devices*, Wiley, New York, 1967; S. M. Sze, *Physics of Semiconductor Devices*, Wiley, New York, 1969; D. E. Hill, *Some Properties of Semiconductors*, (Table), Monsanto Co., St. Peters, Mo., 1971; H. Wolf, *Semiconductors*, Wiley, New York, 1971.

[a] Used in density-of-state calculations.
[b] Used in conductivity calculations.[11]

Table 1.2 Additional Properties of Silicon (Lightly Doped, at 300°K Unless Otherwise Noted)

Property	Symbol	Units	Values
Tetrahedral radius		Å	1.17
Pressure coefficient of energy gap	$\Delta E_g/\Delta p$	eV (atm)$^{-1}$	-1.5×10^{-6}
Product of electron and hole density	n_i^2	cm^{-6} (kT in eV, T in °K)	$1.5 \times 10^{33}T^3 \exp\left(-\dfrac{1.21}{kT}\right)$
Temperature coefficient of lattice mobility			
Electron	$\Delta\mu_n/\Delta T$	cm^2 (V s °K)$^{-1}$	-11.6
Hole	$\Delta\mu_p/\Delta T$	cm^2 (V s °K)$^{-1}$	-4.3
Diffusion constant			
Electron	D_n	cm^2 s^{-1}	34.6
Hole	D_p	cm^2 s^{-1}	12.3
Hardness	H	Mohs	7.0
Elastic constants	c_{11}	dyne cm^{-2}	1.656×10^{12}
	c_{12}		0.639×10^{12}
	c_{44}		0.796×10^{12}
Young's modulus ($\langle 111\rangle$ direction)	Y	dyne cm^{-2}	1.9×10^{12}
Surface tension (at 1412°C)	σ_0	dyne cm^{-2}	720
Latent heat of fusion	H_F	eV	0.41
Expansion on freezing	%		9.0
Cut-off frequency of lattice vibrations	ν_0	Hz	1.39×10^{13}

Main Source. H. F. Wolf, *Semiconductors*, Wiley, New York, 1971, p. 45.

Energy Levels of Elemental Impurities in Si[a]

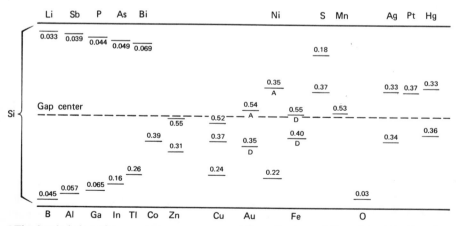

[a] The levels below the gap center are measured from the top of the valence band and are acceptor levels unless indicated by D for donor level. The levels above the gap center are measured from the bottom of the conduction band and are donor levels unless indicated by A for acceptor level.[4]

34

Disadvantage of Germanium over Silicon :
∗ Difficulty in doping (less selectively doped)
∗ Large geometry → Large components and devices → slow respond time
∗ Becomes intrinsic at low temperature due to small bandgap.

1.3 Semiconductor Technology **35**

As technology progressed, however, the limitations of germanium became evident. There was no convenient method for introducing controlled amounts of dopant impurities into small, selected areas of a piece of germanium to form localized regions of p-type and n-type material, and devices were restricted to relatively large geometries. Since the frequency response of a semiconductor device is limited by the time required for carriers to move from one region to another, the device dimensions, as well as the carrier velocities, are important. In addition, the relatively small forbidden gap of germanium causes the intrinsic carrier concentration to approach the dopant concentration at lower temperatures than is the case with silicon. This behavior is apparent in Fig. 1.8. We shall see that most semiconductor devices become useless when the semiconductor behaves intrinsically. Hence, in this respect, silicon is preferable to germanium.

The restriction to large dimensions can also be eliminated by making use of the ease with which the surface of a silicon crystal is oxidized to form a chemically stable oxide. Silicon dioxide, when grown in this manner, is amorphous yet very resistant to most common acids and solvents. Fortunately, however, hydrofluoric acid will attack the silicon-dioxide layer but it will not etch silicon. Equally important, even at very high temperatures, most dopant impurity atoms will not readily penetrate silicon dioxide. It is possible, therefore, to cover selectively a piece of silicon with regions of silicon dioxide. Then, the silicon can be placed in an ambient that will deposit dopant atoms on the surface. The dopant atoms can then enter the regions of exposed silicon but not the areas covered by the silicon dioxide. This selective doping can be controlled by using photo-sensitive polymers that act as shields to an HF etch bath in regions that have been defined by photographic plates. The selective-etch process, developed originally for lithographic printing applications, permits the delineation of very small geometries.

By sequencing the operations just sketched, we may dope as p- or n-type localized regions on a silicon surface having dimensions in the micron range. The ability to dope small regions of silicon selectively is the basis of present-day silicon technology and one of the most important reasons that silicon has replaced germanium as the most commonly used semiconductor material. At an even more fundamental level, the advantage of silicon as a semiconductor material results from the ease with which one precisely controlled technology offers both a semiconductor of excellent quality and an insulator with superb properties.

Crystal Growth[†]

We start our discussion of silicon technology with a brief consideration of the techniques used to obtain the basic single-crystal semiconductor material. The silicon used for many types of semiconductor devices is usually doped to contain between 10^{15} and 10^{20} dopant impurity atoms per cm^3. So that we can control the properties of the material, any unintentional or background concentration of electrically active impurity atoms should be at least two orders of magnitude less than the lowest intentional dopant concentration—that is, about 10^{13} cm^{-3} or less. Since silicon contains roughly 5×10^{22} atoms cm^{-3}, only about one unintentional impurity atom per billion silicon atoms can be tolerated. This purity is well beyond that required in almost any other industry, and indicates that crystal growth is a highly refined technology. In addition to high purity, the semiconductor material must be formed into a nearly perfect single crystal, since grain boundaries and other crystalline defects will generally degrade device performance. Defects may, for example, enhance the recombination of electrons and holes to such a degree that injected excess carriers are lost by recombination before they can travel across the device.

Sophisticated techniques are needed to obtain single crystals of such quality. They are generally grown by what is known as the *Czochralski method*, in which high-purity polycrystalline silicon is first melted in a nonreactive crucible and held at a temperature just above the melting point (Fig. 1.15). A carefully controlled quantity of the desired dopant impurity, such as phosphorus or boron, is added. For the common dopant concentration of 10^{15} cm^{-3}, about 0.1 mg of dopant is added for each kilogram of silicon. A high-quality seed crystal of the desired crystalline orientation is then lowered into the melt while being rotated as indicated in Fig. 1.15a. A portion of this crystal is allowed to dissolve into the molten silicon to remove the strained outer portions and to expose fresh surfaces. The seed crystal then continues rotating and is slowly raised from the melt. As it is raised, it cools and material from the melt adheres to it, thereby forming a larger crystal as shown in Fig. 1.15b. Under the carefully controlled conditions maintained

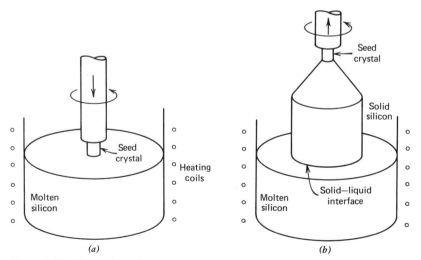

Figure 1.15 Formation of a single-crystal semiconductor ingot by the Czochralski process.

during growth, the new silicon atoms continue the crystal structure of the already solidified material. In this manner cylindrical, single-crystal ingots of silicon several inches in diameter can be fabricated. As crystal-growing technology has developed, the diameter of the crystals has progressively increased from a small fraction of an inch to the three or four inches that are common today.

Once the single-crystal ingot is grown, it is sliced into thin, circular *wafers*. The wafers are then chemically etched to remove the sawing damage and polished with successively finer polishing grits and chemical etchants until a defect-free, mirrorlike surface is obtained. At this point the wafers are ready for device fabrication.

Planar Technology[†]

The silicon *planar process* consists of the selective introduction of dopant atoms into small areas of the silicon from one surface of the wafer in order to form regions of p- and n-type material. The technology is *planar* because fabrication is accomplished by processes carried out from one surface plane. The dopant atoms are typically added by *depositing* them onto the silicon surface and then *diffusing* them into those regions of the silicon that are not protected by a layer of silicon dioxide. Since the dopant atoms are introduced from the surface and typical diffusions are of the order of a few microns* or less, the active regions of the semiconductor device are themselves within a few microns of one surface of the silicon wafer. The remainder of the wafer (generally a few hundred microns) serves simply as a mechanical support for the important surface region.

A crucial advantage of planar technology is that each portion of the fabrication process (prior to lead attachment) is applied to the entire silicon wafer at the same time. In this manner dopant atoms can be introduced simultaneously into several thousand separate, small regions of the wafer. Since the processing cost of a wafer is not changed significantly by its size or by the dimensions of the diffused regions, the use of larger diameter wafers and smaller device dimensions minimizes the processing cost per device. Simultaneous fabrication of several related semiconductor devices and their interconnection in the same piece of semiconductor results in an entire circuit in one silicon chip. This type of element, called an *integrated circuit*, evolved from the capability inherent in the planar process of fabricating many devices at the same time.

The important fabrication steps in the planar process are shown in Fig. 1.16 and include the formation of the protective oxide layer (Fig. 1.16a), its selective removal (Fig. 1.16b), the deposition of dopant atoms on the wafer surface (Fig. 1.16c), and their diffusion into the exposed silicon regions (Fig. 1.16d). These processes determine the surface locations of the dopants and their depths within

* 1 micron (μm) $= 10^{-4}$ cm $= 10^4$ Å.

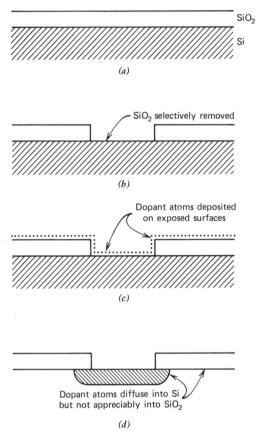

Figure 1.16 Basic fabrication steps in the silicon planar process: (a) oxide formation, (b) selective oxide removal, (c) deposition of dopant atoms on wafer, and (d) diffusion of dopant atoms into exposed regions of silicon.

the top few microns of silicon. The location of the dopants, in turn, determines the characteristics of the devices fabricated.

Oxidation. A silicon dioxide (SiO_2) layer is usually formed on the wafer by the chemical combination of silicon atoms in the wafer with oxygen that is allowed to flow over the semiconductor surfaces while the wafer is heated to a high temperature. During this oxidation process the wafer is placed inside a quartz tube, which is itself set within the cylindrical opening of a resistance-heated furnace (Fig. 1.17). Temperatures in the range 900 to 1200°C are typical, the reaction proceeding more rapidly at higher temperatures. The upper temperature limit is generally determined by softening and degradation of the quartz furnace tube and other fixtures at temperatures much above 1200°C, although the silicon itself

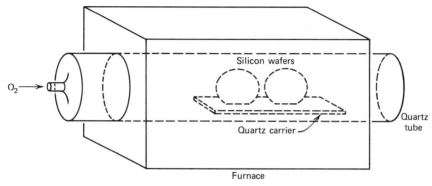

Figure 1.17 An insulating layer of silicon dioxide is grown on silicon wafers by placing them in a high-temperature furnace in a flow of gas containing oxygen.

will not melt until 1412°C. The oxidizing ambient can be either dry or wet oxygen. The latter is generally produced by passing high-purity, dry oxygen through water held close to its boiling point or by admitting hydrogen and oxygen to react directly in the oxidizing furnace. The oxidation proceeds much more rapidly in a wet-oxygen ambient, which is consequently used for the formation of thicker protective layers of silicon dioxide. The growth of a thick oxide in the slower, dry-oxygen ambient may lead to undesirable movement (redistribution) of impurities introduced into the wafer during previous processing.

Oxidation takes place at the Si–SiO$_2$ interface so that the oxidizing species must diffuse through any previously formed oxide and then react with the silicon at this interface. At lower temperatures and for thinner oxides, the surface reaction rate at the Si–SiO$_2$ interface limits the growth rate, and the thickness of the oxide layer is a linear function of the oxidation time. The reaction rate depends on the nature of the surface bonds; therefore, the oxide thickness formed depends on the orientation of the crystal. Silicon with its surface cut normal to the $\langle 100 \rangle$ crystal axis, oxidizes more slowly than does $\langle 111 \rangle$-oriented material.

At higher temperatures and for thicker oxides, the oxidation reaction is limited by diffusion of the oxidizing species through the previously formed oxide. In this case the oxide thickness grown is roughly proportional to the square root of the oxidation time. This square-root dependence is characteristic of diffusion processes and sets a practical upper limit on the thickness that can be conveniently obtained. When oxygen diffusion is rate limiting, the oxidation rate is independent of the orientation of the silicon wafer.

The relation between the oxide thickness x_{ox} formed and the oxidation time t and temperature T is given by[7]

$$x_{ox} = \frac{A}{2}\left[\sqrt{1 + \frac{t + \tau}{A^2/4B}} - 1\right] \tag{1.3.1}$$

where A and B are coefficients that depend on the oxidation temperature. The parameter τ includes the effects of oxide initially present and also an initial stage of oxide growth during which the assumptions leading to Eq. (1.3.1) are not satisfied. For short oxidation times the surface reaction rate limits the oxide growth as discussed above, and Eq. (1.3.1) reduces to a linear relationship

$$x_{ox} = \frac{B}{A}(t + \tau) \tag{1.3.2}$$

For long oxidation times a square-root relationship is obtained:

$$x_{ox} = \sqrt{B(t + \tau)} \approx \sqrt{Bt} \tag{1.3.3}$$

Using experimentally determined values of A, B, and τ, we may plot the oxide thickness as a function of time for several commonly used oxidation temperatures for both dry oxidation (Fig. 1.18) and wet oxidation (oxygen bubbled through 95°C water; Fig. 1.19). These figures apply to $\langle 111 \rangle$-oriented silicon. The oxide thickness will be somewhat less for $\langle 100 \rangle$-oriented silicon in the case of thinner oxides and lower oxidation temperatures where oxide growth is limited by the surface reaction rate for an appreciable fraction of the time. From these curves we see that the formation of a 0.3 μm-thick oxide at 1100°C will take 270 min in a dry-oxygen ambient compared to 15 min in wet oxygen. Oxide thicknesses of a few tenths of a micron are often used, with 1 to 2 μm being the upper practical limit when employing conventional oxidation techniques.

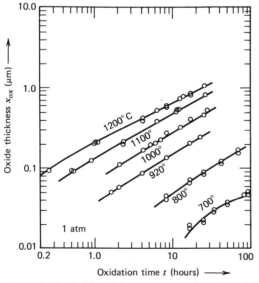

Figure 1.18 Oxide thickness x_{ox} as a function of time in a dry-oxygen ambient for several commonly used oxidation temperatures.[7]

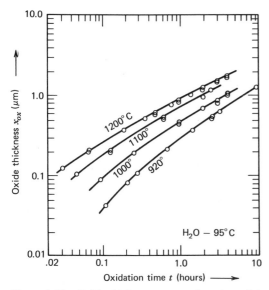

Figure 1.19   Oxide thickness x_{ox} as a function of time in an ambient of wet oxygen obtained by bubbling oxygen through water held at 95°C before introducing it into the oxidation furnace.[7]

Although most oxides are formed by the reaction of the silicon from the wafer with a gaseous oxidizing species, silicon oxides may also be deposited directly on the silicon surface without the use of any silicon from the wafer. Various techniques, such as physical vapor deposition (e.g., evaporation in a vacuum) and chemical vapor deposition may be used. The temperature of the wafer may also vary widely during the deposition. Although the deposited oxides are not usually as high in quality as are thermally grown oxides, they may be used for convenience when high-temperature processing is undesirable—for example, when aluminum, which melts at 660°C, is present on the wafer surface. One commonly used deposited oxide is formed by the chemical reaction of silane and oxygen on a wafer surface heated to about 400°C. The reaction

$$SiH_4 + O_2 \rightarrow SiO_2 + 2H_2\uparrow \qquad (1.3.4)$$

produces an oxide that is less dense and less chemically resistant than thermally grown oxide, but is useful for many applications.

Photomasking.   Once the protective layer of silicon dioxide has been formed on the silicon wafer, it must be selectively removed from those areas into which dopant atoms are to be introduced. Selective removal is usually accomplished by the use of a light-sensitive plastic called *photoresist*. The oxidized wafer is first coated with the liquid photoresist, generally by placing a few drops on a rapidly

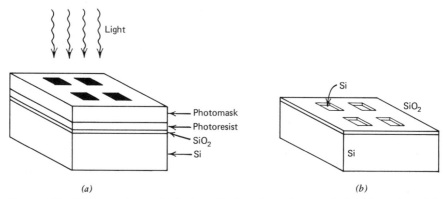

Figure 1.20 The areas from which the oxide is to be etched are defined by polymerizing a light-sensitive photoresist through a photographic negative or mask.

spinning wafer. After drying the photoresist, a photographic negative (called a photomask or mask) is placed over the wafer, as shown in Fig. 1.20*a*, and aligned using a microscope. Then, the photoresist is exposed to ultraviolet or near ultraviolet light, which causes the molecules in the photoresist to become cross-linked (polymerized) in areas that are exposed to the light. The photoresist is not affected in regions that are protected by the opaque portions of the mask.* The photoresist is then developed to remove it from areas where it was not hardened, leaving an acid-resistant protective coating over selected regions of the silicon dioxide.

The unprotected regions of silicon dioxide are chemically etched in a hydrofluoric-acid solution to expose the bare silicon surface, and the photoresist is then removed from the remaining areas. At this point, portions of the wafer are protected by silicon dioxide, and the bare silicon is exposed at *windows* in the oxide (Fig. 1.20*b*). These regions of bare silicon are the locations into which the dopant impurities are diffused.

Diffusion. The diffusion process generally involves two steps. First, the dopant atoms are placed on or near the surface of the wafer by a gaseous *deposition* step or by coating the wafer with a layer containing the desired dopant impurity. This is followed by a *drive-in diffusion*, which moves the dopant atoms further into the wafer. The shape of the resulting dopant distribution is determined primarily by the manner in which the dopant is placed near the surface while the diffusion depth depends primarily on the temperature and time of the drive-in diffusion. Since the characteristics of semiconductor devices depend on the nature of the dopant profile, it is useful to discuss the diffusion process somewhat further.

* Photoresists that polymerize where they are not exposed to light are also available.

When dopant atoms are placed on a piece of silicon and the wafer is heated to a sufficiently high temperature, the dopant atoms migrate in the crystal. This movement occurs because of the concentration gradient of the dopant atoms, with the atoms moving from the region of high concentration near the surface toward regions of lower concentration further into the wafer. The diffusion of dopant atoms arises in the same manner as the diffusion current of free carriers discussed in Section 1.2. The primary difference between the two cases is the temperature necessary before migration will occur. Dopant atoms must typically make their way through the lattice by meeting silicon vacancies, whereas free carriers in the valence or conduction band can move without this taking place. Temperatures of the order of 1000°C are required to supply the energy needed for appreciable diffusion of typical dopant atoms.

The change in the concentration C of the dopant atoms with time in a narrow region of width dx at a depth x from the surface (Fig. 1.21) can be written as the difference between the flux of dopant atoms entering the region from the left and the flux leaving at the right

$$\frac{\partial C(x)}{\partial t} \, dx = F(x) - F(x + dx) \tag{1.3.5}$$

We can approximate the last term by the first two terms of a Taylor-series expansion.

$$F(x + dx) \approx F(x) + \left(\frac{\partial F}{\partial x}\right) dx \tag{1.3.6}$$

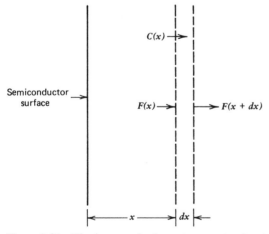

Figure 1.21   The increase in dopant concentration in a region dx is related to the net flux of atoms into the region: $F(x) - F(x + dx)$.

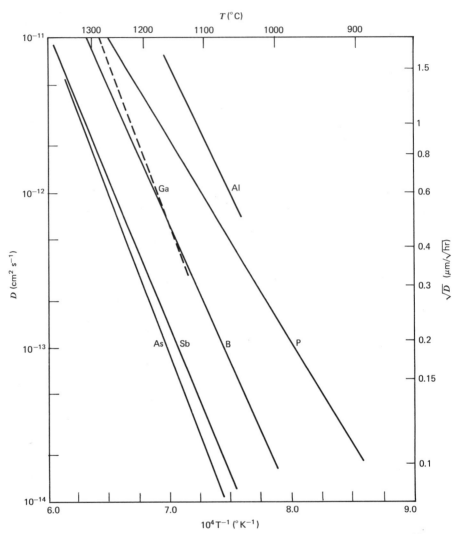

Figure 1.22 The diffusivities of several commonly used dopant impurities in silicon as functions of reciprocal temperature.[8]

to obtain

$$\frac{\partial C(x)}{\partial t} = -\frac{\partial F}{\partial x}$$

(1.3.7)

The first order expression for the flux is proportional to the concentration gradient

$$F = -D\frac{\partial C}{\partial x}$$

(1.3.8)

The parameter D, called the *diffusivity*, describes the ease with which the dopant atoms move in the lattice and is, therefore, a function of temperature. The diffusivities of several common dopant impurities are shown in Fig. 1.22. The simple model considered here neglects many aspects of the diffusion process that become important at higher dopant concentrations. Consequently, the values given in Fig. 1.22 are only valid at low and moderate dopant concentrations and must be used with caution when the dopant densities become high since the diffusivity itself becomes a function of the dopant concentration. Phosphorus is especially troublesome in this respect. Combining Eqs. (1.3.7) and (1.3.8), we find that

$$\frac{\partial C}{\partial t} = D\frac{\partial^2 C}{\partial x^2}$$

(1.3.9)

This is the diffusion equation, which can be solved explicitly for $C(x, t)$ when particular boundary conditions are known. For a given time t in the furnace, therefore, the approximate dopant distribution after a diffusion cycle can be calculated.

When a gaseous deposition source is used to introduce the dopant atoms into the semiconductor, the patterned wafer is placed in a diffusion furnace similar to the furnace used for oxidation, and a gas containing the desired dopant impurity— typically phosphorus or boron— is passed over the wafer. The quantity of dopant that enters the wafer is limited to values less than the solid solubility of the dopant in silicon at the furnace temperature. The solid solubilities in silicon for several common dopants are shown in Fig. 1.23. Because of the low temperatures and short times usually used, the penetration of the dopant atoms into the silicon during this deposition step is generally very small. The drive-in diffusion is then used to spread these atoms over the desired depth.

Since the surface is exposed to the same concentration of dopant atoms during the entire deposition cycle, the solution of the diffusion equation indicates that after the deposition cycle the dopant atoms are distributed according to a complementary-error-function distribution

$$C(x,t) = C_s \, \mathrm{erfc}\left(\frac{x}{2\sqrt{Dt}}\right) = \frac{2C_s}{\sqrt{\pi}} \int_{\frac{x}{2\sqrt{Dt}}}^{\infty} e^{-v^2} \, dv$$

(1.3.10)

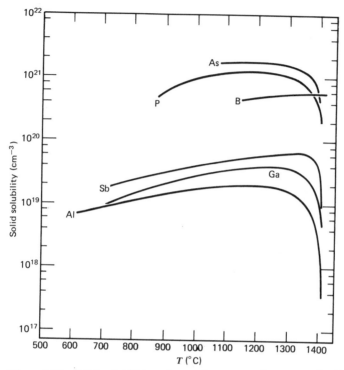

Figure 1.23   Solid solubilities of several common dopant species in silicon as functions of temperature.[9]

where C_s is the surface concentration of dopant atoms. This function is a solution to Eq. (1.3.9) if $C(0,t) = C_s$. Its behavior is illustrated in Fig. 1.24. The combination of parameters $2\sqrt{Dt}$ represents a characteristic diffusion length associated with a particular diffusion cycle and describes the depth of penetration of the dopant. Knowing the surface concentration from the solid solubility and the diffusivity and diffusion time, we can calculate the impurity distribution from Eq. (1.3.10). More important, for the deposition cycle we can find the total density N' of dopant atoms per unit surface area introduced by the diffusion. This is calculated by integrating $C(x,t)$ over all x to find

$$N' = \int_0^\infty C(x,t)\, dx = 2\sqrt{Dt/\pi}\; C_s \qquad (1.3.11)$$

for the complementary-error-function distribution. Thus, from Eq. (1.3.11), the added impurities increase as the square root of the diffusion time t.

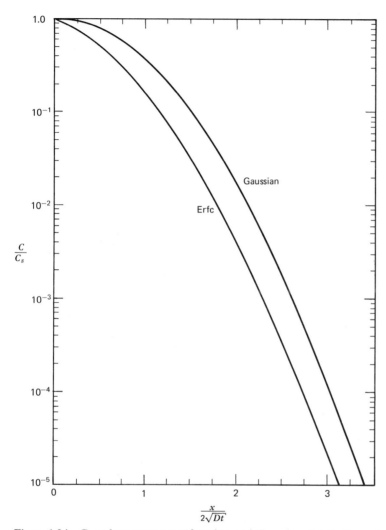

Figure 1.24 Complementary-error-function and Gaussian distributions; the vertical axis is normalized to the surface concentration C_s while the horizontal axis is normalized to the characteristic diffusion length $2\sqrt{Dt}$.

After the deposition is completed, the wafer is placed into another furnace for the drive-in diffusion. Since this drive-in diffusion generally produces an impurity distribution much deeper than that resulting from the deposition step, we often approximate the distribution at the beginning of the drive-in diffusion by a sheet of dopant at the semiconductor surface with a total concentration N' per unit area determined by the deposition cycle. We then assume that the drive-in diffusion simply redistributes this fixed amount of dopant impurity. With this boundary condition the solution of the diffusion equation is a Gaussian distribution

$$C(x,t) = \frac{N'}{\sqrt{\pi Dt}} \exp\left(-\frac{x^2}{4Dt}\right) \tag{1.3.12}$$

where the characteristic diffusion length $2\sqrt{Dt}$ is now determined by the time t and temperature of the drive-in diffusion. The dependence on temperature arises because of the variation in D with temperature. The Gaussian distribution is also shown in Fig. 1.24.

Although gaseous deposition is the most common method for introducing dopant atoms near the surface, another technique is sometimes used. A layer of oxide containing the dopant impurity is deposited on the surface of the wafer, and the dopant atoms are diffused from this glassy layer into the wafer. One convenient method of forming this layer is similar to the chemical vapor deposition of silicon dioxide discussed previously. In this case, however, diborane B_2H_6 or phosphine PH_3 is added to the silane so that the deposited oxide contains a significant amount of boron or phosphorus, which can diffuse into the exposed regions of the silicon wafer during a subsequent drive-in diffusion. When the source of dopant atoms is a doped oxide, the concentration of dopant atoms at the semiconductor surface is generally a fixed fraction of the concentration of dopant in the oxide during the entire drive-in diffusion. Under these conditions solution of the diffusion equation indicates that the impurity profile is a complementary-error-function distribution after the drive-in diffusion.

The two distributions—Gaussian and complementary error function—are the two most frequently predicted by the simple diffusion equation considered above [see Eq. (1.3.9)]. As we have seen, a constant surface concentration leads to a complementary-error-function distribution while the redistribution of a constant number of impurities per unit area causes a Gaussian distribution. More detailed treatments are often necessary—for example, in the case of simultaneous diffusion and oxidation or at high impurity concentrations—but the simplified treatment considered here will allow us to understand the essential characteristics and behavior of devices fabricated by planar technology.

Interconnections and Assembly. Thus far in our discussion of integrated-circuit technology we have considered the basic oxidation, photoresist, and diffusion steps that are combined and repeated to fabricate the desired *p*-type and *n*-type regions comprising a semiconductor device or integrated circuit. After the basic semiconductor structure has been formed by doping the silicon, it is still necessary to provide an electrical path between the various doped regions. The simplest method for accomplishing this is illustrated in Fig. 1.25. First the protective silicon dioxide is removed from areas where we wish to make contact to the semiconductor and a layer of metal is deposited over the surface by evaporation from a hot source in a vacuum. The metal, usually aluminum or an aluminum alloy, is then removed, where it is not desired, by photoresist-masking and etching operations similar to those discussed before.

The wafer is subsequently placed in a low temperature furnace (about 500°C) to alloy the metal to the silicon and to ensure good electrical contact. The silicon wafer is then "diced" into separate pieces each containing one or more semiconductor devices. These chips are soldered to a package, and wires are connected from the leads on the package to the metal on the semiconductor chip (Fig. 1.26). Finally, the package is sealed with a protective ceramic or metal cover or with plastic.

Epitaxy and Polycrystalline Silicon. Although the topics discussed above include the basic components of the planar process, there are two techniques for depositing silicon that are important in integrated-circuit technology. These are *epitaxy* and the deposition of *polycrystalline silicon*.

We have seen that a high concentration of impurities of the same or opposite conductivity type can be added by diffusion to a lightly doped wafer. If, however, a structure that contains a lightly doped layer above a heavily doped region is desired, it cannot be reliably obtained by diffusion techniques. In theory, one can decrease the net carrier concentration by adding an approximately equal concentration of impurities of the opposite conductivity type. However, limited control of the accuracy of the diffusion process generally makes such an approach impractical. In addition, the carrier mobility in such a structure would be degraded since mobility is limited by the total number of ionized impurities $N_d + N_a$ rather than by the net number of impurities $|N_d - N_a|$, which determines the carrier concentration.

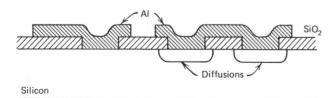

Figure 1.25 A thin layer of aluminum may be used to connect various doped regions of a semiconductor device.

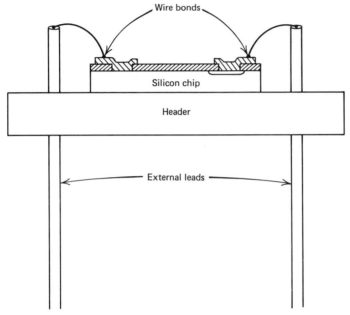

Figure 1.26   The semiconductor chip is mounted on a package, and wires are connected from the leads on the package to the metal on the semiconductor chip.

We can fabricate the desired lightly doped layer above a heavily doped region by the process of *epitaxy* (Fig. 1.27*a*), which is the controlled growth of single-crystal silicon on the host silicon crystal. To deposit an epitaxial layer of silicon on top of a silicon wafer, the wafer is placed in a heated chamber where a gas such as silane SiH_4 or silicon tetrachloride $SiCl_4$ is made to pass over the hot wafer as shown in Fig. 1.27*b*. The gas decomposes on the surface of the wafer leaving a thin layer of solid silicon. If the wafer is heated to a sufficiently high temperature, the deposited silicon atoms may migrate on the silicon surface before they are buried by subsequently arriving atoms. When they reach energetically favorable sites, they are arranged so that the crystal structure of the substrate silicon wafer is continued into the deposited layer. This lightly doped, high-quality, single-crystal layer is of suitable perfection for device fabrication.

To build integrated circuits, it is essential to be able to make conducting layers that are electrically isolated from the substrate wafer. Although aluminum is often used for conducting layers, its low melting point precludes any additional diffusions into the silicon wafer once it is deposited. If, however, a thin layer of silicon is used, further high-temperature processing may be carried out. Silicon is often deposited for this purpose by a similar method to that used for epitaxy. However, since the silicon is not deposited on a silicon crystal but rather on amorphous silicon oxide, it will not be a single crystal. It is usually composed of many small crystallites, and is referred to as *polycrystalline silicon*. Although the electrical

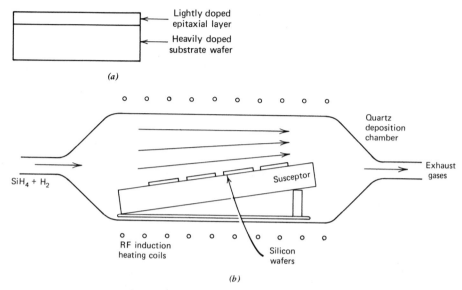

(a)

(b)

Figure 1.27 (*a*) Lightly doped epitaxial layer grown on heavily doped silicon substrate wafer. (*b*) Epitaxial deposition system showing silicon wafers on susceptor which is heated by induction from the *RF* heating coils located outside the quartz deposition chamber.

properties of polycrystalline silicon are markedly inferior to those of single-crystal silicon, it is of great value as a conductor or an electrode in an integrated circuit. In Chapter 8 we discuss the use of polycrystalline silicon as one electrode of a type of transistor called an insulated-gate, field-effect transistor.

Ion Implantation. The technique of *ion implantation* is sometimes used as an alternative to diffusion in order to introduce dopant atoms into selected regions of a semiconductor device. The desired dopant atom is first ionized and then accelerated by an electric field to a high energy (typically from 20 to 200 keV). A beam of these high-energy ions strikes the semiconductor surface (Fig. 1.28) and penetrates exposed regions of the wafer. The penetration is typically less than a micron below

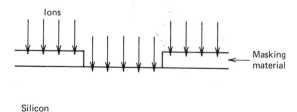

Figure 1.28 In ion implantation, a beam of high-energy ions strikes selected regions of the semiconductor surface, penetrating into these exposed regions.

the surface, and considerable damage is done to the crystal during an implant. Consequently, a subsequent annealing treatment is necessary to restore the lattice quality and to ensure that the implanted dopant atoms are located on substitutional sites where they will act as donors or acceptors. After the ions are implanted, they may be redistributed by a subsequent diffusion if desired.

Various materials, such as photoresist or metal, can be used to prevent ions from entering selected regions of the wafer just as oxide is used to prevent dopant impurities from entering portions of a wafer during a gaseous deposition. Unlike diffusion, however, ion implantation allows precise control of the number of dopant atoms that enter the wafer. Since the dopant atoms strike the wafer as electrically charged ions, the charges may be counted by a relatively simple apparatus during the implantation, and the implantation may be terminated when the desired quantity has been implanted. Besides control of the total number of dopant atoms, ion implantation also offers higher purity of dopant species than does gaseous diffusion. This is because conventional ion implanters make use of mass spectrometers at the source to sort ionic species and allow only the desired dopant in the implant. Ion implantation is, therefore, especially valuable as a replacement for conventional gaseous deposition. In Chapter 8 we discuss the use of ion implantation for very precisely controlled modifications of an electron device.

We have briefly discussed the basic structure used in silicon integrated circuits. Although many variations of the original planar structure are now used in practice, virtually all of them rely on the ability of silicon dioxide or other materials to limit the areas into which impurities enter. In our subsequent discussion we investigate various devices that can be fabricated in silicon integrated circuits and see what advantages and limitations are imposed on device operation by the planar structure.

1.4 DEVICE: INTEGRATED-CIRCUIT RESISTOR

The first specific device for integrated circuits that we consider is an integrated-circuit (IC) resistor. In the introductory section we noted that the resistance of a bar of uniform material is given by the relation

$$R = \frac{\rho L}{A} \tag{1.4.1}$$

where the resistivity ρ is the reciprocal of the conductivity as obtained in Eq. (1.2.7).

$$\sigma = \frac{1}{\rho} = (q\mu_n n + q\mu_p p) \tag{1.4.2}$$

The simplest method for forming an integrated-circuit resistor is to define an opening in a protective silicon-dioxide layer above a uniformly doped semiconductor wafer and to diffuse impurities of the opposite conductivity type into the wafer as shown in Fig. 1.29. In Chapter 3 we see that the junction between two regions of opposite conductivity type presents a barrier to current flow. Therefore, if contacts are made near the two ends of the diffused region and a voltage is applied, a current will flow along the diffused area parallel to the surface.

We have already described the way in which the impurity concentration resulting from a diffusion decreases from the surface into the bulk (Fig. 1.24). Therefore, the resistivity of the diffused region increases as one moves away from the surface into the semiconductor, and the simple relation given by Eq. (1.4.1) for a uniform bar has to be modified.

Conductance

Consider an n-type silicon wafer that has been processed with a p-type diffusion as in Fig. 1.29. The differential conductance dG of a thin layer of the p-type material

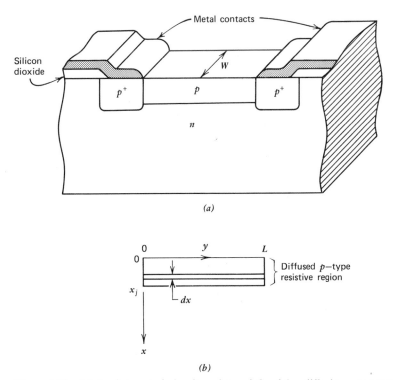

Figure 1.29 (*a*) An integrated-circuit resistor defined by diffusing acceptors into selected regions of an n-type wafer. The p^+ regions are highly doped to assure good contact between the metal electrodes and the resistive p-type region. (*b*) The dimensions of a thin region in the resistor having conductance dG as given by Eq. (1.4.3).

of thickness dx parallel to the surface and at a depth x (shown in Fig. 1.29b) is

$$dG(x) = q\mu_p p(x) \frac{W}{L} dx \tag{1.4.3}$$

We can find the conductance G of the entire diffused region by summing the conductance of each thin slab from the surface down to the bottom of the diffused layer. This sum becomes an integral in the limit of many thin slabs.

$$G = \frac{W}{L} \int_0^{x_j} q\mu_p p(x) \, dx \tag{1.4.4}$$

where x_j is the depth at which the hole concentration becomes negligible (close to the point where $N_a = N_d$).

If the p-region has been formed in a gaseous deposition cycle followed by a drive-in diffusion, we can approximate the impurity profile $N_a(x)$ by a Gaussian distribution [Eq. (1.3.12)]. As shown in Section 1.3, the total density of dopant N' per unit area diffused into the wafer is determined by the nature of the deposition cycle while the characteristic length of the diffused region $2\sqrt{Dt}$ is associated with the drive-in diffusion. The hole concentration at any depth into the wafer is approximately given by the net concentration of p-type dopant atoms in excess of the background n-type dopant concentration of the original starting wafer. In practice most of the current in the diffused area is carried in the regions with the highest dopant concentration, which is usually of the order of 10^{18} cm^{-3} or greater. Since the starting wafer often has a dopant concentration of about 10^{15} cm^{-3}, we can neglect the background concentration and assume that $p(x) = N_a(x)$.* Substituting Eq. (1.3.12) into Eq. (1.4.4), we have

$$G = \frac{N'q}{\sqrt{\pi Dt}} \frac{W}{L} \int_0^{x_j} \mu_p \left[\exp\left(\frac{-x^2}{4Dt} \right) \right] dx \tag{1.4.5}$$

The mobility μ_p in Eq. (1.4.5) is a function of the hole concentration (Fig. 1.12) and, therefore, a function of the distance from the surface. Consequently, we cannot remove it from the integrand.

The most straightforward and accurate means of evaluating Eq. (1.4.5) is numerical integration using a computer. An alternative technique, which is often valuable, is to approximate various regions of the curve of Fig. 1.12 by analytical relationships, which may then be used in Eq. (1.4.5). Both of these techniques produce results of high accuracy within the limits of the original assumptions. However, in practice there are rather severe deviations in the diffused profile from the Gaussian distribution predicted by simple diffusion theory, and accurate numerical

---

* We have assumed here that the holes and the acceptor atoms have the same depth distribution. In Section 3.1 we show that this assumption is slightly inaccurate. In the present case, however, the small difference between N_a and p may be ignored.

results are generally warranted only when the nature of the diffusion profile has been investigated experimentally for the particular dopant under consideration. We can obtain an approximate value of the conductance by using an average value $\bar{\mu}_p$ for the mobility. Since much of the current is carried in a region with a dopant concentration close to the maximum, selecting a value of mobility corresponding to perhaps half the maximum dopant concentration is reasonable and consistent with the use of simplified diffusion theory. The expression for the conductance [Eq. (1.4.5)] then reduces to

$$G = N'q\bar{\mu}_p \frac{W}{L} = g\frac{W}{L} \tag{1.4.6}$$

where $g \equiv N'q\bar{\mu}_p$ is the conductance of a square ($L = W$) resistor pattern. The conductance, in turn, is determined by the product of the average mobility $\bar{\mu}_p$ and the total dopant density per unit surface area N' [Eq. (1.3.11)]. The resistance R is therefore

$$R = \frac{1}{G} = \frac{L}{W}\frac{1}{g} \tag{1.4.7}$$

Generally many resistors in an integrated circuit are fabricated simultaneously by defining different geometric patterns on the same photomask. Since the same diffusion cycle is used for all of these resistors, it is convenient to separate the magnitude of the resistance into two parts: the ratio L/W is determined by the photomask dimensions while $1/g$ is determined by the diffusion process.

Sheet Resistance

Any resistor pattern on the photomask can be divided into squares of dimension W on each side (Fig. 1.30). The number of squares in any pattern is just equal to the ratio L/W. The value of a resistor is thus equal to the product of the number of squares into which it can be divided and the parameter $1/g$, which is usually denoted by the symbol $R_\square$, called the *sheet resistance*. The sheet resistance has units of ohms, but it is conventionally specified in units of ohms per square ($\Omega/\square$) to stress that the value of a resistor is given by the product of the number of squares times the sheet resistance. For example, a resistor 100 μm long and 5 μm wide contains

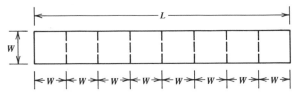

Figure 1.30 The number of squares describing the surface dimensions of a resistor is given by the ratio L/W.

20 squares (20 □). If the diffusion process used produces a diffused layer with a sheet resistance of $200 \, \Omega/\square$, the value of the resistor is $20 \, \square \times 200 \, \Omega/\square = 4.0 \, k\Omega$.

Attainable values of sheet resistance are such that resistors in the kilohm range and higher require patterns containing many squares. Since the width of the pattern is determined by the ability to photomask and etch very narrow lines, the length of the resistor may become very great in order to obtain the required number of squares. The large area needed for high-value resistors is a practical limitation in an integrated circuit, and circuits are usually designed to avoid large-value resistors. When a resistor containing a large number of squares must be used, it is generally designed to minimize area by using a serpentine pattern as shown in Fig. 1.31. The current flow across the corner square of such a pattern is not uniform. The resistance contributed by a corner square can be estimated by taking it to be $\sim 65\%$ of the straight-path value.

For large-area resistors, the dimensions L and W are simply determined by the mask dimensions. However, for very narrow resistors (small W), the effective value of W may differ substantially from the mask dimension because of the nature of the diffusion process. When impurities are diffused into an oxide window of width W (Fig. 1.32), they diffuse laterally under the oxide as well as vertically. If W is much greater than the diffusion depth, we may neglect this effect. However, the width is often made as small as photomasking tolerances allow so that a given number of squares fit into the smallest area. In this case the diffusion depth may become an appreciable fraction of W, and an effective value W' must be used for the width of the resistor to account for the lateral diffusion.

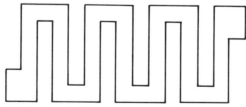

Figure 1.31   *A* serpentine pattern may be used when a long, high-valued resistor must be designed.

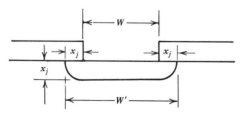

Figure 1.32   Lateral diffusion changes the dimensions of the resistor from the nominal mask dimensions.

Figure 1.33 is a microphotograph of an integrated circuit (Type 741 operational amplifier) in which several diffused-resistor patterns can be identified. The resistance of the pattern indicated by the arrow is nominally 4 kΩ.

Limitations. One additional comment should be made before leaving the subject of diffused resistors. The accuracy of the resistor value is determined by both the accuracy of the pattern definition in the oxide and the control of the diffusion.

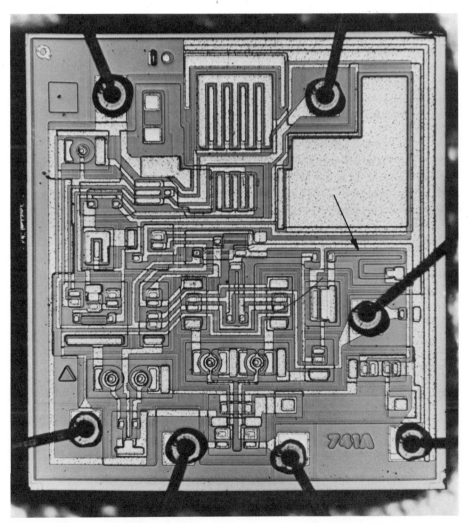

Figure 1.33 Microphotograph of an integrated circuit that makes use of several diffused resistor patterns. A 4 kΩ resistor is indicated by the arrow. (Courtesy of Signetics Corporation.)

Although diffusion processes often cannot be controlled as closely as desired, the pattern definition can be quite accurate if the dimensions are large compared with the limits of the photomasking capability available. The absolute tolerance on the value of a resistor is, therefore, generally limited by the diffusion process. In some critical applications that require well-controlled resistance values, resistors are now produced by ion implantation.

Although the exact value of sheet resistance obtained from a diffusion will vary somewhat in typical processing, it will still be relatively constant over an area of the wafer that is large compared to the size of an integrated circuit. Thus, two nearby resistors can be expected to have the same value of sheet resistance so that the resistance ratio between them will be determined by their relative dimensions. Because these dimensions can be controlled with high precision, significantly greater accuracy can be maintained in the ratios of paired resistors than in their specific resistance values. For this reason, integrated circuits are often designed so that their critical behavior depends on the ratio of two resistors rather than on the absolute value of either resistor.

Diffused resistors are commonly used in silicon integrated circuits because of their compatibility with the remainder of the planar process. They are usually formed at the same time as other circuit elements and, therefore, do not add to the fabrication cost. It is also possible to make resistors in integrated circuits by forming patterns in the *n*-epitaxial material with reverse-biased *pn* junctions. Since the epitaxial material has the highest resistivity of the silicon used to form the circuit, one can obtain sheet resistance values by this method that are roughly five or six times larger ($\sim 1000 \ \Omega/\square$) than those normally attainable in *p*-type diffused resistors ($\sim 200 \ \Omega/\square$) made in integrated circuits. For even higher values of sheet resistance, a double-diffused structure is sometimes used in which an *n*-region is formed over the *p*-region to reduce the vertical dimension of the diffused resistor. Resistors of this type, known as *pinch-resistors*, can increase the sheet resistance by a factor of roughly 40 or 50. It is clearly more complicated to make a pinch-resistor, however, and reproducibility of the sheet resistance is generally not very good.

If a particular circuit requires very large or very accurate values of resistance, alternative techniques may be needed. Instead of forming a resistor by diffusing impurities into the wafer, a resistive film can be deposited on top of the insulating silicon dioxide, which covers most of the circuit. This film is then defined by photomasking and connected to the rest of the circuit by the aluminum interconnections. Although the use of deposited resistors provides flexibility in the design of circuits, the additional processing steps required for their fabrication add to the cost of the circuit. Deposited resistors are, therefore, avoided whenever possible.

In this section we started our discussion of semiconductor devices for integrated circuits by considering diffused resistors. We have applied some of the physical concepts developed earlier in this chapter to investigate the behavior of the device and have seen that particular constraints imposed by technology influence circuit design. We shall see this pattern repeated as we consider other *IC* devices in subsequent chapters.

SUMMARY

One of the cornerstones of solid-state electronics is the band structure of solids. This key concept is related to the quantized energy levels of isolated atoms. Huge differences in the electrical conductivities of metals, semiconductors, and insulators result from basic differences in the band structures of these three classes of materials.

By considering the band structure of semiconductors, we can deduce the existence of two types of current carriers, *holes* and *electrons*. In most cases of practical interest, it is valid to consider holes and electrons as classical free particles inside the semiconductor crystal. The motions of the holes and electrons and the statistical distributions that characterize their energies can be calculated if their effective masses are modified from those of truly free particles. The populations of holes and electrons can be controlled by adding impurities to otherwise pure semiconductor crystals. In this process, known as *doping*, extra holes can be added to single-crystal silicon, for example, by introducing ions with valence three, which are incorporated in the lattice in place of silicon, which has valence four. The resultant crystal is known as *p*-type silicon. To add electrons and thereby obtain *n*-type silicon by doping, impurities with valence five can be substituted for silicon atoms in the lattice.

An understanding of many electronic properties can be gained by applying the laws of statistical physics to the populations of electrons and holes in semiconductors. The *Fermi energy* is an important parameter to describe the populations of free carriers in a crystal at thermal equilibrium. Under usual conditions, the Fermi energy in a semiconductor exists within a *forbidden gap* of energies that is ideally free of any allowed energy states for electrons. Semiconductors in this condition are described as *nondegenerate*. If the Fermi energy exists within a populated band of allowed electronic states, the semiconductor is called *degenerate* because its properties degenerate in the direction of those of a metal.

When free carriers in a semiconductor are subjected to a low or moderate electric field, they move with a constant *drift* velocity. The velocity per unit field is known as the *mobility* and given the symbol μ. The magnitude of the mobility is determined by the nature of the scattering events that are experienced by free carriers as they interact with the lattice. The thermal energies of free carriers in solids cause them to randomize their distribution. The net carrier motion resulting from this process is known as *diffusion*. When the crystal is at a uniform temperature, diffusion occurs from regions where the carriers are more dense to regions where their density is lower. The characteristic parameter for motion by diffusion is called the *diffusion constant*, which is usually denoted by the symbol D. In nondegenerate semiconductors, the diffusion constant D for a given species of carrier is related to the mobility μ of the same species by the *Einstein relationship*: $D/\mu = kT/q$.

From an engineering point of view, there is almost equal importance between the basic properties of a material, on the one hand, and the technological possibilities of that material, on the other. The extraordinary importance of silicon for device applications is seen to result both from excellent material properties and from a very favorable technology; a technology that permits the compatible fabrication of a high-quality semiconductor and a remarkably excellent insulator. The *planar process* for silicon is the basis for the accurate delineation of small-geometry devices. It makes possible the batch fabrication of many devices and is the key to the production of integrated circuits.

Diffused resistors are widely used in integrated circuits. They are typically fabricated by employing the planar process to delineate a pattern in which currents are carried by drifting carriers. Resistance values for diffused resistors are usually calculated from a parameter called the *sheet resistance* ($\Omega/\square$). This parameter is determined by the *IC* processing and is given the symbol $R_\square$. It represents the resistance measured between the opposite edges of a square pattern.

REFERENCES

1. R. B. Adler, A. C. Smith, and R. L. Longini, *SEEC VOL. I: Introduction to Semiconductor Physics*, Wiley, New York, 1964.
2. (a) F. Morin and J. P. Maita, *Phys. Rev.*, *96*, 28, (1954).
   (b) A. S. Grove, *Physics and Technology of Semiconductor Devices*, Wiley, New York, 1967.
3. P. P. Debye and E. M. Conwell, *Phys. Rev.*, *93*, 693 (1954).
4. S. M. Sze and J. C. Irvin, *Solid-State Electron*, *11*, 599 (1968).
5. Reference 1, p. 34.
6. C. Canali, G. Majni, and A. Albergi Quaranta, *J. Phys. Chem. Solids*, *32*, 1707 (1971).
7. B. E. Deal and A. S. Grove, *J. Appl. Phys.*, *36*, 3770 (1965).
8. *Integrated Silicon Device Technology, Vol. IV: Diffusion*, Research Triangle Institute, Durham, North Carolina, 1964 (ASD-TDR-63-316). Reprinted by permission of the publisher.
9. F. A. Trumbore, *Bell Syst. Tech. J.*, *39*, 205 (1960). Reprinted by permission of the publisher.
10. J. L. Moll, *Physics of Semiconductors*, McGraw-Hill, New York, 1964, p. 99.
11. S. Wang, *Solid-State Electronics*, McGraw-Hill, New York, 1966, pp. 263–269.
12. C. Jacoboni et al, *Solid-State Electron*, *20*, 77 (1977).

PROBLEMS

1.1 Phosphorus donor atoms with a concentration of 10^{16} cm^{-3} are added to a pure sample of silicon. Assume that the phosphorus atoms are distributed homogeneously throughout the silicon. The atomic weight of phosphorus is 31.
(a) What is the sample resistivity at 300°K?
(b) What proportion by weight does the donor impurity comprise?

(c) If 10^{17} atoms cm^{-3} of boron are included in addition to the phosphorus, and distributed uniformly, what is the resultant resistivity and conductivity type (i.e., p or n-type material)?

(d) Sketch the energy-band diagram under the condition of (c) and show the position of the Fermi level.

1.2 Find the equilibrium electron and hole concentrations and the location of the Fermi level for silicon at 27°C if the silicon contains the following concentrations of shallow dopant atoms:

(a) 1×10^{16} cm^{-3} boron atoms

(b) 3×10^{16} cm^{-3} arsenic atoms and 2.9×10^{16} cm^{-3} boron atoms

1.3 An n-type sample of silicon has a uniform density $N_d = 10^{16}$ atoms cm^{-3} of arsenic, and a p-type silicon sample has $N_a = 10^{15}$ atoms cm^{-3} of boron. For each semiconductor material determine the following:

(a) The temperature at which half the impurity atoms are ionized. Assume that all mobile electrons and holes come from dopant impurities. [*Hint.* Use Eqs. (1.1.21) and (1.1.22) together with the fact that $f_D(E_f) = \frac{1}{2}$]

(b) The temperature at which the intrinsic concentration n_i exceeds the impurity density by a factor of 10. See Table 1.2 for $n_i(T)$.

(c) The equilibrium minority-carrier concentrations at 300°K. Assume full ionization of impurities.

(d) The Fermi level referred to the valence-band edge E_v in each material at 300°K. The Fermi level if both types of impurities are present in a single sample.

1.4 A piece of n-type silicon has a resistivity of 5 Ω-cm at 27°C. Find the thermal-equilibrium hole concentrations at 27, 100, and 500°C. (Refer to the tables and figures for needed quantities.)

1.5[†] Grain boundaries and other structural defects introduce allowed energy states deep within the forbidden gap of polycrystalline silicon. Assume that each defect introduces two discrete levels: an acceptor level 0.51 eV above the top of the valence band and a donor level 0.27 eV above the top of the valence band. (Note that $E_a > E_d$ and that these are *not* shallow levels.) The ratio of the number of defects with each charge state ($+$, $-$, or neutral) at thermal equilibrium is given by the relation (see reference 10)

$$N_d^+ : N_o : N_a^- = \exp \frac{E_d - E_f}{kT} : 1 : \exp \frac{E_f - E_a}{kT}$$

(a) Sketch the densities of these three species as the Fermi level moves from E_v to E_c. Which species dominates in heavily doped p-type material? In n-type material?

(b) What is the effect of the defects on the majority-carrier concentrations?

(c) Using the above information, determine the charge state of the defect levels and the position of the Fermi level in a silicon crystal containing no shallow dopant atoms. Is the sample p- or n-type?

(d) What are the electron and hole concentrations and the location of the Fermi level in a sample with 2×10^{17} cm^{-3} phosphorus atoms and 5×10^{16} cm^{-3} defects?

1.6 Two scattering mechanisms are important in a piece of semiconductor. If only the first scattering process were present, the mobility would be 800 cm^2(Vs)$^{-1}$. If only the second were present, the mobility would be 200 $cm^2/(Vs)^{-1}$. What is the mobility considering both scattering processes?

1.7 Find the mobility of electrons in aluminum with resistivity 2.8×10^{-6} Ω-cm and density 2.7 g cm^{-3}. The atomic weight of aluminum is 27. Of the three valence electrons in aluminum, on the average 0.9 electrons are free to participate in conduction at room temperature. If $m^* = m_0$, find the mean time between collisions and compare this value to the corresponding value in lightly doped silicon.

1.8 An electron is moving in a piece of lightly doped silicon under an applied field at 27°C so that its drift velocity is one-tenth of its thermal velocity. Calculate the average number of collisions it will experience in traversing by drift a region 1 μm wide. What is the voltage applied across this region?

1.9 In an experiment the voltage across a uniform, 2 μm long region of 1 Ω-cm, n-type silicon is doubled, but the current only increases by 50%. Explain. (The silicon remains neutral at both current levels.)

1.10 The electron concentration in a piece of uniform, lightly-doped, n-type silicon at room temperature varies linearly from 10^{17} cm^{-3} at $x = 0$ to 6×10^{16} cm^{-3} at $x = 2$ μm. Electrons are supplied to keep this concentration constant with time. Calculate the electron current density in the silicon if no electric field is present.

1.11[†] Silicon atoms are added to a piece of gallium arsenide. The silicon can replace either trivalent gallium or pentavalent arsenic atoms. Assume that silicon atoms act as fully ionized dopant atoms and that 5% of the 10^{10} cm^{-3} silicon atoms added replace gallium atoms and 95% replace arsenic atoms. The sample temperature is 300°K.
   (a) Calculate the donor and acceptor concentrations.
   (b) Find the electron and hole concentrations and the location of the Fermi level.
   (c) Find the conductivity of the gallium arsenide assuming that lattice scattering is dominant.
   See Table 1.1 for properties of GaAs.

1.12[†] (Dielectric relaxation in solids.) Consider a homogeneous one-carrier conductor of conductivity σ and permittivity ϵ. Imagine a given distribution of the mobile charge density $\rho(x,y,z; t = 0)$ in space at $t = 0$. We know the following facts from electromagnetism, provided we neglect diffusion current:

$$\nabla \cdot D = \rho; \quad D = \epsilon \mathscr{E}; \quad J = \sigma \mathscr{E}; \quad \nabla \cdot J = \frac{-d\rho}{dt}$$

   (a) Show from these facts that $\rho(x,y,z; t) = \rho(x,y,z; t = 0)\, e^{-t/(\epsilon/\sigma)}$. This result shows that uncompensated charge cannot remain in a uniform conducting material, but must accumulate at discontinuous surfaces or other places of nonuniformity.
   (b) Compute the value of the *dielectric relaxation time* ϵ/σ for intrinsic silicon; for silicon doped with 10^{16} donors cm^{-3}; and for thermal SiO_2 with $\sigma = 10^{-16}$ $(\Omega$-cm$)^{-1}$.[1]

1.13[†] Because of their thermal energies, free carriers are continually moving throughout a crystal lattice. While the net flow of all carriers across any plane is zero at thermal equilibrium, it is useful to consider the directed components that balance to a null. The component values are physically significant in that they measure the quantity of current that can be delivered by diffusion alone. This would be relevant if, for example, one were able to unbalance the thermal equilibrium condition by intercepting all carriers flowing in a given direction.

(a) By considering that $J_x = -qn_0v_x$, show that the current in a solid in any random direction resulting from thermal processes is

$$J = \frac{-qn_0v_{th}}{4}$$

where v_{th} is the mean thermal velocity and n_0 is the free-electron density. (*Hint.* Consider the flux through a solid angle of 2π steradians.)[1]

1.14 Calculate the wavelengths of radiation needed to create hole-electron pairs in intrinsic: germanium, silicon, gallium arsenide, and SiO_2. Identify the spectrum range (e.g., infrared, visible, UV, and X ray) for each case.

1.15[†] The relation between D and μ is given by

$$\frac{D}{\mu} = \frac{1}{q}\frac{dE_f}{d(\ell n\ n)}$$

for a material that may be *degenerate*. (That is, the Fermi-Dirac distribution function must be used since the Fermi level may enter an allowed energy band.) Show that this relation reduces to the simpler Einstein relation $D/\mu = kT/q$ if the material is *non-degenerate* so that Boltzmann statistics can be used.

1.16 A hot probe setup is a useful laboratory apparatus. It is used to determine the conductivity type of a semiconductor sample. It consists of two probes and an ammeter that indicates the direction of current flow. One of the probes is heated (most simply, a soldering iron tip is used), and the other is at room temperature. No voltage is applied, but a current flows when the probes touch the semiconductor. Considering the role of diffusion currents, explain the operation of the apparatus and draw diagrams to indicate the current directions for p- and n-type semiconductor samples, respectively.

1.17[†] A silicon wafer is covered with a 0.2 μm thick oxide layer. What is the additional time required to grow 0.1 μm more oxide in dry oxygen at 1200°C?

1.18[†] Boron atoms in an area concentration of 10^{15} cm^{-2} are introduced into an n-type wafer from a BCl_3 source in a carrier gas. The starting wafer has a uniform donor concentration of 5×10^{15} cm^{-3}. The subsequent drive-in diffusion is carried out at 1100°C in a nitrogen ambient. The desired junction depth is 2 μm.

(a) How long should the drive-in diffusion be continued?

(b) Sketch the resulting boron distribution on log N_a and linear N_a versus x plots.

(c) It is required to lay out a 500 Ω resistor. The minimum mask dimension is 4 μm. What is the length of the resistor as indicated on the mask? What is the silicon surface area used? An exact, closed form solution to this part of the problem is not possible, but the sketches of part (b) should indicate the method of solution. State your approximations.

(d) Qualitatively describe the effect of doubling the time of the drive-in diffusion. Note that the net number of acceptor impurities $N_a - N_d$ can contribute to the conduction. Also note that the mobility decreases as the impurity concentration increases above about 10^{16} cm^{-3} for Si.

1.19 Assume that uniform concentrations of 6.55 μm thickness can be imbedded in silicon of the opposite conductivity type to form an integrated-circuit resistor (Fig. P1.19).

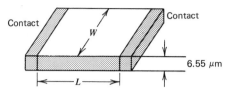

Figure P1.19

Make calculations for two cases: (i) for $N_d = 10^{16}$ cm^{-3} and (ii) for $N_a = 10^{16}$ cm^{-3} to obtain:

(a) The relationships between L and W for resistances of 100 Ω, 1 kΩ, and 10 kΩ between contacts at 25°C.

(b) The actual dimensions if the resistors should dissipate 10 mW of power each and the maximum power dissipation is 1 μW/μm^3.

(c) The temperature coefficients of the resistors (TCR) around 25°C. The TCR for a resistor R is defined by $(1/R)(\partial R/\partial T) \times 100$ (in % per degree).

1.20[†] An acceptor diffusion is carried out, introducing an acceptor profile $N_a = N_s$ erfc (x/λ) with $N_s = 10^{18}$ cm^{-3} and $\lambda = 0.05$ μm into a lightly doped n-type silicon crystal. This diffusion will form a resistor in an integrated circuit.

(a) Show that the resistance across any square pattern on the surface of the resistor is approximately given by:

$$R_\square \approx \left(q\mu \int_0^\infty N_s \, \text{erfc}\left(\frac{x}{\lambda}\right) dx \right)^{-1}$$

State clearly where the approximation is made. This expression can be integrated by parts and the value for $R_\square$ written as:

$$R_\square = \frac{\sqrt{\pi}}{q\mu N_s \lambda}$$

(b) Derive this form.

(c) What is the approximate resistance if the resistor is made 40 μm wide and 2000 μm long?

(d) What is the largest resistance that can be fabricated in a surface area that measures 200 by 70 μm if 10 μm line widths and 10 μm spacings are the minimum dimensions? Sketch the resistor pattern. (Consider each corner square to be 65% effective.)

1.21[†] Consider a simple model for diffusion in one space dimension. Assume that all particles must move only at discrete, regularly spaced times, one time unit apart. When they move, each particle may only jump one space unit either to the left or to the right, with *equal* probability. Start with 1024 particles at $x = 0$, $t = 0$, and build up, step by step, the pattern of particles in space after 10 time units. Plot the distribution in space after each jump and measure the corresponding width W between points of one-half maximum concentration. Then plot W^2 versus time. What does your result suggest about the rate of "spread" of particles by the diffusion process (in *one* space dimension at least)?[1]

CHAPTER 2

METAL-SEMICONDUCTOR CONTACTS

Most of the electronic devices that make up an integrated circuit are connected by means of metal-semiconductor contacts. Moreover, all integrated circuits communicate with the rest of an electrical system via metal-semiconductor contacts. As we shall see, the properties of these contacts can vary considerably, and it is necessary to consider several factors in order to understand them. We shall narrow our discussion whenever necessary to consider only metal contacts to silicon, but first let us consider in general the nature of the thermal equilibrium that is established when a metal and a semiconductor are in intimate contact. The concepts that are developed to understand this equilibrium are very important. They will prove useful many times in our discussion of devices because

65

they underlie the basic properties of semiconductor *pn* junctions as well as the properties of interfaces between semiconductors and insulators and between metals and insulators.

Application of the equilibrium principles to metals and semiconductors provides a simple theory (Schottky theory) of ohmic and rectifying behavior in various metal-semiconductor systems. The Schottky theory is, however, not adequate in many cases: most notably not for metal-silicon systems without further considerations of the real nature of solid interfaces. The important influence of surface states and the origins of these states are therefore discussed. Several applications are then described, but special emphasis is given to the Schottky diode clamp that has been so widely used in fast logic circuits.

2.1 EQUILIBRIUM IN ELECTRONIC SYSTEMS

Metal-Semiconductor System

We shall find it very profitable to construct an energy-band representation for a metal-semiconductor contact. To do so we employ at first a very general viewpoint. We consider the metal and the semiconductor as representing two systems of allowed electronic energy states. Using the concepts developed in Chapter 1, we recognize that these systems of allowed states can be regarded as being almost entirely populated at energies less than the Fermi energy, and nearly vacant at higher energies. When the metal and the semiconductor are remote from one another and therefore do not interact, both the systems of electronic states and their Fermi levels are independent. Let us designate the metal as state-system number 1 and the semiconductor as state-system number 2. Each system has a density of allowed states $g(E)$ per unit energy. Of these $g(E)$ states, $n(E)$ are full and $v(E)$ are empty. (The variables are all functions of energy; hence, the use of E in parentheses.) Our terminology is indicated in Table 2.1 and Fig. 2.1. The Fermi–Dirac distribution functions

$$f_{D1,2} = \frac{1}{1 + \exp\left[(E - E_{f1,2})/kT\right]} \tag{2.1.1}$$

Table 2.1 **Designation Scheme**

Material	Allowed Electron State Density	Filled State Density	Vacant State Density	Fermi–Dirac Distribution Function
1 (Metal)	$g_1(E)$	$n_1(E)$	$v_1(E)$	$f_{D1}(E, E_{f1})$
2 (Semiconductor)	$g_2(E)$	$n_2(E)$	$v_2(E)$	$f_{D2}(E, E_{f2})$

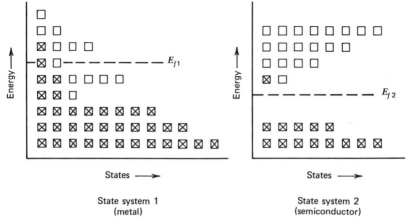

Figure 2.1 Allowed electronic-energy-state systems for two isolated materials. States marked with an x are filled; those unmarked are empty. System 1 is a very qualitative representation for a metal; system 2 qualitatively represents a semiconductor.

allow us to relate $v_{1,2}(E)$, $n_{1,2}(E)$, and $g_{1,2}(E)$.* The filled state density is

$$n_{1,2} = g_{1,2} \times f_{D1,2} \tag{2.1.2}$$

and the vacant state density is

$$v_{1,2} = g_{1,2} \times (1 - f_{D1,2}) \tag{2.1.3}$$

In Eqs. (2.1.1) through (2.1.3) we have omitted the unnecessary parentheses that have served only to emphasize the dependence on energy.

We now consider that the two remote systems are brought into intimate contact. The systems then become interactive and electron transfers take place between them. Equilibrium is reached when there is no net transfer of electrons at any energy. As we noted in Section 1.1, this does not mean a cessation of all processes; rather it means that every process and its inverse are occurring at the same rate. Equilibrium can be expressed mathematically by noting that the transfer probability is proportional to the population of electrons $n(E)$ available for transfer at a given energy and also proportional to the density of available states $v(E)$ to which the electrons may transfer. Because of the Pauli exclusion principle, the available state density is not $g(E)$, the state density, but $v(E)$, the vacant state density. The proportionality factor relating the transfer probability to these two

---

* A more complete discussion of Fermi–Dirac statistics would show that f_D may have slightly altered forms dependent upon specific properties of the allowed electronic states. The alternate forms for f_D typically have factors of 2 or $\frac{1}{2}$ preceding the exponential in Eq. (2.1.1). We shall not be concerned with this refinement.

densities is related to the detailed quantum nature of the states and is the same for transfers from system 1 to system 2 as it is for transfers from system 2 to system 1. Therefore, at thermal equilibrium,

$$n_1 \times v_2 = n_2 \times v_1 \tag{2.1.4}$$

at any given energy.

Using Eqs. (2.1.2) and (2.1.3) in Eq. (2.1.4), we have

$$f_{D1}g_1(1 - f_{D2})g_2 = f_{D2}g_2(1 - f_{D1})g_1$$

or

$$f_{D1}g_1g_2 = f_{D2}g_2g_1 \tag{2.1.5}$$

Equation (2.1.5) can only be true if $f_{D1} = f_{D2}$ or, by Eq. (2.1.1), if $E_{f1} = E_{f2}$. We, therefore, have established the very important property of any two systems in thermal equilibrium: they have the same Fermi energy. Note, from our derivation, that no constraints exist on the density of states functions g_1 or g_2. Regardless of the detailed nature of these functions, a statement that their Fermi levels are equal is equivalent to stating that the two systems are in thermal equilibrium. Generalization of this statement to more than two systems or its reduction to apply to one system of states is straightforward. *At thermal equilibrium the Fermi level is constant throughout a system.*

Inhomogeneously Doped Semiconductor

Before leaving this topic it will be instructive to apply these concepts to one other example. Figure 2.2 is a sketch of the energy-band diagram at thermal equilibrium for an n-type semiconductor doped with N_{d1} donors cm^{-3} for $0 < x < a$ and with N_{d2} donors cm^{-3} for $x > a$. The figure represents what is effectively a single system of states because the dopant concentrations are much less than those of

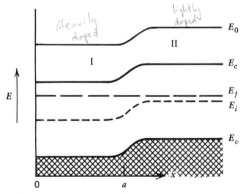

Figure 2.2   Band diagram at thermal equilibrium for an n-type semiconductor having doping N_{d1} for $0 < x < a$ and N_{d2} for $a < x$ ($N_{d1} > N_{d2}$).

the host atoms in the crystal. To sketch the diagram we would first make use of the principle just derived and draw a constant Fermi level. Then, by applying Eq. (1.1.21) to both regions of the crystal together with the knowledge that $n \approx N_{d1}$ for $0 < x < a$ and $n \approx N_{d2}$ for $x > a$, we can locate E_c, the conduction-band edge in each region. As we see in the figure, E_c is not constant but moves higher in energy in the lower-doped region. For a correct representation of the band picture very close to the doping transition at $x = a$, we shall have to develop further theory. A smooth transition will be sketched for the present. Since the band gap is determined by the properties of the silicon lattice, it is a constant of the system. Hence, E_v and E_i are drawn parallel to E_c in Fig. 2.2 and only E_f is constant with position. The increase in the energies associated with the semiconductor band edges properly represents an increased potential energy for electrons in the less heavily doped region. We shall return to consider this and other points about Fig. 2.2 in Chapter 3. For the present our emphasis is on Fermi-level constancy and on the use of this thermal-equilibrium principle to construct an energy-band diagram.

2.2 IDEALIZED METAL-SEMICONDUCTOR JUNCTIONS

Band Diagram

We now turn to the particular features of metal-semiconductor energy-band diagrams. For the purposes of our present discussion, the most distinctive characteristic of the electronic-energy states of the metal and the semiconductor is the relative positions of the Fermi levels within the densities of allowed states $g(E)$. In the metal, the Fermi level is immersed within a continuum of allowed states, while in a semiconductor, under usual circumstances, the density of states is negligible at the Fermi level. Plots of $g(E)$ versus energy for idealized metals and semiconductors are shown in Figs. 2.3a and 2.3b.

These points, which were made in general in our discussion of Fig. 1.3, are apparent in the specific cases of the allowed energy states for gold and those for silicon shown in Figs. 2.4a and 2.4b, respectively. Note that these sketches differ from those in Fig. 2.3 in that they represent allowed energies in the bulk of the material versus position and not the density of allowed energy states versus energy. These two types of sketches are sometimes confused.

New quantities also appear in these figures. First, a convenient reference level for energy is taken, that is, the vacuum or free-electron energy E_0. This represents the energy that an electron would have if it were just free of the influence of the given material. The difference between E_0 and E_f is called the *work function*, usually given the symbol $q\Phi$ in energy units and often listed as Φ in volts for particular materials. In the case of the semiconductor the difference between E_0 and E_f is a function of the dopant concentration of the semiconductor, since E_f

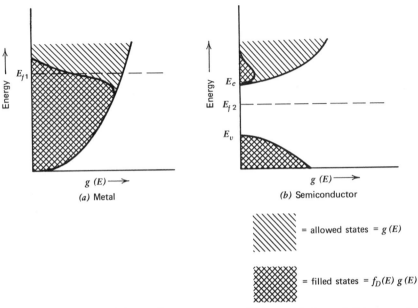

Figure 2.3 (*a*) Allowed electronic-energy states $g(E)$ for an ideal metal. The states indicated by cross-hatching are occupied. Note the Fermi level E_{f1} immersed in the continuum of allowed states. (*b*) Allowed electronic-energy states $g(E)$ for a semiconductor. The Fermi level E_{f2} is at an intermediate energy between that of the conduction-band edge and that of the valence-band edge.

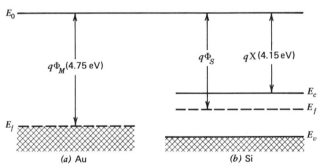

Figure 2.4 Pertinent energy levels for the metal gold and the semiconductor silicon. Only the work function is given for the metal whereas the semiconductor is described by the work function $q\Phi_S$, the electron affinity qX_S, and the band-gap ($E_c - E_v$).

changes position within the gap separating E_v and E_c as the doping is varied. The difference between the vacuum level and the conduction-band edge is, however, a constant of the material. This quantity is called the *electron affinity* and is conventionally denoted by qX in energy units. Tables of X in volts exist for many materials. The symbol X is a Greek capital letter *chi*.

The choice of E_0 as a common energy reference makes clear that if Φ_M is less than Φ_S and the materials are not interacting, the electrons in the metal have, on the average, a total energy that is higher than the average total energy of electrons in the semiconductor. On the contrary, if Φ_M is greater than Φ_S, the average total energy of electrons in the semiconductor is higher than it is in the metal. For the sake of discussion, we consider the latter case where $\Phi_M > \Phi_S$. When an intimate contact is established, the disparity in the average energies can be expected to cause the transfer of electrons from the semiconductor into the metal.

Another way of looking at the establishment of equilibrium for this case where $\Phi_M > \Phi_S$ is to make use of the concepts developed in Section 2.1. The left-hand side of Eq. (2.1.5) in that section is proportional to the flow of electrons from state-system 1 to state-system 2 while the right-hand side of the same equation is proportional to the inverse flow. The net flow is easily seen to be in the direction 2 to 1 if $f_{D2} > f_{D1}$ or, therefore, if $E_{f2} > E_{f1}$. The charge transfer will continue until equilibrium is obtained and a single Fermi level characterizes both the metal and the semiconductor. At equilibrium, the semiconductor, having lost electrons, will be charged positively with respect to the metal.

To construct a proper band diagram for the metal and the semiconductor in thermal equilibrium we need to take note of two additional facts. The first of these is that the vacuum level E_0 must be drawn as a continuous curve. This is because E_0 represents the energy of a "just-free" electron and thus must be a continuous, single-valued function in space. If it were not such a function, one could conceive of means to extract work from an equilibrium situation by emitting electrons and then reabsorbing them an infinitesimal distance away where E_0 had changed value. Second, we note that electron affinity is a property associated with the crystal lattice like the forbidden-gap energy. Hence, it is a constant in a given material. Considering these three factors: constancy of E_f, continuity of E_0, and constancy of X in the semiconductor, we can sketch the general shape of the band diagram for the metal-semiconductor system. The sketch is given in Fig. 2.5a for an *n*-type semiconductor for which $\Phi_M > \Phi_S$.

We see from Fig. 2.5a that there is an abrupt discontinuity of allowed energy states at the interface. The magnitude of this step is $q\phi_B$ electron volts with

$$q\phi_B = q(\Phi_M - X) \tag{2.2.1}$$

Figure 2.5a indicates that electrons at the band edges (E_c and E_v) in the vicinity of the junction in the semiconductor are at higher energies than are those existing

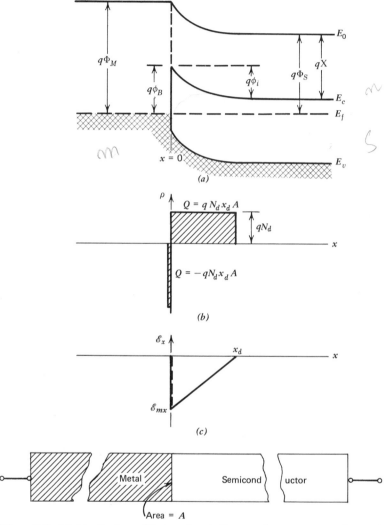

Figure 2.5 (*a*) Idealized equilibrium band diagram (energy versus distance) for a metal-semiconductor rectifying contact (Schottky barrier). The physical junction is at $x = 0$. (*b*) Charge at an idealized metal-semiconductor junction. The negative charge is approximately a delta function at the metallic surface. The positive charge consists entirely of ionized donors (here assumed constant in space) in the depletion approximation. (*c*) Field at an idealized metal-semiconductor junction.

in more remote regions. This is a consequence of the transfer of negative charge from the semiconductor into the metal. Because of the charge exchange there is a field at the junction and a net increase in the potential energies of electrons within the band structure of the semiconductor. The free-electron population is thus depleted near the junction as indicated by the increased separation between E_c and E_f at the surface compared to that in the bulk.

Before considering the electrical properties of the metal-semiconductor junction we should note that our development thus far has relied on an important idealization. The idealization is that the basic band structures of the two materials are unchanged near the surface. We shall gain some useful results by working with this idealized model, but it will be necessary to consider conditions at the surface more carefully later, and then to construct a nonidealized band picture.

Charge, Depletion Region, and Capacitance

The charge and field diagrams for an ideal metal-semiconductor junction are sketched in Figs. 2.5b and 2.5c. To the extent that the metal is a perfect conductor, the charge transferred to it from the semiconductor exists in a plane at the metallic surface. In the idealized n-type semiconductor, positive charge can consist either of ionized donors or of free holes while electrons make up the negative charge. We have made several assumptions about the semiconductor charge in drawing Figs. 2.5b and 2.5c. First, the free-hole population is assumed to be everywhere so small that it need not be considered; second, the electron density is much less than the donor density from the interface to a plane at $x = x_d$. Beyond x_d, the donor density N_d is taken to be equal to n. These assumptions make up what is usually called the *depletion approximation*. Although they are not precisely true, they are generally sufficiently valid to permit the development of very useful relationships. We shall reconsider the space-charge distributions at a junction in equilibrium in Chapter 3 and consider the depletion approximation in more detail.

Under the depletion approximation, the extent of the space-charge region is exactly x_d units, and the field (for this case of a constant doping in the semiconductor) is a decreasing linear function of position (Fig. 2.5c). The maximum field $\mathscr{E}_{mx}$ is located at the interface and, by Gauss' law, its value is given by

$$\mathscr{E}_{mx} = \frac{-qN_dx_d}{\epsilon_s} \tag{2.2.2}$$

where ϵ_s is the permittivity of the semiconductor. The voltage across the space-charge region, which is represented by the negative of the area under the field curve in Fig. 2.5c, is given by

$$\phi_i = -\frac{1}{2}\mathscr{E}_{mx}x_d = \frac{1}{2}\frac{qN_dx_d{}^2}{\epsilon_s} \tag{2.2.3}$$

From Fig. 2.5a, the built-in voltage ϕ_i is also seen to be $\Phi_M - \Phi_S = \Phi_M - X - (E_c - E_f)/q$. The space charge Q_s in the semiconductor is

$$Q_s = qAN_d x_d = A\sqrt{2q\epsilon_s N_d \phi_i} \qquad (2.2.4)$$

where A is the area of the contact.

Applied Bias. Up to this point, we have been considering thermal equilibrium conditions at the metal-semiconductor junction. Now we take up the case of an applied voltage; that is, a nonequilibrium condition. We saw in Fig. 2.5a that there is an abrupt step in allowed electron energies at the metal-semiconductor interface. This step makes it more difficult to cause a net transfer of free electrons from the metal into the semiconductor than it would be to obtain a net flow of electrons in the opposite direction. There is a barrier of $q\phi_B$ electron volts between electrons at the Fermi level in the metal and the conduction-band states in the semiconductor (Fig. 2.5a). To first order this barrier height is independent of bias because no voltage can be sustained across the metal. If we refer to Fig. 2.5c, we see that the voltage drop across the near delta function of space charge in the metal (equivalent to the area between the $\mathscr{E}$-field curve and the axis) is effectively zero in equilibrium. The total voltage drop across the space-charge region (ϕ_i) occurs within the semiconductor, as can be seen in Fig. 2.5a. An applied voltage would similarly be sustained entirely within the semiconductor and would alter the equilibrium-band diagram (Fig. 2.5a) by changing the total curvature of the bands, modifying the potential drop from ϕ_i. Thus, electrons in the bulk of the semiconductor at the conduction-band edge are impeded from transferring to the metal by a barrier that can be changed readily from its equilibrium value $q\phi_i$ by an applied bias. The barrier is reduced when the metal is biased positively with respect to the semiconductor, and it is increased under bias of the opposite polarity. Energy-band diagrams for two cases of bias are shown in Figs. 2.6a and 2.6b. Since these diagrams correspond to nonequilibrium conditions, they are not drawn with a single Fermi level. The Fermi energy in the region from which electrons flow is displaced higher than is the Fermi energy in the region into which electrons flow. Currents, of course, move in the direction opposite to the electron flow.

To investigate bias effects on the barrier, we shall consider the semiconductor to be grounded and take forward bias to correspond to the metal electrode being made positive. The applied voltage is referred to as V_a and the bias sense is indicated in Fig. 2.6c. Under reverse bias the metal is negatively biased ($V_a < 0$). If the metal-semiconductor junction is placed under reverse bias so that the voltage drop across the space-charge zone increases to ($\phi_i - V_a$), then the space charge in the semiconductor increases from the value at equilibrium [Eq. (2.2.4)] to

$$Q_s = A\sqrt{2q\epsilon_s N_d(\phi_i - V_a)} \qquad (2.2.5)$$

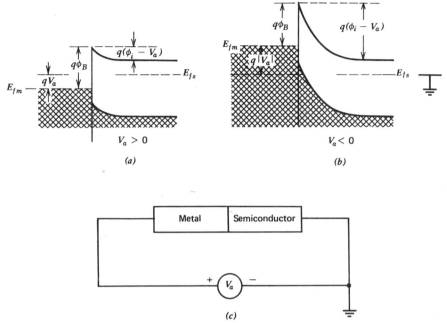

Figure 2.6 Idealized band diagram (energy versus distance) at a metal-semiconductor junction (a) under applied forward bias ($V_a > 0$) and (b) under applied reverse bias ($V_a < 0$). The semiconductor has been taken as the reference (voltage ground) as shown in (c). The vacuum levels for the two cases are not shown to maintain clarity in the diagram.

Under small-signal ac conditions, the junction will show a capacitive behavior that can be calculated by using Eq. (2.2.5):

$$C = \left| \frac{\partial Q_s}{\partial V_a} \right| = A \sqrt{\frac{q \epsilon_s N_d}{2(\phi_i - V_a)}} = \frac{A \epsilon_s}{x_d} \qquad (2.2.6)$$

The last form of Eq. (2.2.6) is a general result for small-signal capacitance C. Since C represents the ratio of a *differential* charge (hence a charge with differential linear extent) to differential voltage, it can always be expressed by the ratio of the permittivity to the total space-charge width. A further discussion of this result is given in Chapter 3.

If Eq. (2.2.6) is solved for the total voltage across the junction, we obtain

$$(\phi_i - V_a) = \frac{A^2 q \epsilon_s N_d}{2C^2} \qquad (2.2.7)$$

The form of Eq. (2.2.7) indicates that a plot of the square of the reciprocal of the small-signal capacitance versus the reverse bias voltage should be a straight line as sketched in Fig. 2.7. The slope of the straight line can be used to obtain the doping in the semiconductor, and the intercept of the straight line with the abscissa should equal ϕ_i. Measurements of small-signal capacitance as a function of dc bias and the subsequent construction of plots similar to Fig. 2.7 from the data are often used to study semiconductors. In the practical case, the most serious inaccuracy comes from obtaining ϕ_i from the intercept with the voltage axis. The slope of the curve, however, usually gives an accurate indication of the semiconductor doping.

Measurements of the small-signal capacitance are useful even in cases where the semiconductor doping varies with distance. In that circumstance, the space-charge picture is altered from the one shown in Fig. 2.5b to the configuration sketched in Fig. 2.8. For a given dc (reverse) bias ($V_a < 0$), the space-charge layer is of width x_d. A small increase in the magnitude of V_a causes a small increase in

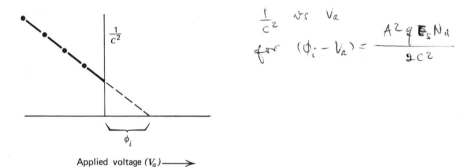

$$\frac{1}{C^2} \ vs \ V_a$$

$$for \ (\phi_i - V_a) = \frac{A^2 q \, \mathcal{E}_s N_d}{2C^2}$$

Applied voltage (V_a) ⟶

Figure 2.7 Plot of $1/C^2$ versus applied voltage for an ideal metal-semiconductor junction.

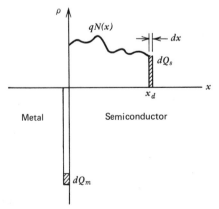

Figure 2.8 Schematic representation of space charge at a metal-semiconductor junction with nonuniform doping in the semiconductor.

Q_s where

$$Q_s = qA \int_0^{x_d} N(x)\, dx \qquad (2.2.8)$$

Consequently, an increment in voltage dV_a results in an increment in Q_s of value

$$dQ_s = qN(x_d) A\, dx = -C\, dV_a$$

or

$$N(x_d) = -C/qA(dx/dV_a) \qquad (2.2.9)$$

where x_d is the depletion-region width that corresponds to an applied dc voltage V_a at which the small-signal capacitance C is to be measured. We can rewrite Eq. (2.2.9) by noting that since the small-signal capacitance C is given by $C = A\epsilon_s/x_d$, the derivative (dx/dV_a) can be written $dx/dV_a = (dx/dC)(dC/dV_a) = -(\epsilon_s A/C^2)\, dC/dV_a$. Thus

$$N(x_d) = \frac{C^3}{q\epsilon_s(dC/dV_a)A^2} \qquad (2.2.10)$$

Equation (2.2.10) can be placed in a more useful form by recognizing that

$$\frac{d(1/C^2)}{dV_a} = -\left(\frac{2}{C^3}\right)\frac{dC}{dV_a}$$

so that

$$N(x_d) = \frac{-2}{q\epsilon_s[d(1/C^2)/dV_a]A^2} \qquad (2.2.11)$$

The result that we have derived in Eq. (2.2.11) shows that the slope of a plot of $1/C^2$ versus reverse bias voltage directly indicates the doping concentration at the edge of the space-charge layer. This slope, when divided into $(2/q\epsilon_s A^2)$ gives $N(x_d)$ directly. Several commercial "semiconductor profilers" make use of this fact; some have direct-reading outputs that convert the slope to its equivalent doping concentration.

Schottky Barrier Lowering.[†] We now reconsider our earlier statement that the barrier to electron flow from the metal into the semiconductor is "to first order" unchanged by an applied bias. The small dependence on applied voltage of this barrier height can be observed particularly under reverse bias. The dependence corresponds to an effect that was explained many years ago by Walter Schottky when electron emission into a vacuum was studied.

In our derivation we consider that within the semiconductor close to the interface it should be possible to approximate the electron energy by using the free-electron theory and by treating the metal as a planar conducting sheet. In this model we take account of the semiconductor in two ways: by assigning the electron an effective mass m_n^*, and by using a relative permittivity that differs from unity ($\epsilon_r = 11.7$ for Si). A sketch of the appropriate energy picture is given in Fig. 2.9 under equilibrium conditions and also for the case of an applied field that tends

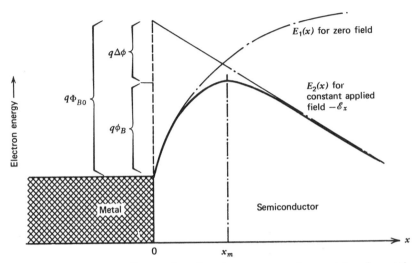

Figure 2.9   Classical energy diagram for a free electron near a plane metal surface at thermal equilibrium $[E_1(x)]$, and with an applied field $-\mathscr{E}_x [E_2(x)]$.

to move electrons away from the metal surface. The function $E_1(x)$ that represents electron energy in the figure is derived by the classical technique; the conducting metal plane has the same effect on the electron as an image charge of opposite sign spaced equidistant behind the plane $x = 0$. In the presence of a field $-\mathscr{E}$ tending to move electrons away from the metal surface, the electron energy $E_2(x)$ is

$$E_2 = \frac{-q^2}{16\pi\epsilon_s x} - q\mathscr{E}x \qquad (2.2.12)$$

From studies of metal-vacuum systems, Eq. (2.2.12) has been shown to be accurate for distances greater than a few tens of angstroms. The plane at which E_2 is a maximum is easily found as is the energy $q\,\Delta\phi$ in Fig. 2.9:

$$q\,\Delta\phi = \sqrt{\frac{q^3 \mathscr{E}}{4\pi\epsilon_s}} \qquad (2.2.13)$$

According to this model the height of the barrier to electron flow out of the metal $q\phi_B$ varies in the same way as does $q\,\Delta\phi$. The current flow is exponentially dependent on this height since only that fraction of the Boltzmann-distributed electrons in the metal with energies above the barrier maximum can pass across it. One therefore expects the emitted current from the metal under reverse bias to vary as

$$J = J_0 \exp \frac{\sqrt{q^3 \mathscr{E}/4\pi\epsilon_s}}{kT} \qquad (2.2.14)$$

Using the depletion approximation, we can relate the field $\mathscr{E}$ to the bias V_a and built-in potential ϕ_i by

$$\mathscr{E} = \sqrt{\frac{2qN_d}{\epsilon_s}} (\phi_i - V_a) \tag{2.2.15}$$

Equations (2.2.14) and (2.2.15) show that because of Schottky barrier lowering, the current emitted over a reverse-biased blocking contact depends exponentially on the fourth root of voltage at higher biases. Although this type of an exponential dependence is sometimes observed in practice, reverse currents that result from the generation of free carriers in the space-charge region may be larger than the component that we have considered here. In that case, the dependence on bias is more gradual. We shall consider generation as a source of reverse current when we discuss current flow in pn junctions in Chapter 4.

2.3 CURRENT-VOLTAGE CHARACTERISTICS

The basic dependence of current on voltage in a Schottky-barrier diode can be deduced from qualitative arguments. These arguments provide fundamental insight into the nature of the equilibrium behavior of the metal-semiconductor system, and we therefore consider them before making a more rigorous calculation of the $I-V$ characteristic.

The starting point for the derivation is to consider the band diagram at thermal equilibrium shown in Fig. 2.5. At equilibrium, the rate at which electrons cross over the barrier into the semiconductor from the metal is balanced by the rate at which electrons cross the barrier into the metal from the semiconductor. From the discussion of diffusion in Chapter 1, we know that free carriers in crystals are constantly in motion because of their thermal energies. In Problem 1.13, for example, this fact was used to deduce that a density n_0 of free carriers in thermal motion can be considered to cause a current density equal to $-qn_0v_{th}/4$ in an arbitrary direction. At thermal equilibrium, of course, this current density is balanced by an equal and opposite flow, and there is zero net current. Applying this concept to the boundary plane of the band diagram of Fig. 2.5, we can consider that there is a tendency for electrons to flow from the semiconductor to the metal and an opposing balanced flux of electrons from the metal into the semiconductor. These currents are proportional to the density of electrons at the boundary. In the semiconductor, this density n_s [from Eq. (1.1.21)] is

$$n_s = N_c \exp\left(-\frac{q\phi_B}{kT}\right) \tag{2.3.1}$$

which can be written in terms of the bulk density $n = N_d$ by applying Eq. (1.1.21) in the bulk of the semiconductor and noting from Fig. 2.5 that $q\phi_B = q\phi_i + E_c - E_f$ in the semiconductor bulk.

$$n_s = N_d \exp\left(-\frac{q\phi_i}{kT}\right) \tag{2.3.2}$$

Thus, equilibrium at the junction corresponds to

$$|I_{MS}| = |I_{SM}| = KN_d \exp\left(-\frac{q\phi_i}{kT}\right) \tag{2.3.3}$$

where I_{MS} and I_{SM} are the thermally induced currents directed from the metal toward the semiconductor and vice versa, and K is a proportionality constant.

When a bias V_a is applied to the junction as in Fig. 2.6, the potential drop within the semiconductor is changed and we can expect the flux of electrons from the semiconductor toward the metal to be modified. If we assume that the surface and the bulk of the semiconductor remain nearly at thermal equilibrium under bias, then the equation for n_s will be modified from Eq. (2.3.2) to

$$n_s = N_d \exp\left[-\frac{q(\phi_i - V_a)}{kT}\right] \tag{2.3.4}$$

The current from the metal to the semiconductor that results from the electron flow out of the semiconductor is therefore altered by the same factor. The flux of electrons from the metal to the semiconductor, however, is not affected by the applied bias because the barrier $(q\phi_B)$ remains at its equilibrium value.

Subtracting these two components, we obtain an expression for the net current from the metal into the semiconductor under applied bias:

$$I = I_{MS} - I_{SM}$$

$$= KN_d \exp\left[-\frac{q(\phi_i - V_a)}{kT}\right] - KN_d \exp\left(\frac{-q\phi_i}{kT}\right) \tag{2.3.5}$$

which can be written

$$I = I_0 \left[\exp\left(qV_a/kT\right) - 1\right] \tag{2.3.6}$$

where $I_0 = KN_d \exp\left(\dfrac{-q\phi_i}{kT}\right)$ is a new constant.

Equation (2.3.6) is often called the ideal diode equation. It arises, as this derivation makes clear, when a barrier to electron flow affects the thermal flux of carriers asymmetrically. Although more detailed analysis will modify the current equation slightly, the essential dependence of current on voltage for metal-semiconductor barrier junctions is contained in Eq. (2.3.6). The ideal diode equation predicts a saturation current $-I_0$ for V_a negative and a very steeply rising current when V_a is positive (Problem 2.10).

More detailed considerations of the $I-V$ characteristics at metal-semiconductor barrier junctions show that the saturation current I_0 is not completely independent of applied voltage. Analyses that lead to this result are carried out in the remainder of this section.

Schottky Barrier[†]

The dependence of current on applied voltage in a metal-semiconductor junction can be obtained by integrating the equations for carrier diffusion and drift across the depletion region near the contact. This approach, which was first taken by Schottky,[1] assumes that the dimensions of this space-charge region are sufficiently large so that the use of a diffusion constant and a mobility value are meaningful— basically that the width of the region is at least a few electronic mean-free paths and the field strength is less than that at which the drift velocity saturates. An alternative physical approach, adopted first by Hans Bethe[2a,b] and based on carrier emission from the metal, is valid even if these constraints are not met, and it leads to the same $I-V$ dependence.[3]

If we refer to the metal-semiconductor contact under bias as shown in Fig. 2.10 and consider a one-dimensional electron flow through the barrier region, then as shown in Chapter 1 [Eq. (1.2.19)], we can write

$$J_x = q \left[n\mu_n \mathscr{E}_x + D_n \frac{dn}{dx} \right] \tag{2.3.7}$$

If we denote potential in the region as ϕ, then by using $\mathscr{E}_x = -d\phi/dx$ and the Einstein relationship [Eq. (1.2.18)], we can write

$$J_x = qD_n \left[\frac{-qn}{kT} \frac{d\phi}{dx} + \frac{dn}{dx} \right] \tag{2.3.8}$$

A sketch of ϕ with the metal taken as grounded is shown in Fig. 2.10b.

We shall rewrite Eq. (2.3.8) in an integrated form that can be evaluated at the two ends of the depletion region ($x = 0$ and $x = x_d$ in Fig. 2.10). Multiplication of both sides of Eq. (2.3.8) by an integrating factor $\exp(-q\phi/kT)$ permits direct

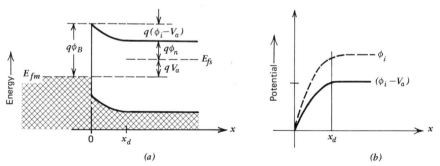

(a)   (b)

Figure 2.10   (a) Band diagram of a rectifying metal-semiconductor junction under forward bias. The applied voltage V_a displaces the Fermi levels: $qV_a = E_{fs} - E_{fm}$. (b) The potential across the surface depletion layer is decreased to $\phi_i - V_a$.

integration of the right-hand side. Using the limits of the depletion region, we obtain

$$J_x \int_0^{x_d} \exp\left(\frac{-q\phi}{kT}\right) dx = qD_n \left\{ n \exp\left(\frac{-q\phi}{kT}\right) \right\}\bigg|_0^{x_d} \tag{2.3.9}$$

In writing Eq. (2.3.9) we have assumed that the particle current represented by J_x is not a function of position and may be placed outside of the integral. This assumption has wide validity. Since the reference for potential is at the metal, the boundary conditions on voltage for use in Eq. (2.3.9) are

$$\phi(0) = 0; \qquad \phi(x_d) = (\phi_i - V_a) \tag{2.3.10}$$

As we see from Fig. 2.10a, $\phi(x_d)$ can also be written as

$$\phi(x_d) = (\phi_B - \phi_n - V_a) \tag{2.3.11}$$

where $q\phi_n$ is $(E_c - E_f)$ in the semiconductor bulk. Boundary conditions for n are also necessary in order to evaluate Eq. (2.3.9). Using Eq. (1.1.21), we can express these as

$$n(0) = N_c \exp\left(\frac{-q\phi_B}{kT}\right)$$

$$n(x_d) = N_d = N_c \exp\left(\frac{-q\phi_n}{kT}\right) \tag{2.3.12}$$

Substituting the boundary values into Eq. (2.3.9), we find

$$J_x = qD_n N_c \exp\left(\frac{-q\phi_B}{kT}\right) \left[\exp\left(\frac{qV_a}{kT}\right) - 1 \right] \bigg/ \int_0^{x_d} \exp\left(\frac{-q\phi(x)}{kT}\right) dx \tag{2.3.13}$$

To obtain current as a function of voltage, the functional dependence of ϕ on x must be inserted into the integrand in the denominator of Eq. (2.3.13) and the integration must be carried out. This functional dependence of ϕ on x is determined by the doping profile in the semiconductor near the contact.

The contact that we have been considering, a barrier to electron flow from a metal into a semiconductor of constant doping, is called a Schottky barrier in recognition of its initial analysis by Walter Schottky.[1] For the Schottky barrier we can use the depletion approximation as in Section 2.2 to derive an expression for the potential in the depletion region:

$$\phi(x) = \frac{qN_d}{\epsilon_s} x \left(x_d - \frac{x}{2} \right) \qquad (0 < x < x_d) \tag{2.3.14}$$

When Eq. (2.3.14) is inserted into Eq. (2.3.13) and the integration is performed, an explicit solution of J_x as a function of V_a is obtained.

$$J_x = J_S \left[\exp\left(\frac{qV_a}{kT}\right) - 1 \right] \tag{2.3.15}$$

where

$$J_S = \frac{q^2 D_n N_c}{kT} \left[\frac{2q(\phi_i - V_a)N_d}{\epsilon_s} \right]^{1/2} \exp \left(\frac{-q\phi_B}{kT} \right) \qquad (2.3.16)$$

Equation (2.3.16) shows that J_S is not independent of voltage; hence, some of the voltage dependence in Eq. (2.3.15) is implicit in J_S. The square-root dependence of J_S on voltage is, however, weak compared to the term that is exponential in voltage in Eq. (2.3.15). It is, therefore, possible to approximate the current-voltage characteristic by writing, instead of Eq. (2.3.15):

$$J_x = J_S' \left[\exp \left(\frac{qV_a}{nkT} \right) - 1 \right] \qquad (2.3.17)$$

where J_S' is independent of voltage and n is taken to be a constant having a value that is usually found experimentally to be between 1.02 and 1.15. (See Problem 2.9). The results of experimental measurements on a forward-biased, aluminum-silicon Schottky barrier are shown in Fig. 2.11. The good fit between the measured data in Fig. 2.11 and Eq. (2.3.17) with n = 1.07 is typical.

The analyses that we have just carried out have made use of a number of assumptions, several of which have been explicitly noted. One important implicit assumption, however, should be mentioned before proceeding because it is very frequently relied upon for first order device analysis. This is the assumption that the system is in *quasi-equilibrium,* that is, almost at thermal equilibrium even though currents are flowing. Quasi-equilibrium was implicitly invoked at a number of points in our analysis; for example, in writing Eqs. (2.3.4) and in using the Einstein relationship to write Eq. (2.3.8). Logically we expect quasi-equilibrium to be more valid under low-bias conditions when currents are small—and, in fact, this is the case. Often, a quasi-equilibrium analysis suffices or else the analysis can be extended to cover all currents within the range of interest by making slight modifications to the theory. The ultimate test of any assumption is, of course, the agreement between measurements and predictions, as in Fig. 2.11.

Mott Barrier[†]

Our derivation thus far has been for a metal-semiconductor junction in which the semiconductor doping is constant throughout the space-charge region; that is, for a Schottky barrier. To derive the current-voltage characteristic for other dopant profiles, we can use the equations through (2.3.13), but we must modify the expression for voltage in the depletion region [Eq. (2.3.14)]. One case of doping that is useful for some metal-semiconductor junctions is known as the Mott barrier after N. F. Mott who analyzed it in connection with studies of diodes composed of oxide compounds.[4]

In the Mott-barrier approximation, the semiconductor is characterized by an abrupt change in doping from a low value near the metal interface to a high value

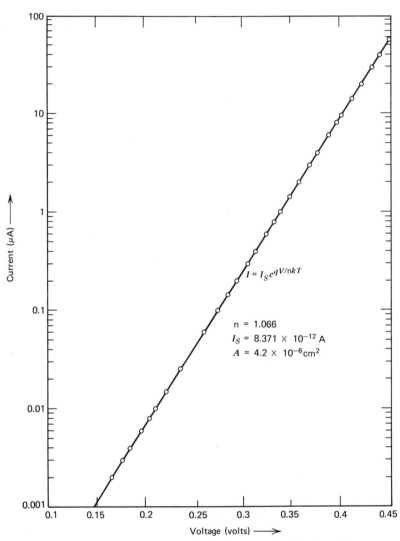

Figure 2.11   Measured values of current (plotted on a logarithmic scale) versus voltage for an aluminum-silicon Schottky barrier. Values for J'_S and n result from an empirical fit of the data to Eq. (2.3.17).

a very short distance from the surface. The distance is short in the sense that essentially no field lines terminate in it. (A short-hand description would be that the distance is much shorter than a Debye length L_D where L_D is discussed in Section 2.4.) An electron-energy diagram for this situation is sketched in Fig. 2.12. Because the length of the lightly doped region is very short, the electric field can be assumed to be constant throughout it. The field lines are assumed not to penetrate into the highly doped region because the donor density there is so high. This condition might, for example, model a case in which the doping near the contact is altered during the fabrication of the junction. It might also represent the situation in which the metal contact is made to a very thin, lightly doped, epitaxial film above a highly doped region in the crystal. We shall see that this latter case is encountered in the design of bipolar integrated circuits.

To obtain the dependence of current on voltage for the Mott barrier, we proceed as with the Schottky barrier, first expressing the potential ϕ as a function of x and then performing the integration indicated in Eq. (2.3.13). From Fig. 2.12 the

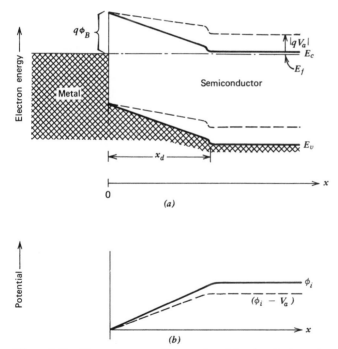

Figure 2.12   Electron-energy diagram for a Mott barrier (near-insulating region at the surface with a sharp transition at $x = x_d$ to a highly conducting region). The solid-line diagram indicates thermal equilibrium; the dotted line, a forward bias of V_a volts with the metal held at ground potential. (b) Potential diagram for the Mott barrier.

dependence of ϕ on x is

$$\phi(x) = (\phi_i - V_a)\frac{x}{x_d} \qquad (0 < x < x_d) \qquad (2.3.18)$$

Using Eq. (2.3.18) in (2.3.13), we obtain a result that may be written in the form of Eq. (2.3.15)

$$J_x = J_M\left[\exp\left(\frac{qV_a}{kT}\right) - 1\right] \qquad (2.3.19)$$

where J_M is more strongly dependent on V_a than was the parameter J_S derived for the Schottky barrier [Eq. (2.3.16)]:

$$J_M = \frac{q^2 D_n N_c(\phi_i - V_a)\exp\left(\dfrac{-q\phi_B}{kT}\right)}{x_d kT\left\{1 - \exp\left[\dfrac{-q(\phi_i - V_a)}{kT}\right]\right\}} \qquad (2.3.20)$$

Mott and Schottky barriers represent idealized metal-semiconductor rectifiers. In many cases, these idealizations suffice. To derive the J-V characteristic in cases for which they are not adequate, an accurate representation for potential throughout the barrier region must be found and used in place of Eqs. (2.3.14) or (2.3.18). In general, the major dependence on voltage is, however, contained in the exponential form that we derived both in Eqs. (2.3.15) and (2.3.19). An empirical fit to measured data is generally possible using a form similar to Eq. (2.3.17) with a value of n that is nearly unity.

Both equations that have been derived for saturation currents, J_S in Eq. (2.3.16) and J_M in Eq. (2.3.20), become physically unreasonable as V_a approaches ϕ_i. In practice, if V_a were to equal ϕ_i, there would be no barrier at the junction and very large currents could then flow. The actual forward voltage applied to the Schottky barrier is virtually never as large as the built-in voltage. A similar consideration will be given greater discussion when currents in pn junctions are described in Chapter 4.

2.4 NONRECTIFYING (OHMIC) CONTACTS

In our discussion of metal-semiconductor contacts, we have thus far considered cases in which the semiconductor near the metal is depleted of majority carriers relative to the bulk density and in which there has been a barrier to electron transfer from the metal. In such cases any applied voltage is dropped mainly across the junction region, and currents are contact limited. _The inverse case, in which the contact itself offers negligible resistance to current flow when compared to the bulk, defines an ohmic contact._ Although this definition of an ohmic contact may sound awkward, it emphasizes this essential aspect: when a voltage is applied across an element, the voltage drop across an ohmic contact is negligible compared to any bulk voltage drop.

Tunnel Contacts

The metal-semiconductor contacts that we have considered in the previous section can, for example, be made ohmic if the effect of the barrier on carrier flow can be made negligible. In practice this is accomplished by heavily doping the semiconductor so that the barrier width x_d is reduced to a very small value. To see this, refer to Eq. (2.2.4) and solve for x_d:

$$x_d = \sqrt{\frac{2\epsilon_s \phi_i}{qN_d}} \qquad (2.4.1)$$

The space-charge region, therefore, narrows as N_d is increased. When the barrier width approaches a few tens of angstroms, a new transport phenomenon, tunneling through the barrier, can take place.

Figure 2.13a is a schematic illustration of the tunneling process through a very thin Schottky barrier. When the barrier is of the order of tens of angstroms and the metal is biased negatively with respect to the semiconductor, electrons in the metal need not be energetic enough to surmount the barrier ($q\phi_B$ units above the Fermi energy) to enter the semiconductor. Instead, they can tunnel through the barrier into the conduction-band states in the semiconductor. Likewise, when the semiconductor is biased negatively with respect to the metal, electrons from the semiconductor can tunnel into electronic states in the metal (Fig. 2.13b). Many electrons are available to take part in these processes and currents rise very steeply as bias is applied. Hence, a metal-semiconductor contact at which tunneling is possible shows a very small resistance. It is virtually always an ohmic contact. Ohmic contacts are frequently made in this way in practice. To assure a very thin barrier, the semiconductor is often doped until it is degenerate (i.e., until the Fermi level enters either the valence or the conduction band).

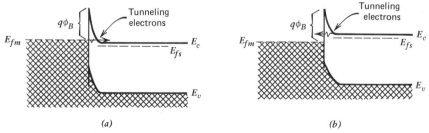

| (a) | (b) |

Figure 2.13 Metal-semiconductor barrier with a very thin space-charge region through which electron tunneling can take place. (a) Tunneling from metal to semiconductor. (b) Tunneling from semiconductor to metal.

Schottky Ohmic Contacts[†]

Another method to obtain an ohmic contact is to cause the majority carriers to be more numerous near the contact than they are in the bulk of the semiconductor. An ohmic contact of this type results if the semiconductor surface is not depleted when it comes into equilibrium with the metal, but rather has an enhanced majority-carrier concentration. Using the ideal Schottky theory that we described in Section 2.2, we see that this condition occurs in a metal-semiconductor junction between a metal and an n-type semiconductor with a larger work function than that of the metal. In this case, electrons are transferred to the surface of the semiconductor and the metal is left with a skin of positive charge.[*] The relevant energy diagram is sketched in Fig. 2.14a, and the corresponding charge and field diagrams are given in Figs. 2.14b and 2.14c. There is a qualitative similarity between these figures and Figs. 2.5a to 2.5c that referred to a rectifying contact; the important distinction between the two situations is that the semiconductor charge consists of free electrons in the case of the ohmic contact, but it is fixed (on positive donor sites) in the barrier case. The charge distribution, field, and potential for such a contact can be calculated using techniques similar to those employed for the rectifying contact.[5] Details of this procedure are considered in Problem 2.11. The results are summarized as follows.

To find a solution for the distribution of the space charge in the semiconductor, we take a reference for potential ϕ in the metal, and we assume that the excess electrons in the semiconductor surface region (denoted as n') are Boltzmann-distributed in energy. Thus, $n' = n_s \exp(-q|\phi|/kT)$ where n_s represents the excess concentration in the semiconductor at the metal-semiconductor interface. This equation describes the net space charge and can be used to write an integrable form of Poisson's equation. We solve Poisson's equation, in turn, to find the space-charge density

$$\rho(x) = -qn_s \bigg/ \left(1 + \frac{x}{\sqrt{2}L_D}\right)^2 \tag{2.4.2}$$

where

$$L_D = \left(\frac{\epsilon_s kT}{q^2 n_s}\right)^{1/2} \tag{2.4.3}$$

is known as the *Debye length* at the surface. We shall discuss the Debye length in more detail after completing our analysis of the contact.

---

[*] For an ohmic contact to a p-type semiconductor, the relative sizes of the work functions in the two regions need to be reversed to achieve a net positive charge in the semiconductor and, thereby, an enhanced hole density near the contact.

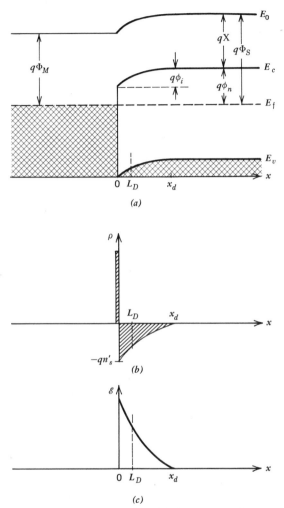

Figure 2.14   (*a*) Idealized equilibrium energy diagram for a Schottky ohmic contact between a metal and an *n*-type semiconductor. (*b*) Charge at an ideal Schottky ohmic contact. A delta function of positive charge at the metal surface couples to distributed excess electrons $n'(x)$ in the semiconductor. (*c*) Field at an idealized Schottky ohmic contact. The Debye length L_D is a characteristic measure of the extent of the charge and field.

The field varies with distance according to the equation

$$\mathscr{E}_x = \frac{\sqrt{2}kT}{L_D q}\left(1 + \frac{x}{\sqrt{2}L_D}\right)^{-1} \tag{2.4.4}$$

and the space-charge layer extends into the semiconductor a distance

$$x_d = \sqrt{2}L_D\left[\exp\left(\frac{q|\phi_i|}{2kT}\right) - 1\right] \tag{2.4.5}$$

where the built-in voltage ϕ_i is seen from Fig. 2.14a to be

$$|\phi_i| = \phi_n - (\Phi_M - X) \tag{2.4.6}$$

where $q\phi_n = (E_c - E_f)$ in the bulk. The condition $\Phi_M - X = \phi_n$ defines what can be called a neutral contact; that is, a contact with no built-in voltage and a surface that has the same density of free electrons as the bulk. In Problem 2.12 we show that a neutral contact can be considered to be "ohmic" for current levels less than $qn_0v_{th}/4$ where n_0 is the electron density and v_{th} is the thermal electron velocity. For $\Phi_M - X \neq \phi_n$, the current limit for ohmic behavior at the contact differs from $qn_0v_{th}/4$ by a factor $\exp\{q[(\Phi_M - X) - \phi_n]/kT\}$. Thus, contacts to n-type material are properly divided into *ohmic* behavior for cases in which the surface bands bend down and into *blocking* behavior for cases in which the bands bend up. The inverse conditions apply to contacts to p-material.

Summing up, we repeat the essential condition for ohmic behavior: an unimpeded transfer of majority carriers between the two materials forming the contact. At ohmic contacts there are generally built-in potentials. Unless penetrable barriers exist, majority carriers must be more numerous at an ohmic contact than they are in the bulk.

Debye Length.   The expressions that we have derived for charge density (Eq. 2.4.2), field (Eq. 2.4.4), and space-charge-layer width (Eq. 2.4.5) are all seen to contain the characteristic length L_D. Figures 2.14a through 2.14c show that L_D is an appropriate qualitative measure of the spatial extent of electrical effects at the boundary. The result of Problem 2.13 confirms this quantitatively by showing that 50% of the space charge in the semiconductor lies within $\sqrt{2}L_D$ units of the surface.

Considering a wider scope than this particular problem, we find that the solution of Poisson's equation in the presence of free charges always leads to a characteristic Debye length. If the charge configuration differs from the example given, the Debye length is still defined by Eq. (2.4.3) except that n_s is replaced by the free-charge density appropriate to that configuration. We shall see an example of this in Chapter 3. The general result that L_D qualitatively measures the spatial extent of space charge is always true. Problem 2.14 shows that a relationship exists between L_D in a given region and the dielectric relaxation time in the same region. This relationship can be interpreted physically in terms of a balance of two forces on free carriers: their diffusion tendency and their field-induced motion.

2.5 SURFACE EFFECTS

When we began our consideration of metal-semiconductor contacts in Section 2.2 and made use of Fermi-level equality to infer an equilibrium energy-band picture, we made an important idealization. We considered that both the semiconductor and the metal had allowed energy states (energy bands) at the surface that were no different from those in the bulk. The real situation is more complicated, and we will need to develop the theory further to derive a more practical physical model. The most important correction needed is to take account of the effects of surface states. Although these effects modify some conclusions, we shall be able to retain most of the ideas that were developed in previous sections.

Surface States

To begin, we should clarify the term *surface states*. These are extra allowed states for electrons that are present at the semiconductor surface, but not within the bulk. These extra states arise from a number of sources. First, let us consider an absolutely clean surface, that is, one composed purely of atoms appropriate to the host lattice. There will be extra states on this surface because the energy field of the crystal is one-sided, that is, electrons in the surface region are bonded only from the side directed toward the bulk (Fig. 2.15). We would expect the characteristic energies for electrons at such sites to differ from the characteristic energies in the bulk. Surface states of this type are called Tamm or Shockley states after original investigations by these scientists.[6,7] The density of the Shockley-Tamm states in a particular semiconductor is of the order of the density of atoms at the surface, or roughly $N_0^{2/3}$ (cm^{-2}) where N_0 is the bulk atomic density (atoms cm^{-3}). For

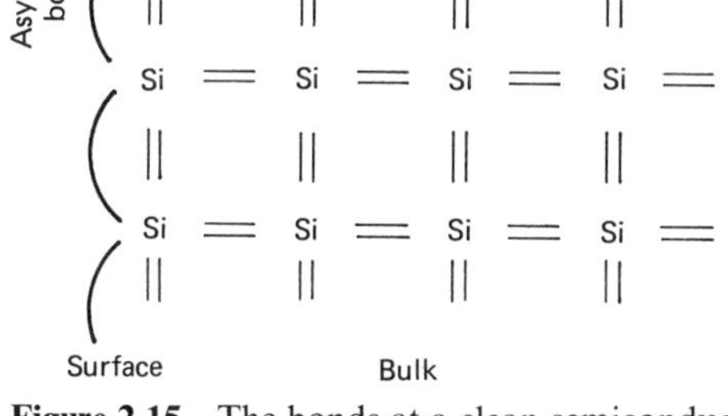

Figure 2.15 The bonds at a clean semiconductor surface are anisotropic and, consequently, differ from those in the bulk.

silicon, N_0 is 5×10^{22} cm^{-3} (cf Table 1.1) and the density of the Tamm-Shockley states is about 10^{15} cm^{-2}. The energetic distribution of these states is not well established although studies on the diamond lattice[8] indicate their density to be peaked about one-third of the forbidden-gap energy above the valence band as sketched in Fig. 2.16.

Bonded foreign atoms at the surface are sources for other types of surface states. An example is oxygen, which is virtually always found at a silicon surface. Oxygen can produce surface states spread over a range of energies depending upon the nature of the specific bonds it shares with silicon atoms. Other surface complexes of metals, hydroxyl ions, etc., can also be expected on any processed silicon surface. The net effect of all of these sources is a density of available electronic surface states that is not zero at any energy, although there may be pronounced peaks at specific energy values.

Besides varying in energy, surface states also vary in type and may be classified according to the charge they carry at equilibrium. For example, states that are neutral when occupied by electrons and positively charged when unoccupied are classified as donor states. States that are negative when occupied but neutral when empty are classified as acceptor states in a manner analogous to bulk dopant atoms, as discussed in Section 1.1.

One other classification of surface states arises from the properties of real interfaces between solids. On an atomic scale (a few angstroms), such interfaces are not plane but consist rather of zones of intermediate materials and impurities that are several to tens of atomic layers in thickness. Within these intermediate zones some surface states are physically close to the bulk semiconductor, and they remain in thermal equilibrium with bulk states even under fairly rapid changes in potential. These are called *fast surface states* because the electrons occupying them come into equilibrium quickly. In contrast to these are so-called *slow states*, which are states

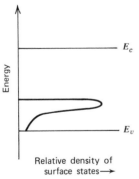

Figure 2.16 Approximate distribution of Tamm-Shockley states in the diamond lattice.[8] The distribution appears to peak sharply at an energy roughly one-third of the band gap above E_v.

situated more remote from the bulk semiconductor within the intermediate layer. These states take relatively longer times to reach thermal-equilibrium conditions with the bulk states. Although the demarcation between these categories of states is vague, general usage sets the boundary between them at a response time corresponding to about 1 kHz.

Surface Effects on Metal-Semiconductor Contacts[†]

The presence of surface states can modify the contact theory that we have presented. If the semiconductor surface states are not neutral or if they change their charge state when the contact is formed, then the charge configurations that we obtain by applying Schottky theory are not valid. A theory for metal-semiconductor systems with surface states has been presented.[9] We shall not discuss this theory in detail, but rather give a summary of the viewpoints adopted and show the correspondence of theory with practical results.

To account for surface effects, the metal-semiconductor contact is treated as if it contained an intermediate region sandwiched between the two crystals. The intermediate region is thought to consist of an interfacial layer ranging from several to the order of ten atomic dimensions in width. This layer contains the impurities and added surface states. It is too thin to be an effective barrier to electron transfers (which can occur by tunneling), but it can sustain a voltage drop. Extra allowed electronic states are postulated to be distributed in energy at the assumed planar boundary between the interfacial layer and the semiconductor. Figure 2.17 is a

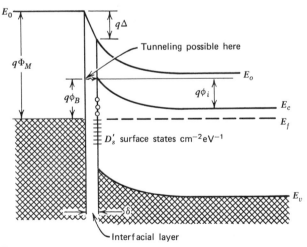

Figure 2.17 Band structure near a metal-semiconductor contact according to the model of Cowley and Sze.[9] The model presumes a thin interfacial layer of thickness δ that sustains a voltage Δ at equilibrium. Acceptor-type surface states distributed in energy are assumed to be described by a distribution function D_s' states cm^{-2} eV^{-1}.

sketch showing the band structure and the surface states. In the figure the states are assumed to be acceptor type and to have a density D'_s states cm^{-2} eV^{-1}. Note in Fig. 2.17 the thin layer, δ units in width, which sustains a voltage drop of Δ units of voltage. Because this layer is thin enough for electron tunneling, the metal-semiconductor barrier height is measured between the Fermi level and the conduction band at the semiconductor surface. A surface layer should be associated with the semiconductor even without the metal being present (Fig. 2.18). When acceptor surface states are present as in Fig. 2.18, the n-type semiconductor is depleted of electrons near its surface, and negative charge exists in the surface acceptor states.

Consider the formation of a blocking contact to a semiconductor with surface states such as that shown in Fig. 2.18. Such a contact would have an energy diagram similar to that in Fig. 2.17. To draw this diagram we note that from Schottky theory a blocking contact would be formed if contact were made to a metal having Φ_M greater than Φ_S, because electrons would be transferred from the semiconductor into the metal. Transfer of electrons from the semiconductor bends the conduction band away from the Fermi level. For the semiconductor in Fig. 2.18 this would remove charge from the D'_s surface states by lifting them above E_f. The larger is D'_s, the greater is the charge removed for each incremental energy increase in E_c near to the contact. If D'_s is very large, therefore, a negligible movement of the Fermi level at the semiconductor surface would transfer sufficient charge to equalize the Fermi levels. Under this condition, the Fermi level is said to be *pinned* by the high density of states. Note that pinning of the Fermi level is not exclusively associated with the surface states being acceptor type. Any electronic states clustered near the Fermi energy will cause the Fermi level to be pinned when the state density becomes very large because slight changes in Fermi energy would result in very sizable charge transfer.

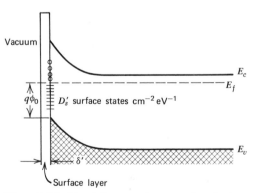

Figure 2.18 Band diagram for a semiconductor surface showing a thin surface layer containing acceptor-type surface states distributed in energy. A surface-depletion region is present because of charge in the surface states.

Table 2.2   **Schottky Barriers to Silicon**[10]
        (**qX for Silicon = 4.15 eV**)

Silicon Type	Metal	$q\Phi_M$ (eV)	$q\phi_B$ (eV)
n	Al	4.1	0.69
p	Al	—	0.38
n	Pt[a]	5.3	0.85
p	Pt[a]	—	0.25
n	W	4.5	0.65
n	Au	4.75	0.79
p	Au	—	0.25

[a] Pt is known to react with silicon to form a thin layer of platinum silicide.

When the Fermi level is pinned, it has been shown[9] that the barrier height $q\phi_B$ becomes

$$q\phi_B = (E_g - q\phi_0) \qquad (2.5.1)$$

where $q\phi_0$ is $(E_f - E_v)$ at the surface when the semiconductor is not covered by metal as shown in Fig. 2.18. As D'_s approaches zero, the barrier height $q\phi_B$ approaches the height predicted by basic Schottky theory as given in Eq. (2.2.1) and repeated here for reference.

$$q\phi_B = q(\Phi_M - X) \qquad (2.5.2)$$

Whether a metal-semiconductor barrier height will be predicted by Eq. (2.5.1) or by Eq. (2.5.2) or by an intermediate result depends on the value of D'_s at energies near to the Fermi level and also on specific properties ascribed to the boundary layer, such as its precise thickness and permittivity.[9]

In practice most Schottky barrier heights for important semiconductors are predicted more accurately by Eq. (2.5.1) than by Eq. (2.5.2). There is thus only small dependence on the metal work function. This is true for silicon, germanium, and gallium arsenide although cadmium sulfide is an exception. For silicon, as with germanium, gallium arsenide, and gallium phosphide, the quantity $q\phi_0$ is found experimentally to be about equal to $\frac{1}{3}E_g$ so that the barrier height $q\phi_B$ from Eq. (2.5.1) is typically close to $\frac{2}{3}$ of the band gap or roughly 0.75 eV for silicon. The reason for this consistency may be the characteristic very high density of states, which appears to be common to the diamond-type lattice and which pins the Fermi level at this energy (Fig. 2.16). Some measured barriers for various metals contacting n and p-type silicon crystals are shown in Table 2.2.

2.6 METAL-SEMICONDUCTOR DEVICES: SCHOTTKY DIODES

The industrial use of metal-semiconductor barrier devices, generally designated as Schottky diodes, is very widespread. The area of greatest application is that of

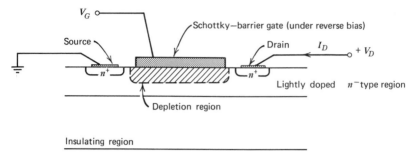

Figure 2.19 Schottky-barrier-gate, field-effect transistor. Current I_D flowing from drain to source is modulated by gate voltage V_G that controls the dimensions of the space-charge region. This, in turn, affects the conducting area for I_D. The source and drain contacts are ohmic because they are made to highly doped material.

digital logic circuits, that is, circuits that perform binary arithmetic. Schottky diodes are often used in these circuits as fast switches that can be made on integrated-circuit chips within very small dimensions on the surface. There is also an increasing interest in Schottky diode power rectifiers since large-area devices with an excellent thermal path through the contact metal allow high currents to flow. These high currents are sustained at comparatively lower voltage drops than would exist in diffused *pn*-junction diodes.

Schottky diodes are also used as variable capacitors that can be operated efficiently in the microwave region of the spectrum. For variable capacitance applications, the diode is kept under reverse bias continuously. Voltage changes across it then are able to modulate the depletion-region width and capacitance in the manner discussed in Section 2.2.

Another commercial application that makes use of modulation of the depletion width is the Schottky-barrier, field-effect transistor.* This device, pictured in Fig. 2.19, consists of a barrier junction at the input side that acts as a control electrode, (or *gate*) and two ohmic junctions through which output current flows (described as the *source* and *drain* electrodes). The output current varies when the cross section of the conducting path beneath the gate electrodes is changed. This device is a special form of a junction field-effect transistor (JFET), an amplifying device in which the control electrode is most usually made from a reverse-biased *pn* junction. Substitution of a Schottky barrier for the *pn* junction in JFETs has proven especially useful with semiconductor materials in which technological difficulties make

---

* Since this is the first point at which we meet the term *transistor*, it is appropriate to point out that the word has its origin in an amalgamation of the descriptive terms *transfer resistor*. It was coined by W. Shockley and co-workers at the Bell Telephone Laboratories where the junction field-effect transistor and the bipolar transistor, to be discussed in Chapter 5, were both invented.

the fabrication of *pn* junctions unreliable. The major application of Schottky barriers to JFETs has been to make high-frequency, gallium-arsenide devices. We shall discuss the operation of JFETs in more detail after we have considered *pn*-junction operation in Chapter 3.

Schottky Diodes in Integrated Circuits. Two fortuitous designer's choices, made for reasons having nothing to do with Schottky barriers, combine to allow a very simple technology for Schottky diodes in silicon digital integrated circuits. These choices are the use of high resistivity *n*-type silicon in which to build *npn* transistors and the use of evaporated aluminum metal to form the "wire" interconnections in integrated circuits. The aluminum forms a blocking contact to the lightly doped, *n*-type silicon provided that the silicon surface has been very thoroughly cleaned. It is only necessary that the doping of the silicon be sufficiently low so that the barrier may not be penetrated by tunneling electrons, as was described in Section 2.4. Practically, this limits the doping to less than about 10^{17} cm^{-3} (Problem 2.7).

The barrier height between *n*-type silicon and aluminum is about 0.70 eV, and diodes made by evaporation of aluminum onto the silicon surface in a vacuum have characteristics that approximate theoretical predictions quite well (Fig. 2.11). Because of the concentration of electric-field lines near the corners, however, reverse breakdown is not an abrupt function of voltage and occurs at a relatively low bias (~ 15 V). Several techniques, such as the use of diffused guard rings (Fig. 2.20a) or field plates (Fig. 2.20b), have been developed to improve the reverse characteristics. Because they complicate circuit processing, however, these techniques are avoided unless especially needed. Excellent Schottky diodes with high barriers can be made using refractory metals; platinum, in particular, has been employed in many instances. A technique of integrated-circuit metallization called *beam-leading* uses sputtered* platinum. If sputtering is included in the manufacturing process, an advantageous means of cleaning the silicon surface very

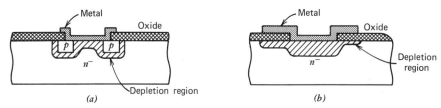

(a)   (b)

Figure 2.20 Special processing techniques improve the performance of Schottky diodes. (*a*) The diffused guard ring leads to a uniform electric field and eliminates breakdown at the junction edge. (*b*) The field plate is an alternative means for achieving the same effect.

---

* Sputtering is a technique for laying down thin films of materials in which a source, or target, is bombarded with high-energy gaseous ions. Sputtering is especially useful with refractory materials that are not easily evaporated.

thoroughly by bombarding it with high-energy ions or *sputter-etching* can be employed. After sputter-etching, the subsequent sputter-deposition of platinum forms an excellent Schottky diode. Although Schottky diodes can be made to *p*-type silicon, the barrier heights are inherently smaller (roughly $\frac{1}{3}$ of the the band gap or 0.36 V), and the yield and electrical performance are correspondingly poorer.

Designers who make use of Schottky diodes for digital logic circuits often call them *clamps* because they fix or clamp a condition on the circuit causing it to function desirably. We shall see the reasons in Chapter 6, but at this point we merely assert that substantial improvement in the speed of a digital logic circuit is gained if clamps are put between the collector and base of a switching transistor to keep its collector-base junction from being forward biased. Schottky diodes are nearly ideal for this purpose.

To place this topic in context, we first discuss the concept of a *turn-on voltage*. This term refers to the forward drop that must be placed across a diode to "turn-it-on"; that is, to cause it to pass substantial current. The current-voltage equations for metal-semiconductor barriers [Eqs. (2.3.17) and (2.3.19)] show, however, a continuous dependence of current on voltage with no abrupt change of characteristics to identify as a turn-on voltage. From an engineering point of view, however, there is a threshold for conduction that becomes apparent when we plot the current-voltage characteristics for a Schottky diode on linear scales (Fig. 2.21). The linear scale permits only a small range of current to be plotted meaningfully. The very strong dependence of current on voltage allows the data to be fit fairly well by two straight lines, one nearly horizontal at $J = 0$ and one nearly vertical. The intersection of the nearly vertical line with the voltage axis defines a turn-on voltage V_o. When digital circuits are designed, V_o is a good approximation to the voltage drop across any diode that is conducting. From Eq. (2.3.17), at a given

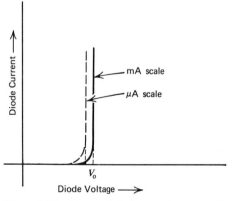

Figure 2.21   Linear plot of current versus voltage for a Schottky diode illustrating the concept of a turn-on voltage.

forward current density J_F, V_o is given by

$$V_o = \frac{nkT}{q} \ln\left(\frac{J_F}{J'_S} + 1\right)$$ (2.6.1)

Thus, from the designer's point of view, V_o is most strongly dependent on J'_S. For diodes designed to pass currents in the milliamp range, aluminum Schottky diodes fabricated on n-type silicon typically have a V_o of about 450 mV. Because this number is about 200 mV smaller than V_o for a comparable pn junction (Chapter 4), a Schottky diode placed in parallel with a pn-junction diode will not permit the forward bias to rise sufficiently for the junction diode to conduct. In the case of a Schottky-clamped collector junction, the bipolar transistor is held out of a condition called *saturation*, in which switching transients are inherently slowed. Thus, digital circuits using Schottky clamps are faster by several nanoseconds than unclamped circuits.

The only penalty paid in electrical performance for using the Schottky diode is the small amount of extra capacitance with which the reverse-biased Schottky diode loads the circuit. The clamped transistor takes up only slightly more surface area on the silicon chip than does the unclamped transistor, a big advantage for an integrated-circuit device. There is, however, generally some loss in fabrication yield when Schottky processing is used because metallization and surface preparation are more critical than in circuits that do not make use of Schottky barriers.

SUMMARY

An ensemble of electrons at thermal equilibrium is characterized by a single *Fermi energy*. The Fermi energy is therefore the appropriate reference for a sketch of allowed energy states versus position. It is particularly useful in drawing the appropriate thermal-equilibrium diagram for an inhomogeneous material and for systems of materials in intimate contact. If two materials, characterized by different Fermi energies (and therefore not in thermal equilibrium with each other) are brought sufficiently close to interact with one another, electrons will be transferred from the material with the higher Fermi energy to the material with the lower Fermi energy. Application of these principles to metals and semiconductors, under the idealized conditions that the bulk energy-state configuration continues to the surface and that the semiconductor is homogeneous, is the basis for Schottky, metal-semiconductor contact theory. This theory predicts blocking contacts and rectifying behavior for n-type semiconductors if the metal work function Φ_M exceeds the semiconductor work function Φ_S and ohmic behavior if Φ_S is greater than Φ_M. The inverse is true for metal contacts to p-type semiconductors. If the space-charge regions at a Schottky barrier become thin enough for substantial electron tunneling (which results from doping the semiconductor very highly), the contacts are also ohmic.

To develop a theory of metal-semiconductor contacts, it is necessary to make use of Poisson's equation in conjunction with the thermal-equilibrium band diagram. Simplifying assumptions such as the *depletion approximation* and *quasi-equilibrium* make the theory tractable. Basic Schottky theory predicts a number of observed properties successfully. Among these are the major dependences of the current-voltage characteristics and the reverse-bias capacitance behavior for the Schottky barrier. Contacts to a semiconductor having a high-resistivity region on top of a low-resistivity bulk are analyzed by the same techniques to give results for a *Mott barrier*. A similar analysis scheme applied to the Schottky ohmic contact gives rise to the *Debye length*, a characteristic measure of the extent of electric-field penetration in a region having significant free charge. Although basic Schottky theory provides much useful information about metal-semiconductor contacts, it is unsuccessful in predicting barrier heights to silicon. A major inadequacy of the theory lies in the treatment of surface effects. Surface states originate from the termination of the lattice and from imperfect and impure surfaces. Theory for metallic blocking contacts to real silicon surfaces is based upon the assumption of a thin interfacial layer of imprecise composition, but of well-specified electronic behavior. The applications of metal-semiconductor contacts in the form of Schottky diodes are widespread and growing. Schottky diodes have special relevance to integrated circuits because they are relatively easy to obtain using standard silicon planar technology.

REFERENCES

1. W. Schottky, *Naturwissenschaften, 26*, 843 (1938).
2. (a) H. A. Bethe, *Theory of the Boundary Layer of Crystal Rectifiers*, MIT Radiation Laboratory Report 43–12 (1943).
   (b) S. M. Sze, *Physics of Semiconductor Devices*, Wiley, New York, (1969), pp. 378–381.
3. H. K. Henisch, *Rectifying Semiconductor Contacts*, Oxford at the Clarendon Press, (1957), p. 172.
4. N. F. Mott, *Proceedings Cambridge Philosophical Society, 34*, 568 (1938).
5. A. Rose, *Concepts in Photoconductivity and Allied Problems*, Interscience at Wiley, New York, (1963).
6. I. Tamm, *Phys. Z. Sowjetunion 1, 733* (1933).
7. W. Shockley, *Phys. Rev., 56*, 317 (1939).
8. D. Pugh, *Phys. Rev. Lett., 12*, 390 (1964).
9. A. M. Cowley and S. M. Sze, *J. Appl. Phys., 36*, 3212 (1965).
10. S. M. Sze, *Physics of Semiconductor Devices*, Interscience at Wiley, New York, (1969).

PROBLEMS

2.1 Draw a diagram similar to Fig. 2.2 for silicon in which the doping changes abruptly from 5×10^{18} donors cm^{-3} to 8×10^{15} donors cm^{-3}.

    (a) What are the work functions associated with the two regions of the crystal?

    (b) What is the potential difference between the two regions of silicon?

2.2 The energy-band diagrams for a metal and a semiconductor are sketched to the same scales in Fig. P2.2a. Three possible band diagrams for a contact between these materials are sketched in Figs. P2.2b, P2.2c, and P2.2d. Assume thermal equilibrium.
(a) Explain why each figure is incorrect.
(b) Draw a correct diagram, assuming that basic Schottky theory applies.

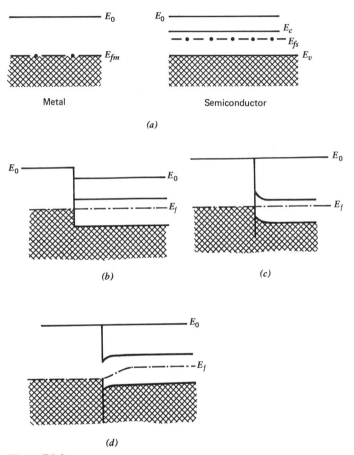

(a)

(b)

(c)

(d)

Figure P2.2

2.3 Consider metal-semiconductor junctions that behave according to simple Schottky theory.
(a) Draw the theoretical energy-band diagram for copper (work function 4.5 eV) in contact with silicon having a work function of 4.25 eV.
(b) If light were to shine on this junction and create hole-electron pairs:
(i) Which way would current flow within the device when the junction were connected into a circuit?

(ii) What would be the maximum voltage that could be measured across the junction (zero output current)?

(c) Draw the energy-band diagram for copper in contact with silicon having a work function of 4.9 eV.

(d) Compare the electrical behavior of the metal-semiconductor systems described in (a) and (c).

2.4   The accompanying data were obtained on metal contacts to silicon of equal area (Fig. P2.4). If Schottky theory applies, which metal probably has the higher work function? Which data were taken on 1 Ω-cm silicon and which on 5 Ω-cm silicon? Justify your answers and explain the use of "probably" (Consider Section 2.5.)

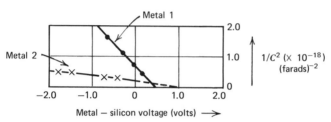

Figure P2.4

2.5 (a) Calculate the small-signal capacitance at zero dc bias and at 300°K for an ideal Schottky barrier between platinum (work function 5.3 eV) and silicon doped with $N_d = 10^{16}$ cm^{-3}. The area of the Schottky diode is 10^{-5} cm^2.

(b) Calculate the reverse bias at which the capacitance is reduced by 25% from its zero-bias value.

2.6† (a) Find the location x_m of the plane at which the barrier to emitted electrons [$E_2(x)$ in Fig. 2.9] is a maximum and prove Eq. (2.2.13).

(b) For an applied field of 10^5 V cm^{-1}, calculate x_m and $q\,\Delta\phi$.

2.7   Consider an aluminum Schottky barrier made to silicon having a constant donor density N_d. The barrier height $q\,\phi_B$ is 0.65 eV. The junction will pass high currents under reverse bias by tunneling from the metal if the barrier presented to the electrons is thin enough, as described in the following. We assume that the onset of efficient tunneling occurs when the Fermi level in the metal is equal to the edge of the conduction band (E_c) at a distance of 100 Å into the semiconductor.

(a) If this condition is reached at a total junction bias ($\phi_i - V_a$) of 5 V, what is the maximum value of N_d?

(b) What limit does this place on the resistivity of the epitaxial layers used in Schottky-clamped circuits?

(c) Draw a sketch of the energy-band diagram under the condition of efficient tunneling.

2.8† Carry through the steps needed to derive Eq. (2.3.13).

2.9† Consider Eq. (2.3.16) under conditions of low forward bias. Show that Eq. (2.3.17) can be derived by using $(1 - V_a/\phi_i)^{1/2} = \exp\left[\frac{1}{2}\ln(1 - V_a/\phi_i)\right]$ and by approximating the resultant expression for J_S. This approach leads to Eq. (2.3.17) with n = (1 + $kT/2q\phi_i$), which is generally smaller than observed values. Other effects such as rounding of the barrier contribute to values for n in Eq. (2.3.17) that are somewhat higher than those found from the expression derived in this problem.

2.10 Using linear scales, plot I versus V_a for a diode that obeys the ideal-diode law [Eq. (2.3.6)] under the condition that:
(a) $I_0 = 1$ pA and $T = 150°K$.
(b) $I_0 = 1$ nA and $T = 300°K$.
(c) $I_0 = 1$ μA and $T = 450°K$.
(d) Considering the discussion in Section 2.6, state an appropriate value for the turn-on voltage V_o for each plot of I versus V_a.
For clarity in the diagrams, use a scale change at $V = 0$. Show the forward characteristic through 5 mA.

2.11† Use the equations in Section 2.4 to represent the space charge in a Schottky ohmic contact between a metal and an n-type semiconductor and set up Poisson's equation. The equation form will be $d^2\phi/dx^2 = K \exp(\phi/V_t)$, where $V_t = kT/q$. It is convenient to convert this function of voltage ϕ and position x to a function of field $\mathcal{E}$ and voltage ϕ. This can be done by making use of $d^2\phi/dx^2 = \mathcal{E} \, d\mathcal{E}/d\phi$. The resultant equation can be solved to find $\mathcal{E} = \sqrt{2n_s kT/\epsilon_s} \exp(\phi/2V_t)$.
(a) Carry through the steps which have been outlined in this problem, (b) derive Eqs. (2.4.2), (2.4.4), and (2.4.5) by continuing with this analysis.

2.12† Draw the band diagram for a "neutral" contact, as described in Section 2.4. Consider the results of Problem 1.13, which express the random thermal flux of free electrons in a semiconductor as $qn_0 v_{th}/4$ where n_0 is the electron density and v_{th} is the thermal velocity. If currents drawn from the metal into the semiconductor are less than this value, the contact will not limit the flow and can be regarded as ohmic.
(a) Show that a contact is ohmic for fields in the semiconductor less than $v_{th}/4\mu$.
(b) Calculate the limiting ohmic current in a neutral contact made to a semiconductor having $N_d = 10^{16}$ cm^{-3} and $A = 10^{-5}$ cm^2. Take $v_{th} = 10^7$ cm s^{-1}.
(c) What is the limiting ohmic current if the bands are bent such that $q(\Phi_M - X - \phi_n) = 0.65$ eV?

2.13† Using Eqs. (2.4.2) and (2.4.5) show that half of the space charge in the Schottky ohmic contact exists within $\sqrt{2}L_D$ of the surface.

2.14 Show that the dielectric relaxation time $\tau_r = \epsilon_s/\sigma$ (discussed in Problem 1.12) can be related to the Debye length L_D by $L_D = (D\tau_r)^{1/2}$ where D is the diffusion constant in the material.

2.15 Assuming that basic Schottky theory applies, sketch the energy-band diagram for:
(a) an ohmic contact between p-type silicon and a metal at equilibrium, (b) a blocking contact between p-type silicon and a metal under 2 V reverse bias.

2.16 Both Schottky-barrier diodes and ohmic contacts are to be formed by depositing a metal on a silicon-integrated circuit. The metal has a work function of 4.6 eV. For ideal Schottky behavior, find the allowable doping range for each type of contact. Consider both p and n-doped regions and comment on the practicality of processing the integrated circuit with the required doping.

2.17[†] A back-biased Schottky diode made to silicon is to be used as a tuning element for a broadcast-band radio receiver (550 to 1650 kHz). For ease of operation, it is desirable to have the resonant frequency $(1/2\pi\sqrt{LC})$ of a tuned circuit change linearly with voltage when a dc voltage applied to the circuit changes from 0 to 5 V. If the tuning inductance L were 2 mH, it is readily calculated that the capacitance at the two extremes of bias to achieve this behavior should be 41.8 and 4.65 pF, respectively. Consider that the diode area is 10^{-3} cm^2 and find the desired dopant variation for N_d. (Calculate the numerical values and sketch a semilogarithmic plot of the results.)

Hint. To attack this problem note that

$$\frac{df}{dV} = -\frac{1}{4\pi\sqrt{LC}}\frac{1}{C}\frac{dC}{dV} = 0.22 \text{ MHz/V}$$

from the problem information. Use Eq. (2.2.10) together with $C = A\epsilon_s/x_d$ to find $N(x_d)$.

Equilibrium in electronic system :
- filled states : $n = g(E)f(E)$
- vacant states : $v = g(E) \not\!\!\!\!f\!\!\!\!E (1 - f(E))$
- at T.E. $\bar{E}_{f,m} = \bar{E}_{f,s}$.

CHAPTER 3

pn JUNCTIONS

We have seen in Chapter 2 that a system of electrons is characterized by a constant Fermi level at thermal equilibrium. This principle was used initially to deduce the energy-band diagram of a semiconductor having two doping levels. Systems that are initially not in thermal equilibrium will approach equilibrium as electrons are transferred from regions with higher Fermi energy to regions of lower Fermi energy. The transferred charge causes the buildup of barriers against further electron flow, and the potential drop across these barriers builds to a value that just equalizes the Fermi levels. These concepts formed the bases for an extensive analysis of metal-semiconductor contacts.

In this chapter we consider similar phenomena in a single crystal of semiconductor material containing regions having different dopant concentrations. We shall find it useful to employ two important approximations that are frequently encountered in device analysis. One, the *depletion approximation*, has already been introduced in Chapter 2. The second, the *quasi-neutrality approximation*, serves the same simplifying purpose as the depletion approximation; it makes very complicated problems tractable by focusing on the dominant physical effects in a given region of a device. The quasi-neutrality approximation is employed in a region of a semiconductor containing a slowly varying dopant concentration, while the depletion approximation is useful in the important case of a semiconductor containing adjacent *p*-type and *n*-type regions.

105

We discuss in some detail the transition region at a *pn* junction and the barrier associated with this transition region. We next consider the influence of an applied reverse voltage on the transition region and show that changes in this applied voltage lead to capacitive behavior. Limitations on the magnitude of the reverse bias imposed by two important breakdown mechanisms are then discussed. Since most semiconductor devices contain one or more *pn* junctions, our results will be directly applicable to the analysis of practical devices. As an example of the use of the concepts developed for a reverse-biased *pn* junction, we conclude this chapter by discussing junction field-effect transistors.

Our focus in this chapter is on *pn* junctions at equilibrium and under reverse bias. We consider currents to be negligible except when junction fields are large enough to lead to avalanche or Zener breakdown. Current flow in *pn* junctions is discussed in Chapter 4.

3.1 GRADED IMPURITY DISTRIBUTIONS

In this section we consider equilibrium in a semiconductor with a dopant concentration that varies in an arbitrary manner with position (Fig. 3.1a). We assume that initially the majority-carrier concentration equals the dopant concentration at every point in the material—a nonequilibrium condition. Then, we investigate the means by which the system approaches thermal equilibrium. From Eq. (1.2.15)

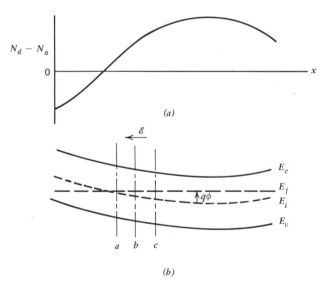

Figure 3.1 (*a*) Net dopant concentration as a function of position in an arbitrarily doped semiconductor. (*b*) Corresponding energy-band diagram versus position indicating the potential ϕ. Locations a, b, and c are referred to in the text.

we note that a gradient in the mobile carrier concentration leads to diffusion of carriers from regions of higher concentration to regions of lower concentration. As the carriers move from their initial locations, they leave behind uncompensated, oppositely charged dopant ions. This separation of positive and negative charges creates a field that opposes the diffusion flow. Equilibrium is eventually reached when the tendency of the carriers to diffuse to regions of lower density is exactly balanced by the tendency to move in the opposite direction because of the electric field created by the charge separation. Thus, the equilibrium situation is characterized by a mobile carrier distribution that does not coincide exactly with the fixed dopant distribution, and also by a *built-in* electric field that keeps the two charge distributions from separating further. The space charge resulting from the mechanism just described is typically a very small fraction of the dopant density (shown in Problem 4.3), but the field arising from it has important influences on device behavior.

We can consider the implications of the built-in electric field in terms of the energy-band diagram. Since the dopant density and carrier concentration vary with position along the semiconductor, the separation between the Fermi level and the valence and conduction-band edges also varies with position, while the Fermi level itself remains constant throughout the system at thermal equilibrium. Figure 3.1*b* shows the energy-band diagram corresponding to the dopant distribution of Fig. 3.1*a*. The separation between the Fermi level and the band edge is less in regions of high carrier density than in regions of lower density, and the intrinsic Fermi level E_i crosses the Fermi level E_f where the net dopant concentration $N_d - N_a$ is zero.

Potential. The presence of an electric field may be implied directly from this energy-band diagram, as well as from the particle model discussed above. Since Fig. 3.1*b* represents the energy-band diagram of an electron, the energy of an electron is measured by its distance above the Fermi level on the band diagram. The separation of the Fermi level from the conduction-band edge may be taken to represent the potential energy of an electron while the energy interval above the conduction-band edge represents kinetic energy. Since the electric potential ϕ is related to the potential energy by the charge $-q$, the potential may be written

$$\phi = -\frac{1}{q}(E_c - E_f) = \frac{1}{q}(E_f - E_c) \tag{3.1.1}$$

The reference point for potential energy is arbitrary, however, and we may shift our reference from E_c to E_i. The potential is then given by

$$\phi = -\frac{1}{q}(E_i - E_f) = \frac{1}{q}(E_f - E_i) \tag{3.1.2}$$

as shown in Fig. 3.1*b*. According to this definition the potential is positive for an *n*-type semiconductor and negative for *p*-type material.

Field. Since the electric field is the negative of the spatial gradient of the potential, the field $\mathscr{E}_x$ is found from Eq. (3.1.2) to be

$$\mathscr{E}_x = \frac{-d\phi}{dx} = \frac{1}{q}\frac{dE_i}{dx} \tag{3.1.3}$$

Thus, a spatial variation of the band edges (hence, of the intrinsic Fermi level) implies the existence of an electric field in the semiconductor. At point *b* of Fig. 3.1*b*, dE_i/dx is negative, and the field is directed toward the left. The resulting force on negatively charged electrons is toward the right; consequently, the field provides a force that opposes the tendency of electrons to diffuse from a high-concentration region at *c* to the low-concentration region at *a*. The situation in a *p*-type semiconductor is analogous, with the proper changes in signs and notation.

We now look at the problem quantitatively and find an expression for the electric field in terms of the graded impurity distribution. Once the system is in thermal equilibrium, there can be no current flowing at any point in the semiconductor. In particular, since thermal equilibrium requires that every process and its inverse are in balance, the electron current and the hole current must separately be zero at thermal equilibrium. In Section 1.2 we found that the total electron current is given by

$$J_n = q\mu_n n\mathscr{E}_x + qD_n\frac{dn}{dx} \tag{3.1.4}$$

This expression is applicable both in *n*-type material, where the electrons are majority carriers, and in a *p*-type semiconductor, where they are minority carriers.

The first term of Eq. (3.1.4) represents the drift current, and the second, the diffusion current. When the total electron current is zero, the two terms are exactly balanced. No current actually flows, rather the two tendencies, drift and diffusion, are in equilibrium. Since $J_n = 0$, we may solve for the field in terms of the electron concentration and its gradient

$$\mathscr{E}_x = -\frac{D_n}{\mu_n}\frac{1}{n}\frac{dn}{dx} = -\frac{kT}{q}\frac{1}{n}\frac{dn}{dx} \tag{3.1.5}$$

where we have used the Einstein relation defined in Eq. (1.2.18). Similarly, the field may be expressed in terms of the hole concentration by either considering the expression for hole current [Eq. (1.2.20)] or using the mass-action law [Eq. (1.1.13)] in Eq. (3.1.5):

$$\mathscr{E}_x = \frac{kT}{q}\frac{1}{p}\frac{dp}{dx} \tag{3.1.6}$$

Equations (3.1.5) and (3.1.6) show that, if we could find the mobile carrier concentrations and their spatial derivatives, we could specify the fields in the semiconductor.

In developing this concept we can gain some physical insight by considering the relation between the electron density and the position of the band edge (or equivalently the intrinsic Fermi level) with respect to the Fermi level. Consider an electron at x_2 in Fig. 3.2 with an energy E. A portion $E - E_c$ of this energy is kinetic energy; the remainder is potential energy. The electron can move freely in the region between x_1 and x_4 because it has energy greater than the potential energy associated with this region. The electron would require more potential energy than its total energy to enter the region to the left of x_1 or to the right of x_4. Consequently, it is forbidden to enter these regions, and the spatial variation of the energy bands indicates a *potential barrier* to the motion of the electrons.

Now let us relate the number of carriers at any two points in the material to the energy-band structure. We know that the electron density at x_2 is less than that at x_3 since the separation between the conduction-band edge and the Fermi level is greater at x_2. We may relate the carrier densities to the potential ϕ through the use of Eq. (3.1.5). Since $\mathscr{E}_x = -d\phi/dx$, we have

$$d\phi = \frac{kT}{q} \frac{dn}{n} \tag{3.1.7}$$

at any point in the semiconductor. If we integrate this equation between any two points—for example, from x_2 to x_3—we obtain

$$\phi_3 - \phi_2 = \frac{kT}{q} \ell n \frac{n_3}{n_2} \tag{3.1.8}$$

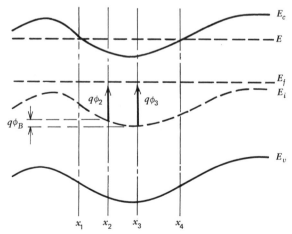

Figure 3.2   Energy-band diagram of an arbitrarily doped semiconductor, showing that an electron with energy E is constrained to remain in the region between x_1 and x_4 where $E > E_c$.

Rewriting Eq. (3.1.8) in exponential form, we find

$$\frac{n_3}{n_2} = \exp\left[\frac{q}{kT}(\phi_3 - \phi_2)\right] \tag{3.1.9}$$

As might be expected, the carrier densities depend on the potential difference $\phi_B = \phi_3 - \phi_2$ between the two points. A physical interpretation of this result is that a fraction $\exp(-q\phi_B/kT)$ of the electrons at x_3 has enough energy to reach x_2.

Poisson's Equation. Equation (3.1.9) is often useful when the variation of carrier concentrations with position must be found. As is often the case, we start such an analysis by writing Poisson's equation:*

$$\frac{d^2\phi}{dx^2} = -\frac{\rho}{\epsilon_s} = -\frac{q}{\epsilon_s}(p - n + N_d - N_a) \tag{3.1.10}$$

where ρ is the space-charge density. Using our definition of potential [Eq. (3.1.2)] in Eq. (1.1.26), we may relate the carrier concentration n to the potential function ϕ:

$$n = n_i \exp\left(\frac{q\phi}{kT}\right) \tag{3.1.11}$$

Poisson's equation can then be rewritten in the form

$$\frac{d^2\phi}{dx^2} = \frac{q}{\epsilon_s}\left(2n_i \sinh\frac{q\phi}{kT} + N_a - N_d\right) \tag{3.1.12}$$

Equation (3.1.12) is the differential equation for the potential distribution in an arbitrarily doped semiconductor material. Unfortunately, this equation cannot be solved in the general case, and approximations must be made to obtain solutions appropriate to specific situations.

To proceed, we consider two special cases. In the first, the dopant concentration varies gradually with position (small gradient case) as, for example, the donor distribution within a diffused *n*-type region. The second case is just the opposite and involves abrupt spatial variations of dopant concentration (large gradient case), as found, for example, in the junction between *p*-type and *n*-type semiconductor regions.

---

* Since much of semiconductor device analysis is concerned with the spatial variations of carriers and potentials from one region of a device to another, Poisson's equation is often encountered, as we have already seen in Chapter 2. Together with approximations that place it in mathematically tractable form, it is one of the most useful principles in semiconductor device analysis.

Quasi-Neutrality. For the small gradient case we consider n-type silicon in which the dopant concentration may vary from about 10^{18} to 10^{16} cm^{-3} in several thousand angstroms after a typical dopant diffusion. This change in dopant concentration corresponds to a potential difference of about 0.1 V and a field [Eq. (3.1.5)] of the order of 10^4 V cm^{-1} or less. If we take a specific case in which the field varies from zero to 10^4 V cm^{-1} in 0.5 μm, the average field gradient is 2×10^8 V cm^{-2}. Using this value in Poisson's equation [Eq. (3.1.10)] and neglecting the minority-carrier density p, we find that the difference between n and N_d must be less than about 10^{15} cm^{-3}. Since this number is only a small fraction of the donor concentration over most of the region under consideration, it is reasonable to approximate n by N_d in order to proceed with the analysis. In essence the approximation means that the semiconductor region under consideration is nearly neutral or *quasi-neutral*; that is, the majority-carrier distribution does not differ much from the donor distribution. The *quasi-neutrality approximation* is the more valid the smaller is the gradient in dopant density. Under the assumption of quasi-neutrality, the field in the n-type semiconductor is found directly from the donor concentration by using Eq. (3.1.5):

$$\mathscr{E}_x = -\frac{kT}{q}\frac{1}{N_d}\frac{dN_d}{dx} \tag{3.1.13}$$

In a p-type semiconductor under quasi-neutral conditions, the field is similarly

$$\mathscr{E}_x = \frac{kT}{q}\frac{1}{N_a}\frac{dN_a}{dx} \tag{3.1.14}$$

Since Eqs. (3.1.13) and (3.1.14) depend on the quasi-neutrality approximation, they are not valid if the dopant concentration has a very steep gradient.

One frequently considered case in which the quasi-neutrality approximation is useful is that of an exponential dopant distribution

$$N_d = N_0 \exp\left(\frac{-x}{\lambda}\right) \tag{3.1.15}$$

where λ is the characteristic length describing the decrease of donor atoms away from the semiconductor surface at $x = 0$. Typical Gaussian or complementary-error-function diffusion profiles (Section 1.3) are often approximated by exponential distributions for mathematical simplicity. Because of the relationship of an exponential function and its derivative, the field has a constant value $kT/q\lambda$ throughout the region in which exponential doping may be assumed. As we shall see in Chapter 5, the approximation of a constant field simplifies some useful cases of device analysis. The exponential approximation may, however, obscure important implications of real diffusions that become apparent when more detailed dopant distributions are considered.

3.2 THE *pn* JUNCTION

In the analysis of the previous section we restricted our discussion to material of one conductivity type with carrier and dopant concentrations that had a gradual variation with position. These limitations permitted a solution for the electric field in a quasi-neutral region having a spatially varying dopant concentration. Now, we consider the other extreme: a semiconductor with a dopant concentration that has a very steep gradient. In this case there can be significant departures from charge neutrality in localized regions of the semiconductor. In particular, we consider the junction between a *p*-type and an *n*-type semiconductor and find that we may treat the transition region as if it were depleted of mobile carriers. This *depletion approximation* is the opposite extreme of the *quasi-neutrality* approximation. When analyzing device structures, it is frequently very useful to divide them into regions that are assumed to be quasi-neutral and other regions that are considered to be completely depleted of mobile carriers. Although an idealization, this simplification is adequate for many calculations.

To build a model for the *pn* junction, we begin by considering initially separated *n*- and *p*-type semiconductor crystals (Fig. 3.3*a*). When these are brought into intimate contact as shown in Fig. 3.3*b*, the large difference in electron concentrations between the two materials causes electrons to flow from the *n*-type semiconductor into the *p*-type semiconductor. Similarly, holes flow from the *p*-type region into the *n*-type region. As these mobile carriers move into the oppositely doped material, they leave behind uncompensated dopant atoms near the junction so that an electric field is built up. The field lines extend from the donor ions on the *n*-type side of the junction to the acceptor ions on the *p*-type side (Fig. 3.3*c*). The presence of this field causes a potential barrier between the two types of material. When equilibrium is reached, the magnitude of the field is such that the tendency of electrons to diffuse from the *n*-type region into the *p*-type region is exactly balanced by the tendency of the electrons to drift in the opposite direction under the influence of the built-in field.

Potential Barrier. The magnitude of the potential barrier associated with this field may be found by considering the difference in the Fermi levels of the initially separated materials (Fig. 3.3*a*) as was done for the metal-semiconductor system in Chapter 2. When the two sections of semiconductor are at equilibrium, the Fermi level must be constant throughout the entire system. Consequently, the potential barrier that forms between the two materials must be just equal to the initial difference between the Fermi levels in the separated pieces of semiconductor. This is equivalent to the difference in work functions of the separated semiconductor

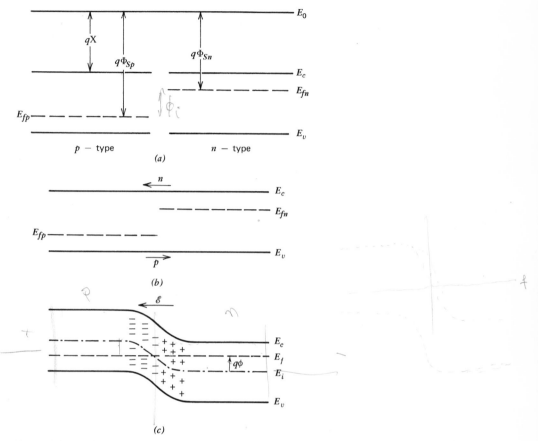

Figure 3.3 (*a*) *n*-type and *p*-type semiconductor regions separated and not in thermal equilibrium. (*b*) The two regions brought into intimate contact allowing diffusion of holes from the *p*-region and electrons from the *n*-region. (*c*) Transfer of free carriers leaves uncompensated dopant ions; these cause a field that opposes and balances the diffusion tendencies of holes and electrons.

regions, since the work function of a semiconductor is defined as

$$q\Phi_s \equiv qX + (E_c - E_f) \tag{3.2.1}$$

where qX is the electron affinity.

Far from the junction, the carrier concentrations remain at the values they had in the isolated semiconductor crystals. The electron concentration in the *n*-type material is equal to the donor concentration, and the hole density is given by

n_i^2/N_d. Similarly, in the p-type material far from the junction, the hole concentration is equal to N_a and the electron concentration is given by n_i^2/N_a. Since the carrier concentrations are known far from the junction, we can solve for the potential defined in Eq. (3.1.2) by making use of Eq. (3.1.11). Close to the junction, we cannot give specifically the free-carrier concentration without further analysis. In this region, however, we can see qualitatively from considering the sketch of potential shown in Fig. 3.3c together with Eq. (3.1.11) that the carrier concentrations are much less than the concentrations in the respective neutral materials since $|\phi|$ is small. Thus, the transition region is often referred to as a depletion region in which the space charge is overwhelmingly made up of dopant ions.

In the *depletion approximation* we assume that the semiconductor may be divided into distinct zones that are either neutral or completely depleted of mobile carriers. These zones join each other at the edges of the depletion or space-charge region, where the majority-carrier density is assumed to change abruptly from the dopant concentration to zero. The depletion approximation will appreciably simplify the solution of Poisson's equation [Eq. (3.1.10)].

Since the carrier concentrations are assumed to be much less than the net ionized dopant density in the depletion region, the second derivative of the potential is proportional to the net dopant concentration in the depletion region:

$$\frac{d^2\phi}{dx^2} = \frac{-q}{\epsilon_s}(N_d - N_a) \tag{3.2.2}$$

In the general case N_d and N_a may be functions of position, and we cannot solve Eq. (3.2.2) explicitly.

Step Junction

When considering silicon planar technology, two important dopant configurations immediately come to mind. In the first, one uniformly doped semiconductor layer is grown on another uniformly doped piece of silicon by epitaxial deposition. By proper control of this deposition process, the change in concentration from one conductivity type to the other may occur very abruptly. This type of junction may be approximated by a step function change in dopant concentration and is the easiest to handle mathematically. This *step* or *abrupt junction* may also serve as a useful approximation to a diffused junction in some situations.

Approximate Analysis. We can solve Eq. (3.2.2) for the step junction shown in Fig. 3.4, where the dopant concentration changes abruptly from N_a to N_d at $x = 0$. Using the depletion approximation, we assume that the region between $-x_p$ and x_n is totally depleted of mobile carriers, as shown in Fig. 3.4b, and that the mobile majority-carrier densities abruptly become equal to the respective dopant concentrations at the edges of the depletion region. The charge density is, therefore, zero everywhere except in the depletion region where it takes the value of the

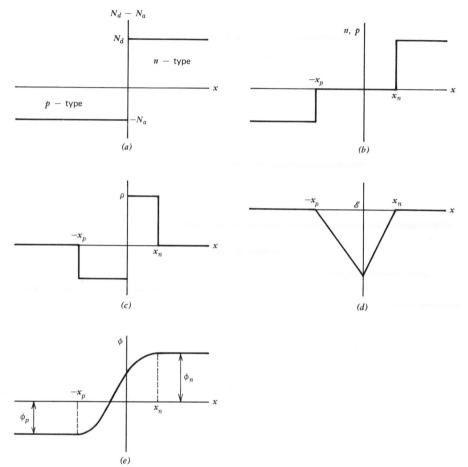

Figure 3.4 Properties of a step junction as functions of position using the complete-depletion approximation: (*a*) net dopant concentration, (*b*) carrier densities, (*c*) space charge used in Poisson's equation, (*d*) electric field found from first integration of Poisson's equation, and (*e*) potential obtained from second integration.

ionized dopant concentration (Fig. 3.4*c*). In the *n*-type material ($x > 0$) Eq. (3.2.2) becomes

$$\frac{d^2\phi}{dx^2} = -\frac{d\mathscr{E}}{dx} = -\frac{qN_d}{\epsilon_s} \tag{3.2.3}$$

which may easily be integrated from an arbitrary point in the *n*-type depletion region to the edge of the depletion region at x_n, where the material becomes neutral and the field vanishes. Carrying through this integration, we find the field to be

$$\mathscr{E}(x) = -\frac{qN_d}{\epsilon_s}(x_n - x) \qquad 0 < x < x_n \tag{3.2.4}$$

The field is negative throughout the depletion region and varies linearly with x, having its maximum magnitude at $x = 0$ (Fig. 3.4d). The direction of the field toward the left is physically reasonable since the force it exerts must balance the tendency of the negatively charged electrons to diffuse toward the left out of the neutral n-type material. The field in the p-type region is similarly found to be

$$\mathscr{E}(x) = -\frac{q}{\epsilon_s}N_a(x + x_p) \qquad -x_p < x < 0 \tag{3.2.5}$$

The field in the p-type region is also negative in order to oppose the tendency of the positively charged holes to diffuse toward the right.

At $x = 0$, the field must be continuous, so that

$$N_a x_p = N_d x_n \tag{3.2.6}$$

Thus, the width of the depleted region on each side of the junction varies inversely with the magnitude of the dopant concentration; the higher the dopant concentration, the narrower the space-charge region. In a highly asymmetrical junction where the dopant concentration on one side of the junction is much higher than that on the other side, the depletion region penetrates primarily into the lightly doped material, and the width of the depletion region in the heavily doped material may often be neglected. The charge, field, and potential at this type of junction, called a *one-sided step junction*, are identical to the results obtained for the ideal Schottky barrier, and the sketches of these quantities shown in Fig. 3.4 reduce to the equivalent sketches in Fig. 2.5. The properties of one-sided, abrupt, planar junctions are so frequently referred to that it is worthwhile to collect them in the useful nomograph form of Table 3.1.

The expressions for the field may be integrated again to obtain the potential variation across the junction. In the n-type material

$$\phi(x) = \phi_n - \frac{qN_d}{2\epsilon_s}(x_n - x)^2 \qquad 0 < x < x_n \tag{3.2.7}$$

as shown in Fig. 3.4e, where ϕ_n is the potential at the neutral edge of the depletion region as obtained by solving Eq. (3.1.11)

$$\phi_n = \frac{kT}{q}\ln\frac{N_d}{n_i} \tag{3.2.8}$$

Similarly, in the p-type material

$$\phi(x) = \phi_p + \frac{qN_a}{2\epsilon_s}(x + x_p)^2 \qquad -x_p < x < 0 \tag{3.2.9}$$

Table 3.1   **Nomograph for Uniformly Doped, One-sided, Abrupt Planar Diodes (300°K)**
(Courtesy Bell Telephone Laboratories)

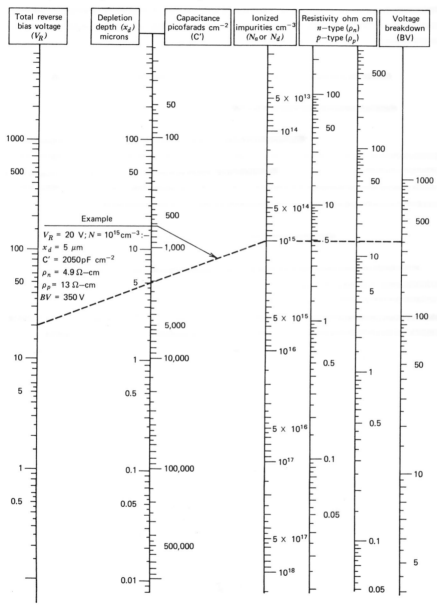

where $\phi_p < 0$ is the potential at the neutral edge of the depletion region in the
p-type material.

The total potential change ϕ_i from the neutral p-type region to the neutral
n-type region is $\phi_n - \phi_p$ ($\phi_p < 0$) and is dependent on the dopant concentration
in each region:

$$\phi_i = \phi_n - \phi_p = \frac{kT}{q} \ln \frac{N_d}{n_i} + \frac{kT}{q} \ln \frac{N_a}{n_i} = \frac{kT}{q} \ln \frac{N_d N_a}{n_i^2} \qquad (3.2.10)$$

Note that the *built-in potential* ϕ_i is positive, that is, the n-side is at a higher potential
than the p-side, which is proper to obtain a balance between drift and diffusion
across the junction. The major portion of the potential change occurs in the
region with the lower dopant concentration, and the depletion region is wider
in the same region. Note that the potential at the junction plane ($x = 0$) is not
exactly zero unless the junction is symmetrical (i.e., $N_a = N_d$). The total depletion-
region width is found from Eqs. (3.2.6) through (3.2.10) to be

$$x_n + x_p = \left[\frac{2\epsilon_s}{q} \phi_i \left(\frac{1}{N_a} + \frac{1}{N_d} \right) \right]^{1/2} \qquad (3.2.11)$$

Thus, the depletion-region width depends most strongly on the material with the
lighter doping and varies approximately as the inverse square root of the smaller
dopant concentration.

The analysis of the abrupt *pn* junction has employed the depletion approximation to solve Poisson's equation and to specify the boundary conditions. Before
proceeding further we discuss briefly the more exact solution and consider the
consequences of the depletion approximation. We again look at a junction with a
step-function change in dopant concentration at $x = 0$ (Fig. 3.5a). However, we
no longer assume an abrupt change from neutral regions to completely depleted regions at x_n and $-x_p$; instead, we consider transition regions that are only partially
depleted near these boundaries (Fig. 3.5b). Since the net charge densities in the
transition regions (Fig. 3.5c) are less than the dopant concentrations, the fields
change more gradually than predicted by the depletion approximation. The solid
line of Fig. 3.5d indicates the field found from the more exact solution, while the
dashed line represents the field found using the depletion approximation. In the
more exact analysis the field extends further into the semiconductor interior and
the space-charge zone is widened.

Debye Length.† Although an exact solution of the step junction can be obtained
without making use of the depletion approximation, we will not carry that out
at this point, but rather consider an analytical approximation for the potential
near the edges of the space-charge region (near x_n and $-x_p$) in order to investigate
the validity of the depletion approximation. If we consider only small variations
of potential from ϕ_n near $x = x_n$, we may rewrite Eq. (3.1.10) by neglecting the

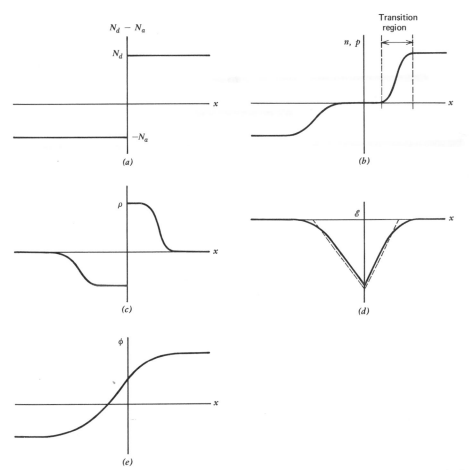

Figure 3.5 Properties of a step junction as functions of position considering a gradual transition between neutral and depleted regions: (*a*) net dopant concentration, (*b*) carrier densities, (*c*) space charge, (*d*) electric field, and (*e*) potential. Compare with Fig. 3.4.

minority-carrier concentration p and letting $\phi' = \phi_n - \phi$ in Eq. (3.1.11)

$$\frac{d^2\phi'}{dx^2} = \frac{q}{\epsilon_s}(N_d - n) = \frac{q}{\epsilon_s}\left[N_d - n_i \exp\left(\frac{q(\phi_n - \phi')}{kT}\right)\right]$$

$$= \frac{q}{\epsilon_s}N_d\left[1 - \exp\left(-\frac{q\phi'}{kT}\right)\right] \tag{3.2.12}$$

Since we are restricting ϕ' to be small, we may expand the exponential term in Eq. (3.2.12) in a Taylor series, retaining only the first two terms so that the equation reduces to

$$\frac{d^2\phi'}{dx^2} = \frac{q}{\epsilon_s} N_d \frac{q\phi'}{kT} = \frac{\phi'}{L_D^2} \qquad (3.2.13)$$

where L_D, the extrinsic *Debye length* given by

$$L_D = \left[\frac{\epsilon_s kT}{q^2 N_d}\right]^{1/2} \qquad (3.2.14)$$

is a characteristic length associated with the spatial variations of potential.* We met a similar form for the Debye length in Eq. (2.4.3) in our discussion of Schottky ohmic contacts. From the solution of Eq. (3.2.13), we see that the potential varies exponentially with distance near the edges of the space-charge region with a characteristic length equal to the extrinsic Debye length. Since the carrier concentration itself depends exponentially on the potential, the carrier concentration changes rapidly from the dopant concentration to essentially zero within a few Debye lengths. Therefore, the depletion approximation is questionable only within a few extrinsic Debye lengths of the edges of the space-charge region, x_n and $-x_p$. For symmetric junctions with typical dopant densities of 10^{16} cm^{-3}, L_D equals 400 Å while x_n is found from simultaneous solutions of Eqs. (3.2.6) and (3.2.11) to be approximately 2100 Å. Consequently, for this case the depletion approximation is reasonable, but clearly still an approximation.

Linearly Graded Junction

Most usually the step or abrupt junction is not an adequate representation for a *pn* junction made by diffusion. It is especially inapplicable to most practical cases of *double-diffused junctions*, that is, junctions formed by two successive diffusions of opposite type dopant atoms. A general analytical solution of Poisson's equation for double-diffused junctions [Eq. (3.1.12)] is not possible, and specific cases are usually considered only approximately. If more accurate results are needed, numerical techniques are used.

---

* More rigorously, the Debye length L_D describes the screening of potentials by the rearrangement of mobile carriers and depends on both hole and electron concentrations:

$$L_D = \left[\frac{\epsilon_s kT}{q^2(n+p)}\right]^{1/2}$$

For an extrinsic n-type semiconductor, the minority-carrier density is negligible and L_D takes the form of Eq. (3.2.14). The Debye length increases as carrier density decreases. The maximum length is the intrinsic Debye length L_{Di} given by the expression

$$L_{Di} = \left[\frac{\epsilon_s kT}{2q^2 n_i}\right]^{1/2}$$

In addition to the abrupt junction there is another doping profile that can be treated exactly and that gives useful results for approximating real *pn* junctions. In a *linearly graded* junction the net dopant concentration varies linearly from the *p*-type material to the *n*-type material. This type of junction is characterized by a constant *a*, which is the gradient of the net dopant concentration and thus has units of cm^{-4}. The net dopant concentration can be written as

$$N_d - N_a = ax \tag{3.2.15}$$

throughout the space-charge region (Fig. 3.6*a*). The field and potential are readily found from Poisson's equation by using the depletion approximation. Since the space charge varies linearly with position in the depletion region, the field varies quadratically and the potential varies as the third power of position in the space-charge region (Fig. 3.6). (The details of the analysis are considered in Problem 3.5.)

Although linearly graded junctions are not realized physically, many practical cases can be approximated by a linearly graded junction over at least a limited voltage range. If an abrupt junction is heated so that the dopant atoms diffuse across the junction, the junction becomes less abrupt and may be approximated

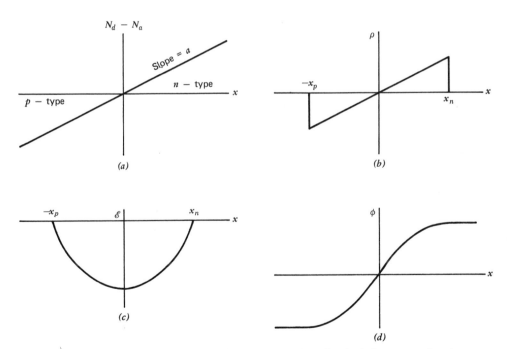

Figure 3.6 Properties of a linearly graded junction using the depletion approximation: (*a*) net dopant concentration: $N_d - N_a = ax$, (*b*) space charge, (*c*) electric field, and (*d*) potential.

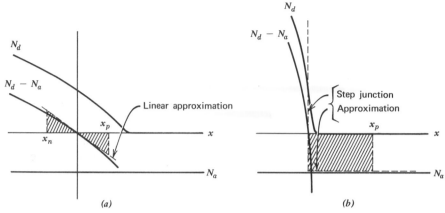

Figure 3.7 (*a*) A diffused junction may be approximated by a linearly graded junction if the diffusion length $2\sqrt{Dt}$ is much longer than the space-charge region. (*b*) A "one-sided" step junction is more appropriate if the space-charge region is much greater than the diffusion length $2\sqrt{Dt}$.

by a linearly graded junction as long as the space-charge region is narrow com-pared to the diffusion length of the impurity atoms. Even diffused junctions are sometimes approximated by linearly graded junctions over a limited distance as shown in Fig. 3.7*a* for an *n*-type diffusion into a *p*-type wafer. On the other hand, a diffusion into a uniformly doped wafer may be approximated by a one-sided step junction if the diffusion length is short compared to the width of the space-charge region (Fig. 3.7*b*).

Exponential Doping. Although more realistic junctions cannot be treated analyti-cally, some qualitative comments about their behavior may be made. We may approximate either a complementary-error-function or Gaussian diffusion profile by an exponential function over a considerable distance. Within this approximation the net dopant concentration after an *n*-type diffusion into a uniformly doped *p*-type wafer may be written as

$$N_d - N_a = N_0 e^{-x/\lambda} - N_a \qquad (3.2.16)$$

as shown in Fig. 3.8*a*. In Eq. (3.2.16), λ is the characteristic length associated with the diffusion, and N_a is the dopant concentration in the *p*-type wafer. As we saw in Section 3.1, there is an electric field in the exponentially doped, quasi-neutral, *n*-type side of the junction to balance the tendency of carriers to diffuse to regions of lower carrier density. There is no field in the uniformly doped *p*-type neutral region. The absence of mobile carriers in the depletion region on the *n*-type side of the junction requires that the field there be higher than in the quasi-neutral

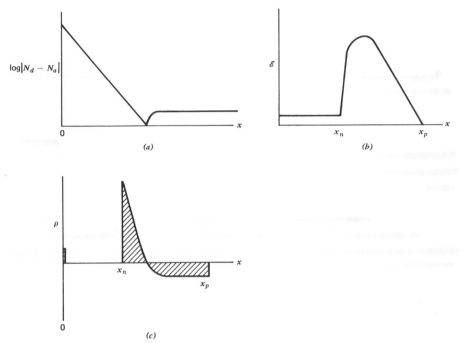

Figure 3.8   Properties of an exponential junction as functions of position: (*a*) net dopant concentration (semilogarithmic scale), (*b*) electric field, and (*c*) space charge.

n-type region; the field throughout the structure is indicated schematically in Fig. 3.8*b*. The space charge is found by differentiating the field. Since the field is constant in the quasi-neutral region, the space charge there is zero, although there is a sheet of charge at the surface (Fig. 3.8*c*). The remainder of the space charge is confined to the depletion region.

3.3 REVERSE-BIASED *pn* JUNCTIONS

In Section 3.2, we used the depletion approximation to simplify our consideration of electrical effects at *pn* junctions in thermal equilibrium. To discuss the junction under bias, we again make use of the depletion approximation together with several assumptions about the applied bias. We assume that ohmic contacts connect the *p* and *n* regions to the external voltage source so that negligible voltage is dropped at the contacts. We also consider that small currents flow through the neutral regions and cause only very low voltage drops. We consider the *n*-region to be grounded and voltage V_a to be applied to the *p*-region. With these assumptions, the entire applied voltage appears across the junction. Furthermore, under these assumptions the solutions of Poisson's equation that were found for thermal-equilibrium conditions in the last section apply also to the junction under bias. Only the total potential across the junction changes from the built-in value ϕ_i to $\phi_i - V_a$.

If V_a is positive, the barrier to majority carriers at the junction is reduced and the depletion region is narrowed. A conceptually useful way of visualizing the reduction in depletion-layer width is that the applied voltage moves majority carriers toward the edges of the depletion region where they neutralize some of the space charge. This reduces the overall depletion-region width. The total potential across the junction is $\phi_i - V_a$ with $V_a > 0$, and the junction is *forward biased*. Under forward bias, appreciable currents can flow even for small values of V_a. We shall defer further consideration of forward bias until Chapter 4.

If negative voltage is applied to the p-region, the barrier to majority flow increases. Again the total potential drop may be expressed as $\phi_i - V_a$ except that now V_a is negative and the junction is under *reverse bias*. Under reverse bias, majority carriers are pulled away from the edges of the depletion region, which therefore widens. There is very little current flow since the bias polarity aids the transfer of electrons from the p-side to the n-side and holes from the n-side to the p-side. Since these are minority carriers in each region, they are very low in density. We shall discuss fully the currents that do flow in a reverse-biased *pn* junction in Chapter 4.

Depletion Width, Maximum Field. For an abrupt *pn* junction we may find the depletion-region width as a function of voltage by replacing the built-in potential ϕ_i in Eq. (3.2.11) by $\phi_i - V_a$ so that

$$x_d = x_n + x_p = \left[\frac{2\epsilon_s}{q} \left(\frac{1}{N_a} + \frac{1}{N_d} \right) (\phi_i - V_a) \right]^{1/2} \tag{3.3.1}$$

As V_a becomes considerably larger than ϕ_i, the depletion region begins to vary as the square root of the reverse-bias voltage in a step junction. Expressions for x_d for other dopant profiles are modified from the thermal-equilibrium result similarly. For example, the depletion-region width is given by

$$x_d = \left[\frac{12\epsilon_s(\phi_i - V_a)}{qa} \right]^{1/3} \tag{3.3.2}$$

in a linearly graded junction with a doping gradient a [Eq. (3.2.15)].

It is often important to know the maximum field at the junction $\mathscr{E}_{mx}$ and its relationship to applied voltage. The step-junction case is particularly simple because the field varies linearly with distance [Eqs. (3.2.4) and (3.2.5)]. Hence, the area under the field curve that represents the potential may be written by inspection as one-half the maximum field times the depletion-layer width. Formally

$$\tfrac{1}{2}\mathscr{E}_{mx}x_d = (\phi_i - V_a)$$

so that

$$\mathscr{E}_{mx} = \frac{2(\phi_i - V_a)}{x_d} \tag{3.3.3}$$

in a step junction. For the linearly graded junction, the results of Problem 3.5 can be used to find that the maximum field is given by

$$\mathscr{E}_{mx} = \frac{3(\phi_i - V_a)}{2x_d} \tag{3.3.4}$$

Capacitance. In Section 2.2 we analyzed capacitive behavior in a metal-semiconductor junction resulting from modulation of the stored charge in the junction depletion region as a function of a varying applied voltage. Analogous behavior occurs in *pn* junctions except that the depletion region extends with bias in two directions. As the dopant concentration on one side of a *pn* junction increases, the depletion region on that side of the junction narrows. Ultimately, the width of the depletion region on the heavily doped side will become a negligible part of the total space-charge-layer width. Under this condition the *pn* junction becomes identical to the blocking metal-semiconductor junction.*

For a step junction with dopant concentrations N_a and N_d on its two sides, we may find the small-signal capacitance per unit area by using the expression for the charge Q_s in the depletion region on either side of the junction

$$\frac{Q_s}{A} = Q_s' = qN_d x_n = qN_a x_p \tag{3.3.5}$$

in the definition of the small-signal capacitance C' *per unit area*:

$$C' = \frac{dQ'}{dV_a} = qN_d \frac{dx_n}{dV_a} = qN_a \frac{dx_p}{dV_a} \tag{3.3.6}$$

Since $x_p = (N_d/N_a)x_n$ and $x_d = x_n + x_p$, we find from Eq. (3.3.1) that

$$\frac{dx_n}{dV_a} = \frac{1}{N_d} \left[\frac{\epsilon_s}{2q\left(\frac{1}{N_a} + \frac{1}{N_d}\right)(\phi_i - V_a)} \right]^{1/2} \tag{3.3.7}$$

and

$$C' = \left[\frac{q\epsilon_s}{2\left(\frac{1}{N_a} + \frac{1}{N_d}\right)(\phi_i - V_a)} \right]^{1/2} \tag{3.3.8}$$

---

* Looked at another way, as charge becomes more concentrated and finally approaches a δ function in space, fields are terminated more and more abruptly. Thus, the heavily doped semiconductor assumes the properties associated with idealized metals; in particular, all charge is confined to the surface.

Thus, for V_a greater than ϕ_i, the capacitance of a step *pn* junction will decrease approximately inversely with the square root of the reverse bias. Using Eq. (3.3.1) in Eq. (3.3.8), we obtain the expression $C' = \epsilon_s/x_d$, the general relationship for small-signal capacitance.

Although we presented a heuristic proof for the general equation $C' = \epsilon_s/x_d$ in Chapter 2, the meaning of the result is important enough to be reinforced at this point by making a careful derivation. We consider an arbitrarily doped *pn* junction having a depletion region that extends from $-x_p$ to x_n (Fig. 3.9a). The charge per unit area Q' stored between a point x and the edge of the depletion region at x_n is just

$$Q' = q \int_x^{x_n} N \, dx \tag{3.3.9}$$

where $N = N_d - N_a$ is the net dopant density. Since $\mathscr{E}_x(x_n) = 0$, the electric

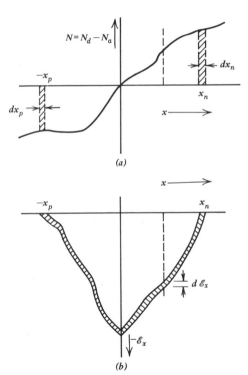

(a)

(b)

Figure 3.9 (a) Dopant concentration in an arbitrarily doped junction showing modulation of the carrier densities at the edges of the space-charge region by an applied voltage. (b) Electric field distributions for two slightly different applied voltages.

field at x is found from Gauss' law to be

$$-\mathscr{E}_x(x) = \frac{1}{\epsilon_s}\int_x^{x_n} qN\,dx = \frac{Q'}{\epsilon_s} \qquad (3.3.10)$$

and is illustrated in Fig. 3.9b. As the applied voltage V_a is changed by a small amount dV_a, the width of the n-type side of the depletion region changes by dx_n, and the charge stored between x and the edge of the depletion region changes by

$$dQ' = qN(x_n)\,dx_n \qquad (3.3.11)$$

Consequently, the field at x changes by

$$-d\mathscr{E}_x = \frac{dQ'}{\epsilon_s} = \frac{q}{\epsilon_s}N(x_n)\,dx_n \qquad (3.3.12)$$

as the voltage changes. Since the area under the $\mathscr{E}_x$ versus x curve corresponds to the total potential $\phi_i - V_a$, the differential change in voltage corresponds to the change in area under the curve (the shaded region of Fig. 3.9b):

$$dV_a \approx -x_d\,d\mathscr{E}_x = \frac{x_d}{\epsilon_s}\,dQ' \qquad (3.3.13)$$

Using Eq. (3.3.13) in the definition of the small-signal capacitance [Eq. (3.3.6)], we find

$$C' = \frac{dQ'}{dV_a} = \frac{\epsilon_s}{x_d} \qquad (3.3.14)$$

Thus, we have proven this simple relationship for an arbitrarily doped junction. The voltage dependence of the depletion-region width x_d and of the capacitance depends, of course, on the actual dopant profile.

As in the case of the metal-semiconductor junction, measurements of the variation with applied voltage of the small-signal capacitance of a pn junction may be used to determine the dopant concentration as a function of position. However, an additional complication arises in the pn junction since the depletion region extends in both directions from the junction plane, that is, the plane where $N = N_d - N_a = 0$. To investigate this further, we differentiate Eq. (3.3.14) and write $x_d = x_n + x_p$ to find

$$\frac{dC'}{dx_n} = -\frac{\epsilon_s}{(x_n + x_p)^2}\left(1 + \frac{dx_p}{dx_n}\right) \qquad (3.3.15)$$

Since the changes in charge on either side of the junction are equal in magnitude,

$$|dQ'| = |qN(-x_p)dx_p| = qN(x_n)dx_n \qquad (3.3.16)$$

and

$$\frac{dC'}{dx_n} = -\frac{(C')^2}{\epsilon_s}\left(1 + \frac{N(x_n)}{|N(-x_p)|}\right) \qquad (3.3.17)$$

Combining this expression with Eq. (3.3.11) and the definition of the capacitance, we obtain

$$N(x_n) = -\frac{(C')^3}{\epsilon_s q\,(dC'/dV_a)}\left(1 + \frac{N(x_n)}{|N(-x_p)|}\right) \qquad (3.3.18)$$

If the *p*-type side of the junction is much more heavily doped than the *n*-type side, the factor on the right of Eq. (3.3.18) reduces to unity, and the equation simplifies to the same form as Eq. (2.2.10), which we obtained for the metal-semiconductor junction. Physically, this is reasonable since a depletion region extends very little into a highly doped semiconductor, just as it extends a negligible amount into the metal side of a metal-semiconductor junction. In the more general case, however, the presence of the factor $[1+N(x_n)/N(-x_p)]$ in the expression for $N(x_n)$ complicates the interpretation of capacitance measurements in terms of dopant concentrations. The use of iterative techniques and measurements of other quantities are then required to obtain the dopant distributions.

3.4 JUNCTION BREAKDOWN

We have seen that the depletion-region width and the maximum electric field in a *pn* junction increase as reverse bias increases. Intuitively, we know that there must be physical limitations to these increases. At very high voltages, some of the materials making up the device structure, such as the insulating layers of silicon dioxide or packaging materials, may rupture, or else the current through the *pn* junction itself may increase rapidly. In the first case, irreversible damage usually occurs, and the device is destroyed. The second case—breakdown of the barrier to current flow within the junction itself—is generally not destructive (unless the high currents involved melt a portion of the junction) and is of very practical importance. The voltage at which breakdown occurs depends on the structure of the junction and the dopant concentrations in a reasonably well-defined manner and junctions can be constructed with predictable breakdown characteristics.

At high fields in a semiconductor, one of two breakdown processes can occur. In one process, the field may be so high that it exerts sufficient force on a covalently bound electron to free it. This creates two carriers, a hole and an electron to contribute to the current. In terms of an energy-band picture, in this breakdown process an electron makes a transition from the valence band to the conduction band without the interaction of any other particles. This type of breakdown is called *Zener breakdown* and involves electron tunneling through energy barriers, a phenomenon that was introduced in our discussion of ohmic contacts in Section

2.4. In the second breakdown process, free carriers are able to gain enough energy from the field between collisions for them to break covalent bonds in the lattice when they collide with it. In this process, called *avalanche breakdown*, every carrier interacting with the lattice as described above creates two additional carriers. All three carriers can then participate in further avalanching collisions, leading to a sudden multiplication of carriers in the space-charge region when the maximum field becomes large enough to cause avalanche.

One or the other of these two breakdown processes predominates in a given *pn* junction. The factors that determine which process occurs will be more evident after details of the breakdown processes have been considered. We discuss first avalanche breakdown, which is most frequently observed. Then we consider Zener breakdown and, finally, we contrast these two processes.

Avalanche Breakdown[†]

We consider an electron traveling in the space-charge region of a reverse-biased *pn* junction. The electron travels, on the average, a distance l, its mean free path, before interacting with an atom in the lattice and losing energy. The energy ΔE gained from the field $\mathscr{E}$ by the moving electron between collisions is

$$\Delta E = q \int_0^l \mathscr{E} \cdot \mathbf{dx} \qquad (3.4.1)$$

The boldface letters denote a vector product. If the electron gains sufficient energy from the field before colliding with an atom, it can break the bond between the atom core and one of the bound electrons during a collision so that three carriers—the initial electron and the hole and electron created by the collision—are free to leave the region of the collision. This process is indicated schematically in Fig. 3.10. If all three carriers are assumed to have equal mass, conservation of energy and momentum requires that the original electron possess kinetic energy of at least $\frac{3}{2}E_g$ in order to break the bond (Problem 3.10).

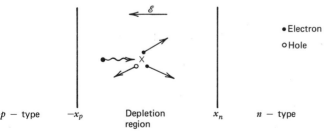

Figure 3.10 Schematic representation of the avalanche process. An incident electron (wavy arrow) gains enough energy from the field to excite an electron out of a silicon-silicon bond during a lattice collision. This creates an additional electron-hole pair.

For simplicity, we assume an abrupt junction in which the *n*-type region is much more heavily doped than the *p*-type region. In this case, as will be shown in Chapter 4, most of the carriers entering the depletion region under moderate reverse bias are electrons from the *p*-type region. The few holes entering from the *n*-type region may be neglected in our analysis. Near the edges of the space-charge region, the electric field is low and essentially no carriers can gain enough energy from the field to create a hole-electron pair before they lose their kinetic energy in a collision with the lattice. Consequently, avalanche is confined to the central portion of the space-charge region where the field is sizable. This central region has been labeled x_1 in the plot of field versus distance for this case shown in Fig. 3.11*a*.

We consider the number of carriers created by avalanche in a small volume of width dx located within x_1 at x (Fig. 3.11*b*). Let the density of electrons entering x_1 from the left at x_a be n_0. Avalanching will increase this density between x_a and x so that the electrons entering the volume dx at x from the left have a density $n_0 + n_1$. The probability that electrons create electron-hole pairs while traveling through dx is given by the product of a proportionality factor α_n, called the *ionization coefficient*, times the length dx. Since the electrons gain energy more rapidly when the field is higher, we expect the ionization coefficient to be a function of electric field and, hence, of position. The added density of electrons (and, therefore, the added density of holes) created in dx by the electrons entering from the left is

$$dn = dp = \alpha_n n \, dx = \alpha_n(n_0 + n_1) \, dx \qquad (3.4.2)$$

Since we have assumed that no holes enter the ionization region at x_c, any holes entering the infinitesimal volume at x from the right were created between x and x_c. We designate the density of these holes as p_2. The holes will also be avalanche-multiplied within dx to create added hole and electron densities.

$$dn = dp = \alpha_p p \, dx = \alpha_p p_2 \, dx \qquad (3.4.3)$$

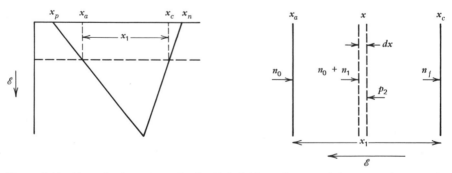

Figure 3.11 (*a*) Ionization occurs in the high-field portion x_1 of the space-charge region. (*b*) Carrier pairs are created in the region dx at x by electrons flowing from the left and by holes flowing from the right.

where α_p is the ionization coefficient for holes. The total increase in electron density created within dx is equal to the sum of Eqs. (3.4.2) and (3.4.3)

$$dn = \alpha_n(n_0 + n_1) \, dx + \alpha_p p_2 \, dx \tag{3.4.4}$$

If we let n_f be the density of electrons that reaches x_c,

$$n_f = n_0 + n_1 + n_2 \tag{3.4.5}$$

where n_2 is the density of electrons created between x and x_c, and $n_2 = p_2$ since electrons and holes are created in pairs. We may then write

$$\frac{dn}{dx} = (\alpha_n - \alpha_p)(n_0 + n_1) + \alpha_p n_f \tag{3.4.6}$$

To proceed further, it is necessary to have additional information about the ionization coefficients α_n and α_p. We can gain some useful perspective about avalanche breakdown by taking $\alpha_n = \alpha_p$ although this is only approximately true. Making that assumption and defining $\alpha \equiv \alpha_n = \alpha_p$ we can integrate Eq. (3.4.6) subject to the boundary conditions $n(x_a) = n_0$ and $n(x_c) = n_f$. The result of this integration is

$$n_f - n_0 = n_f \int_{x_a}^{x_c} \alpha \, dx \tag{3.4.7}$$

We denote the ratio of the density of electrons n_f leaving the space-charge region to the density n_0 entering as the multiplication factor M:

$$M = \frac{n_f}{n_0} = \frac{1}{1 - \int_{x_a}^{x_c} \alpha \, dx} \tag{3.4.8}$$

As the integral in the denominator of Eq. (3.4.8) approaches unity, the multiplication factor increases without bound. Hence, avalanche breakdown can be defined to occur when

$$\int_{x_a}^{x_c} \alpha \, dx = 1 \tag{3.4.9}$$

Had we not taken $\alpha_n = \alpha_p$, there would be a more complicated integral than that in Eq. (3.4.9), but a form similar to Eq. (3.4.8) in which the denominator vanishes could be derived.

As stated earlier, the ionization coefficient is a strong function of the electric field since the energy necessary for an ionizing collision is imparted to the carrier by the field. The exact field dependence of the ionization coefficient is complicated, but an expression of the form

$$\alpha = A\mathscr{E} \exp\left(-\frac{B}{\mathscr{E}}\right) \tag{3.4.10}$$

is often used. That Eq. (3.4.10) is a plausible expression for α may be seen by the following arguments. The density of ionizing collisions at x is proportional to n^*, the density of excited electrons arriving at x with sufficient energy to create electron-hole pairs. The density n^*, in turn, is just the total electron density n times the probability that an electron has not collided in a distance d necessary to gain adequate energy, that is,

$$n^* = n \exp\left(-\frac{d}{l}\right) \tag{3.4.11}$$

where l is the mean-free path. The length d may be found from Eq. (3.4.1) by letting E_1 be the minimum energy necessary for an ionizing collision and $\mathscr{E}$ be the average field that accelerates the electron:

$$d = \frac{E_1}{q\mathscr{E}} \tag{3.4.12}$$

The number of ionizing collisions in the distance dx is also proportional to dx/d if we assume that the electron collides soon after gaining sufficient energy to ionize an atom. Then

$$dn = A'n^* \frac{dx}{d} = \frac{A'q\mathscr{E}}{E_1} \left[\exp\left(-\frac{E_1}{q\mathscr{E}}\right)\right] n \, dx \tag{3.4.13}$$

where A' is a proportionality constant. Comparing Eqs. (3.4.2) and (3.4.13), we see that the functional dependence of α on field in (3.4.10) is at least reasonable. Since the ionization coefficient depends very strongly on the electric field, the multiplication factor also increases rapidly with field. Thus, a small increase in field as the voltage approaches breakdown causes a sharp increase in current, a behavior that is dramatically confirmed in practical diodes.

Not only does the ionization coefficient vary with field, and consequently with position, in the space-charge region, but the width of the space-charge region also varies with voltage. Therefore, the evaluation of M in Eq. (3.4.8) is difficult, and an empirical approximation of the form

$$M = \frac{1}{1 - (|V_R|/BV)^n} \qquad (2 < n < 6) \tag{3.4.14}$$

is often used, where $V_R < 0$ is the applied (reverse) bias and BV is the breakdown voltage, at which the current is observed to increase rapidly.

To relate the breakdown voltage to the material parameters, we consider a one-sided step junction with $N_a \ll N_d$ and assume that breakdown occurs when the maximum field in the junction reaches a critical value $\mathscr{E}_1$ that causes the integral in Eq. (3.4.8) to approach unity. Since the maximum field is approximately given by

$$\mathscr{E}_{mx} = \left(\frac{2qN_a|V_R|}{\epsilon_s}\right)^{1/2} \tag{3.4.15}$$

the breakdown voltage has the approximate form

$$BV = \frac{\epsilon_s \mathscr{E}_1^2}{2qN_a} \tag{3.4.16}$$

This equation illustrates the important fact that the breakdown voltage decreases for more heavily doped material, although the decrease is not quite as rapid as indicated. In practical diodes the breakdown voltage generally varies with doping as $N^{-2/3}$. The more gradual variation with doping in the practical case is a consequence of the more efficient avalanche multiplication in junctions having wider depletion zones. At higher dopant concentrations, a slightly higher critical field $\mathscr{E}_1$ is needed (Fig. 3.12).

All comments thus far have been in terms of one-dimensional geometry. In a practical planar junction, an important consideration arises because of the finite lateral dimension of the junction. In Section 1.3 we considered a planar junction formed by diffusion through an opening in a silicon dioxide layer. The impurities diffused laterally beneath the Si–SiO$_2$ interface as well as vertically into the silicon, creating a rounded region of the junction beneath the edge of the SiO$_2$ (Fig. 1.16d). The field in this corner region can be markedly higher than the field in the remainder of the junction, causing breakdown to occur there at unexpectedly low voltages. The reduction in breakdown voltage is especially severe for a shallow junction with a small radius of curvature. It becomes less serious as the junction depth x_j increases. The influence of junction curvature on the breakdown voltage is shown in Fig. 3.13 for a one-sided silicon planar step junction.

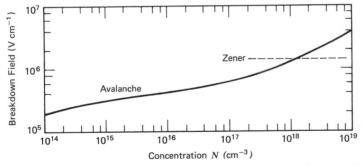

Figure 3.12 The critical electric fields for avalanche and Zener breakdown in silicon as functions of dopant concentration.[1,2,3]

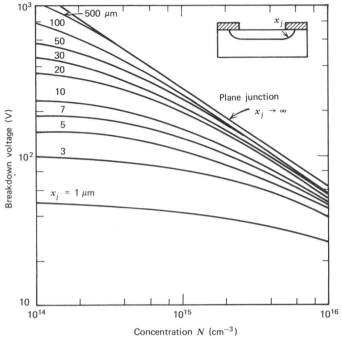

Figure 3.13 Breakdown voltage of one-sided, planar, silicon step junction showing the effect of junction curvature.[4,5]

Zener Breakdown[†]

As the dopant concentration increases, the width of the depletion region decreases and the critical field at which avalanche occurs also increases. At very high dopant concentrations, the field required for avalanche breakdown to occur exceeds the field necessary for Zener breakdown and the latter becomes more probable. As stated earlier, Zener breakdown occurs when the force exerted by the applied field is strong enough to rip an electron from its covalent bond to create an electron-hole pair directly. Figure 3.14 is an energy-band picture that shows Zener breakdown schematically.* As the figure indicates, a large number of electrons in the valence band on the *p*-type side of the junction are separated by the narrow depletion region from empty allowed states at the same energy in the conduction band of the *n*-type material. As the dopant concentration in the semiconductor increases, the width of the depletion region at a given reverse bias decreases, and

---

* For this discussion it is useful to consider electrons in both the conduction band and the valence band rather than employing the concept that only holes are important in the valence band.

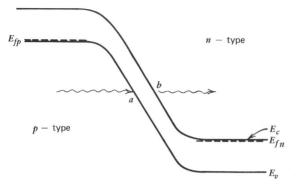

Figure 3.14 Energy-band diagram of a reverse-biased junction that has a high dopant concentration on both sides. Tunneling or Zener breakdown is likely in this type of junction.

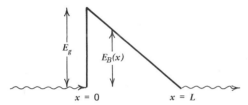

Figure 3.15 The probability of tunneling across the junction may be approximated by considering tunneling through a triangular barrier.

the energy bands in the depletion region are bent more steeply. Because of the wave nature of the electron, there is a finite probability that an electron in the valence band of the p-type semiconductor approaching the forbidden gap can *tunnel* through the forbidden region and appear at the same energy in the conduction band of the n-type semiconductor. Since the probability of transmission of an electron through a barrier is a strong function of the thickness of the barrier, tunneling is only significant in very highly doped material in which the fields are high and the depletion region is narrow.

To investigate the probability of tunneling from the valence band to the conduction band, we may approximate the barrier that the electron sees by a triangular barrier (Fig. 3.15). The height of the energy barrier E_B decreases linearly from E_g at $x = 0$ to 0 at $x = L$, and the average field is $\mathscr{E} = E_g/qL$. The probability of tunneling Θ can be approximated by using the barrier height in the equation*

$$\Theta \approx \exp\left[-2 \int_0^L \sqrt{\frac{2m^* E_B}{\hbar^2}}\, dx \right] \qquad (3.4.17)$$

* This "WKB" approximation is discussed in most quantum mechanics text books: see, for example, reference 6.

Carrying through the integration, we find the tunneling probability to be

$$\Theta = \exp\left(-\frac{B}{\mathscr{E}}\right) = \exp\left(-\frac{qBL}{E_g}\right) \tag{3.4.18}$$

where

$$B = \frac{4\sqrt{2m^*}E_g^{3/2}}{3qh} \tag{3.4.19}$$

Thus, the probability of tunneling decreases rapidly as the electric field decreases or the tunneling distance increases.

We can roughly estimate the field necessary to obtain appreciable tunneling from the above approximations. The current is the product of the area, the electron charge, the number of valence-band electrons in the *p*-region arriving at the barrier per second that "see" empty states across the barrier, and the probability that each electron tunnels through the barrier. The number of electrons arriving at the barrier can be expressed as the density $\mathscr{N}$ of electrons in the valence band times their velocity, and the current may be written as

$$I = qA\mathscr{N}v\Theta \tag{3.4.20}$$

For appreciable tunneling a current of, for example, 10 mA may flow across a junction 10^{-5} cm^2 in area. The density of electrons in the valence band that are at energies corresponding to empty allowed states in the conduction band across the barrier is comparable to the atomic density of about 10^{22} cm^{-3}. We assume that the electrons are moving with their thermal velocity of about 10^7 cm s^{-1} so that 10^{29} electrons cm^{-2} s^{-1} strike the barrier. The tunneling probability corresponding to this current is found from Eq. (3.4.18) to be about 10^{-7}. Using this value and a semiconductor band gap of 1 eV, we find the corresponding tunneling distance and electric field to be about 40 Å and 10^6 V cm^{-1}, respectively. That is, for appreciable current to flow by tunneling or Zener breakdown, the barrier that the electrons "see" must be less than about 40 Å wide and the electric field in the depletion region must be greater than about 10^6 V cm^{-1}. These values are consistent in order-of-magnitude with observations.

As the dopant concentrations decrease, the width of the space-charge region increases and the probability of tunneling decreases rapidly. Avalanche breakdown then becomes more likely than Zener breakdown. Thus, Zener breakdown is only expected for the most heavily doped junctions, while more lightly doped junctions break down by the avalanche mechanism.

Devices exhibiting Zener breakdown generally have breakdown voltages lower than those that break down by avalanching. In silicon, pure Zener breakdown is usually found in diodes having $BV < 5$ V. At higher voltages, avalanche breakdown predominates most often. Commercially available diodes with well-defined breakdown characteristics are generally termed *Zener diodes* regardless of their breakdown mechanism.

It is possible to determine whether avalanche or Zener breakdown is occurring in a junction by noting the temperature sensitivity of BV, the breakdown voltage. Although not large, the temperature variation of the two types of breakdown is of opposite sign. In the case of Zener breakdown, BV decreases with temperature because the flux of valence-band electrons available for tunneling increases as temperature rises. The effect of temperature on BV for avalanching junctions is just the opposite. The breakdown voltage increases as temperature increases because the mean-free path of energetic electrons [l in Eq. (3.4.11)] decreases.

3.5 DEVICE: THE JUNCTION FIELD-EFFECT TRANSISTOR

We have seen that the depletion-layer width of a pn junction can be varied by modulating a reverse-bias voltage applied to the junction. In this section, we consider a device that makes use of this mechanism to control the current through a region bounded by one or more pn junctions. Since little current flows into a reverse-biased pn junction, only a small amount of power is consumed at the control electrode, while substantially more power can be delivered by the controlled current. The device, which is called a *junction field-effect transistor* (often abbreviated JFET) can therefore be used as a power amplifier.

Consider the structure shown in Fig. 3.16, which consists of a lightly doped n-type layer on top of a p-type substrate. For the reasons outlined in Section 1.3 it is usual to obtain this structure by growing an n-type epitaxial layer on a p-type substrate. The n-type region is then relatively uniform and the dopant concentration is well controlled. Alternatively, a well-controlled n-type layer can be introduced by ion-implantation techniques.

After the uniform lightly doped n-type layer is formed, two heavily doped n-type regions (denoted n^+) are added by diffusion as shown in Fig. 3.16 so that good ohmic contact can be obtained (as discussed in Section 2.4) to the lightly doped region known as the *channel*. Note that the channel region is similar to the resistor

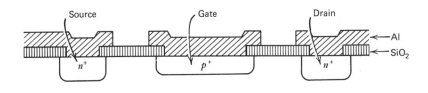

Figure 3.16 Basic structure of an n-channel, junction field-effect transistor. A lightly-doped n-type region is bounded by the p-type substrate and a p-type gate diffusion.

structure discussed in Section 1.4 except for the presence of the p^+ layer. The n^+ electrodes are called the *source* and *drain* electrodes. The source is the electrode that supplies majority carriers to the channel. Consequently, conventional current flows from drain to source in an *n*-channel JFET. If the doping type were changed in all regions of the structure shown in Fig. 3.16, the sketch would represent a *p*-channel JFET. In the *p*-channel device, conventional current flows from source to drain.

The *pn* junction above the channel in Fig. 3.16 serves as the control element when reverse bias is applied to it and is called the *gate*. The channel is defined from above by the depletion zone at the gate and from below by the depletion zone at the substrate *pn* junction. The substrate is usually at ground potential. If the drain is biased positively, current flows from it to the source through the channel. If we now ground the source and apply a negative voltage to the *p*-electrode, the junction depletion region widens and the channel narrows. As the channel narrows, its resistance increases, and less current flows from drain to source. Consequently, a signal applied to the gate controls the current flowing through the channel.

Having seen qualitatively the basis for JFET operation, we will find it straightforward to develop a quantitative theory for the device. We shall see that many of the ideas that we develop will be useful in our later discussion of the MOS or insulated-gate, field-effect transistor (MOSFET or IGFET).

Device Analysis. To analyze the JFET, we consider first a very small bias V_D applied to the drain electrode while the source is grounded. Under this condition the gate-channel bias and therefore the width of the gate depletion region is uniform along the entire channel. The voltage at the gate is V_G. An expanded view of the channel region is shown in Fig. 3.17. We assume a one-dimensional structure with a gate length L between the source and drain regions and a width W perpendicular to the plane of the paper. (Usually $W \gg L$.) Drain current flows along the dimension L. We assume a one-sided step junction at the gate with N_a in the *p*-region much greater than N_d in the channel. Therefore, the depletion layer extends

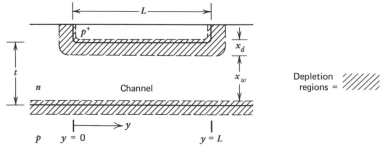

Figure 3.17 Channel region of a JFET with gate length L showing depletion regions. Typical dimensions might be $L \approx 10 \ \mu m$ and $t \approx 1.5 \ \mu m$.

primarily into the n-channel. The distance between the p-type gate and the substrate is t, the thickness of the gate depletion region in the n-type channel is x_d, and the thickness of the neutral portion of the channel is x_w. In order to focus on the role of the gate we assume that the depletion zone at the substrate junction extends primarily into the substrate so that $x_w \simeq (t - x_d)$. This is approximately true in practice.

The resistance of the channel region can be written

$$R = \frac{\rho L}{x_w W} \tag{3.5.1}$$

where $\rho = (q\mu N_d)^{-1}$ is the resistivity of the channel. Hence, the drain current is

$$I_D = \frac{V_D}{R} = \left(\frac{W}{L}\right)(q\mu_n N_d x_w V_D) \tag{3.5.2}$$

The dependence on gate voltage is incorporated in Eq. (3.5.2) by expressing $x_w = t - x_d$, where x_d from Eq. (3.3.1) is

$$x_d = \left[\frac{2\epsilon_s}{qN_d}(\phi_i - V_G)\right]^{1/2} \tag{3.5.3}$$

and ϕ_i is the built-in potential. The current can now be written as a function of the gate and drain voltages.

$$I_D = \frac{W}{L} q\mu_n N_d t \left\{1 - \left[\frac{2\epsilon_s}{qN_d t^2}(\phi_i - V_G)\right]^{1/2}\right\} V_D \tag{3.5.4}$$

The factors in front of the bracketed terms represent the conductance G_0 of the n-region if it were completely undepleted (the so-called metallurgical channel) so that Eq. (3.5.4) may be rewritten as

$$I_D = G_0 \left\{1 - \left[\frac{2\epsilon_s}{qN_d t^2}(\phi_i - V_G)\right]^{1/2}\right\} V_D \tag{3.5.5}$$

Hence, at a given gate voltage, we find a linear relationship between I_D and V_D. This is a consequence of our having assumed small applied drain voltages. The square-root dependence on gate voltage in Eq. (3.5.5) arises from our assumption of an abrupt gate-channel junction. From Eq. (3.5.5) we see that the current is maximum at zero applied gate voltage and decreases as $|V_G|$ increases. The equation predicts zero current when the gate voltage is large enough to deplete the entire channel region.

Now that we can see the physics behind the device, we remove the restriction of small drain voltages and consider the problem for arbitrary V_D and V_G values (with the restriction that the gate must always remain reverse biased). With V_D

arbitrary, the voltage between the channel and the gate is a function of position y. Consequently, the depletion-region width and therefore the channel cross section also vary with position. The voltage across the depletion region is higher near the drain than near the source in this n-channel device. Therefore, the depletion region is wider near the drain as shown in Fig. 3.18.

We now make use of what has been called the *gradual-channel approximation*. This approximation assumes that the channel- and depletion-layer widths vary slowly from source to drain so that the depletion region is influenced only by fields in the vertical dimension and not by fields extending from drain to source. In other words, the field in the y direction is much less than that in the x direction in the depletion regions, and we may find the depletion-region width from a one-dimensional analysis.

Within this approximation, we can write an expression for the increment of voltage across a small section of the channel of length dy at y as

$$d\phi = I_D \, dR = \frac{I_D \, dy}{W q \mu_n N_d (t - x_d)} \tag{3.5.6}$$

The width x_d of the depletion region is now controlled by the voltage $\phi_i - V_G + \phi(y)$, where $\phi(y)$ is the potential in the channel at point y, so that

$$x_d = \left[\frac{2\epsilon_s}{q N_d} (\phi_i - V_G + \phi(y)) \right]^{1/2} \tag{3.5.7}$$

This expression may be used in Eq. (3.5.6), which is then integrated from source to drain to obtain the current-voltage relationship for the JFET

$$\frac{I_D \int_0^L dy}{W q \mu_n N_d} = \int_0^{V_D} \left\{ t - \left[\frac{2\epsilon_s}{q N_d} (\phi_i - V_G + \phi) \right]^{1/2} \right\} d\phi \tag{3.5.8}$$

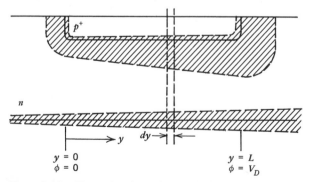

Figure 3.18 Channel region of a JFET showing variation of the width of depletion regions along the channel when the drain voltage is significantly higher than the source voltage.

After integrating and rearranging, we find

$$I_D = G_0 \left\{ V_D - \frac{2}{3} \left(\frac{2\epsilon_s}{qN_d t^2} \right)^{1/2} \left[(\phi_i - V_G + V_D)^{3/2} - (\phi_i - V_G)^{3/2} \right] \right\} \quad (3.5.9)$$

At low drain voltages Eq. (3.5.9) reduces to the simpler expression of Eq. (3.5.5), and the current increases linearly with drain voltage. However, as the drain voltage increases, the current increases more gradually.

At large enough drain voltages, Eq. (3.5.9) indicates that the current reaches a maximum and begins decreasing with increasing drain voltage, but this maximum corresponds to the limit of validity of our analysis. Indeed, from Fig. 3.19 we see

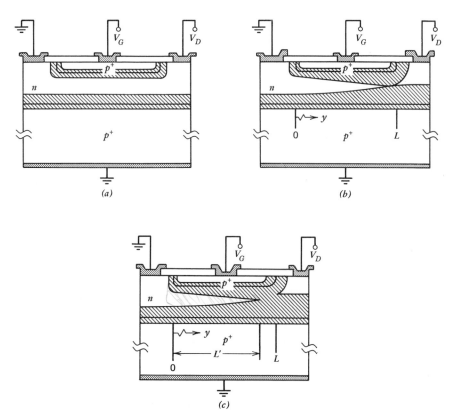

Figure 3.19 Behavior of the depletion regions in a JFET. (*a*) For very small drain voltage, the channel is nearly an equipotential and the dimensions of the depletion regions are uniform. (*b*) When V_D is increased to $V_{D\,\text{sat}}$, the depletion regions on both sides of the channel meet at the pinch-off point at $y = L$. (*c*) When $V_D > V_{D\,\text{sat}}$, the pinch-off point (at $y = L'$) moves slightly closer to the source.[7]

that, as the drain voltage increases, the width of the conducting channel near the drain decreases, until finally the channel is completely depleted in this region (Fig. 3.19*b*). When this occurs, Eq. (3.5.6) becomes indeterminate ($x_d \rightarrow t$). The equations are, therefore, only valid for V_D below the drain voltage that *pinches off* the channel. Current continues to flow when the channel has been pinched off because there is no barrier to the transfer of electrons traveling down the channel toward the drain. As they arrive at the edge of the pinched-off zone, they are pulled across it by the field directed from the drain toward the source. If the drain bias is increased further, any additional voltage is dropped across a depleted, high-field region near the drain electrode, and the point at which the channel is entirely depleted moves slightly toward the source (Fig. 3.19*c*). If this slight movement is neglected, the drain current remains constant (saturates) as the drain voltage is increased further, and the bias condition is referred to as *saturation*. The drain voltage at which the channel is entirely depleted near the drain electrode is found from Eq. (3.5.7) to be

$$V_{D\,\text{sat}} = \frac{qN_dt^2}{2\epsilon_s} - (\phi_i - V_G) \tag{3.5.10}$$

and the corresponding drain current is

$$I_{D\,\text{sat}} = G_0\left[\frac{qN_dt^2}{6\epsilon_s} - (\phi_i - V_G)\left\{1 - \frac{2}{3}\left[\frac{2\epsilon_s(\phi_i - V_G)}{qN_dt^2}\right]^{1/2}\right\}\right] \tag{3.5.11}$$

Based on our analysis, we may divide the drain current-drain voltage characteristic into three regions (Fig. 3.20): (1) the linear region at low drain voltages, (2) a region with less than linear increase of current with drain voltage, and (3) a

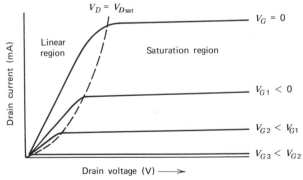

Figure 3.20 The output (drain-current, drain-voltage) characteristics of a JFET as a function of the gate voltage. Operation in the "linear" region (to the left of the $V_D = V_{D\,\text{sat}}$ curve) corresponds to Fig. 3.19*a*. In saturation, (to the right of $V_D = V_{D\,\text{sat}}$) Fig. 3.19*c* applies.

saturation region where the current remains relatively constant as the drain voltage is further increased. As expected from the physics of the device, Eq. (3.5.11) predicts the current to be maximum for zero gate bias and to decrease as negative gate voltage is applied. As the gate voltage becomes more negative, the drain saturation voltage and the corresponding current decrease so that a family of curves may be generated (Fig. 3.20), each curve showing the drain current versus drain voltage characteristic for a particular value of gate voltage. At a sufficiently negative value of gate voltage, the saturation drain current becomes zero. This turn-off voltage V_T is found from Eq. (3.5.11) to be

$$V_T = \phi_i - \frac{qN_d t^2}{2\epsilon_s} \tag{3.5.12}$$

The drain current actually increases slightly as the drain voltage is increased beyond $V_{D\,sat}$ because the end point for the integration in Eq. (3.5.8) now becomes L' rather than L, where L' is the point at which the channel becomes completely depleted $[\phi(L') = V_{D\,sat}]$ (Fig. 3.19). For $V_D > V_{D\,sat}$ the expression for $I_{D\,sat}$ in Eq. (3.5.11) is multiplied by the ratio L/L' (which is greater than unity).

Although most JFETs exhibit characteristics with well-defined saturation regions as shown in Fig. 3.20, significant departures can be seen in devices with short channel lengths. In a device with a donor density of 10^{16} cm^{-3} in the channel region, an excess of 5 V beyond $V_{D\,sat}$ depletes an additional 1.0 μm. Thus, for a typical minimum channel length of 8 μm, the ratio $L/L' \approx \frac{8}{7}$ and the deviation from simple theory is not severe. For experimental devices with channel lengths of the order of 2 μm, however, important deviations occur.*

Field-effect transistors are often operated in the saturation region where the output current is not appreciably affected by the output (drain) voltage but only by the input (gate) voltage. For this bias condition, the JFET is almost an ideal current source controlled by an input voltage. The transconductance g_m of the transistor expresses the effectiveness of the control of the drain current by the gate voltage. It is defined by the relation

$$g_m \equiv \frac{\partial I_D}{\partial V_G}\bigg|_{V_D=\text{const}} \tag{3.5.13}$$

and may be found by differentiating Eq. (3.5.9):

$$g_m = G_0 \left(\frac{2\epsilon_s}{qN_d t^2}\right)^{1/2} [(\phi_i - V_G + V_D)^{1/2} - (\phi_i - V_G)^{1/2}] \tag{3.5.14}$$

---

* Channel-length modulation is discussed further in connection with insulated-gate field-effect transistors in Chapter 8 along with other short-channel effects.

In the saturation region, g_m reaches a maximum value, which is found from Eq. (3.5.14):

$$g_{m\,\text{sat}} = G_0 \left[1 - \left(\frac{2\epsilon_s}{qN_d t^2} (\phi_i - V_G) \right)^{1/2} \right] \qquad (3.5.15)$$

Our analysis has included several simplifying assumptions. In practical devices, however, some of these assumptions may not be sufficiently valid to obtain a good match between theory and experiment. One such assumption is that the depletion-layer width is controlled by the gate-channel junction and not by the channel-substrate junction. There will be a variation of the potential across the channel-substrate junction along the channel, with the maximum potential and depletion-layer thickness near the drain. Consequently, the channel becomes completely depleted at lower drain voltage than is indicated by Eq. (3.5.10). A bias applied between the source and substrate is sometimes gainfully used to control the characteristics of a JFET. The effect of such a bias can be readily calculated by including the effect of substrate bias in the expression for the undepleted channel thickness in Eq. (3.5.6).

The presence of lightly doped regions between the active channel and the heavily doped source and drain contacts can be more troublesome. Because of photo-masking limitations and breakdown-voltage requirements, the n^+ contact diffusions generally are separated from the p-type gate diffusion. The series resistance of the intermediate regions may cause deviations from the ideal characteristics, especially at high current levels, and must be considered when analyzing or designing practical devices.

We have seen that the characteristics of the JFET are very sensitive to the thickness of the channel region t and to its dopant concentration. The n-type region in which the channel is formed can be made with excellent control by epitaxial deposition. The more crucial fabrication step is diffusion of the p-type gate. This diffusion can introduce troublesome variations in the effective channel thickness. For critical applications, ion implantation can be used to introduce the gate impurities since the quantity and location of the dopant added are better controlled by implantation than by gaseous deposition and diffusion. Introduction of both the n-type channel dopant and the p-type gate dopant by ion implantation is a further refinement of the fabrication technique and eliminates the necessity for epitaxial deposition in cases where it is not required for other devices in the same integrated circuit.

Figure 3.21 shows an integrated circuit in which several JFETs are used (type 355 operational amplifer). The interdigitated structures on the lower edge of the circuit are large input JFETs. The sources and drains are made in a comblike pattern. The gate electrode snakes back and forth between the comb electrodes that contact the source and drain.

Figure 3.21 An integrated circuit that employs JFETs to obtain high input resistance. (Type 355 Operational Amplifier, National Semiconductor Corp.)

SUMMARY

As was the case for metal-semiconductor contacts, a basic understanding of some important properties of inhomogeneously doped semiconductors can be gained by applying the principles of thermal equilibrium. If the doping in a semiconductor is nonuniform but of one type, there will be a built-in electric field that balances the diffusion tendency of free carriers with an opposing drift tendency. Often, the space charge associated with this field is small and the semiconductor can be treated as quasi-neutral so that the net dopant concentration equals the majority-carrier concentration. The *quasi-neutral approximation* becomes less valid as the gradient of the dopant concentration becomes larger. At *pn* junctions, dopant gradients are generally very large and quasi-neutrality does not apply. Effects at a *pn* junction are usually analyzed by making use of the *depletion approximation*. In the depletion approximation the space charge is assumed to terminate abruptly and is made up of uncompensated dopant ions. The use of this approximation leads to predictions for field and potential that are inaccurate within a region that is roughly an extrinsic Debye length from each respective neutral boundary. There is a built-in voltage at a *pn* junction in equilibrium just as there is in a metal-semiconductor junction. The built-in voltage equals the difference that the Fermi levels of the *p*- and *n*-regions would have if they were isolated. Very little current flows when the voltage applied to the junction has a polarity that increases the built-in potential, that is, if a positive voltage is connected to the *n*-type region with the *p*-type region grounded. This bias polarity, called reverse bias, causes the junction space-charge zone to widen. Depletion-zone widening can be detected readily by small-signal capacitance measurements made at various dc reverse biases. The specific behavior of a series of capacitance-voltage measurements can provide useful information about the dopant concentration in the junction region. At high fields, semiconductors can suddenly become highly conductive because of the internal generation of extra free carriers. When this occurs in a reverse-biased junction there is a sudden increase in current. Therefore, the phenomenon is referred to as breakdown. One of two mechanisms may be responsible for breakdown: (1) *avalanche* or (2) *tunneling* (*Zener breakdown*). An important use of reverse-biased *pn* junctions in devices is for the gate of a *junction field-effect transistor* (JFET). Operation of a JFET depends directly on the modulation of the depletion-layer width x_d in a reverse-biased *pn* junction. This reverse-bias voltage can modulate a current if that current is made to flow through a region having a cross section that depends on x_d.

REFERENCES

1. A. S. Grove, *Physics and Technology of Semiconductor Devices*, Wiley, New York, 1967, p. 193. Reprinted by permission of the publisher.
2. A. G. Chynoweth, W. L. Feldman, C. A. Lee, R. A. Logan, G. L. Pearson, and P. Aigrain, *Phys. Rev.*, *118*, 425 (1960).
3. S. L. Miller, *Phys. Rev.*, *105*, 1246 (1957).

4. A. S. Grove, *ibid.*, p. 197.

5. H. L. Armstrong, *IRE Trans. Electron Devices ED-4*, 15 (1957). Reprinted by permission of the publisher.

6. E. Merzbacher, *Quantum Mechanics*, Second Edition, Wiley, New York, 1970.

7. A. S. Grove, *ibid*, p. 244.

PROBLEMS

3.1 An abrupt silicon *pn* junction has dopant concentrations of $N_a = 1 \times 10^{15}$ cm^{-3} and $N_d = 2 \times 10^{17}$ cm^{-3}.

   (a) Evaluate the built-in potential ϕ_i at room temperature.

   (b) Using the depletion approximation, calculate the width of the space-charge layer and the peak electric field for junction voltages V_a equal to 0 V and -10 V.

3.2 Consider abrupt silicon *pn* junctions that are very heavily doped on one side and have dopant concentrations of (a) 10^{15} cm^{-3}, (b) 10^{16} cm^{-3}, (c) 10^{17} cm^{-3}, and (d) 10^{18} cm^{-3} on the less heavily doped side. Find as a function of dopant concentration the length that can be depleted of mobile carriers before the maximum electric field reaches the breakdown field shown in Fig. 3.12. What are the corresponding applied voltages?

3.3 Calculate the magnitude of the built-in field in the quasi-neutral region of an exponential impurity distribution:

$$N = N_0 \exp\left(-\frac{x}{\lambda}\right)$$

Let the surface dopant concentration be 10^{18} cm^{-3} and $\lambda = 0.4$ μm. Compare this field to the maximum field in the depletion region of an abrupt *pn* junction with acceptor and donor concentrations of 10^{18} cm^{-3} and 10^{15} cm^{-3}, respectively, on the two sides of the junction.

✓ 3.4 (a) Find and sketch the built-in field and potential for a silicon *pin* junction with the doping profile shown in Fig. P3.4. Indicate the length of each depletion region. (The symbol *i* represents a very lightly doped or nearly intrinsic region.)

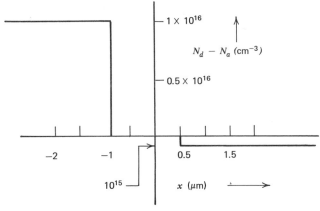

Figure P3.4

(b) Compare the maximum field to the field in a *pn* junction that contains no lightly doped intermediate region, but has the same dopant concentrations as in part (a) in the other regions.

(c) Explain physically what is happening in the intrinsic region. (That is, what does the depletion approximation mean here?)

(d) Discuss how the depletion capacitance for this structure varies with voltage, comparing it with the depletion capacitance of a structure with no intrinsic region but with the same dopant concentrations in the other regions. Sketch $1/C^2$ versus applied reverse bias for the two cases; use the same axes so that the two cases may be directly compared.

✓ 3.5 Use the *depletion approximation* to study a linearly graded junction with $(N_d - N_a) = ax$ throughout the depletion region. Assume that the equilibrium space-charge region is x_{d0} units wide. In terms of the parameters given, derive expressions for (a) the built-in potential, (b) the electric field as a function of applied voltage and distance, and (c) the depletion capacitance as a function of voltage.

3.6† We know that the capacitance of an abrupt *pn* junction varies as $V_a^{-1/2}$ for $V_a \gg \phi_i$, where ϕ_i is the built-in potential and V_a is the reverse bias applied across the junction. The capacitance of a linearly graded junction varies as $V_a^{-1/3}$. In a TV tuning circuit we need a capacitance that varies as V_a^{-1} for $V_a \gg \phi_i$. Qualitatively, discuss the general form of doping profile needed, indicating in each of the three cases the variation of the depletion-region width with voltage.

3.7† Assume that the dopant distributions in a piece of silicon are as indicated in Fig. P3.7.

(a) If a *pn* junction is desired at $x_0 = 1\ \mu m$, what should be the value of the surface-dopant density N_{d0}?

(b) Assume the depletion approximation and make a sketch of the space charge near to the junction. Approximate this space charge as if this were a linearly graded junction. Choose an appropriate value for the doping gradient a.

(c) Under the approximation of part (b), take $\phi_i = 0.7$ V and use Eqs. (3.3.2) and (3.3.4) to calculate $\mathscr{E}_{mx}$ at thermal equilibrium. Then, sketch plots of the fields at thermal equilibrium throughout the regions. (Take $N_{a0} = 10^{18}\ cm^{-3}$, $x_0 = 10^{-4}$ cm, $\lambda_a = 10^{-4}$ cm, and $\lambda_d = 2 \times 10^{-4}$ cm.)

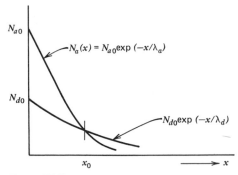

Figure P3.7

3.8 The small-signal capacitance at a *pn*-junction diode of area 10^{-5} cm^2 is measured under a total dc reverse-bias junction voltage ($\phi_i - V_a$) of 1.0 V and found to be 1.3 pF. A sketch showing $(1/C^2)$ as measured for the diode versus V_a is given in Fig. P3.8.
   (a) If the diode is considered as a one-sided step junction, what is the indicated doping level on the lower conductivity side?
   (b) Sketch the doping density on the low conductivity side of the junction. Calculate the location of any point at which the dopant density changes.
   (c) Use the intercept on the $(1/C^2)$ plot to find the doping density on the highly doped side.

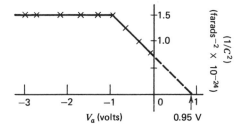

Figure P3.8

3.9 All parts of the following question refer to the system shown in Fig. P3.9. The sketch shows the cross section of a silicon wafer fabricated by a planar process. It consists of a *p* substrate of resistivity 10 Ω-cm and an epitaxial layer 2.5 μm thick having 5×10^{15} donors cm^{-3}. At *A*, a platinum (Pt) contact is made directly to the epitaxial Si surface. At *B*, an n^+ contact is diffused 1.5 μm deep into the epitaxial material. The donor density throughout the n^+ region can be assumed to be 3×10^{18} cm^{-3}. Ignore all edge effects in the following considerations.
   (a) What is the energy interval between the Fermi level and the mid-gap energy (E_i):
       (i) In the epitaxial *n*-region? (ii) In the substrate *p*-region?
   (b) The barrier height (energy difference between the conduction-band edge and the Fermi level) at the Pt-Si junction is measured to be 0.85 eV.

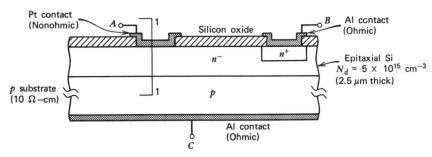

Figure P3.9

(i)   What is the built-in voltage for the Pt-Si junction?

(ii) Is the 0.85 eV barrier consistent with idealized Schottky theory? Comment intelligently.

(c) Show whether or not it would be possible to deplete fully the n^- layer below the Pt contact without reaching a breakdown field of 3×10^5 V cm^{-1}. (Consider that contact B is shorted to contact C and that both are held at ground while voltage is applied to contact A.) What voltage would be required to accomplish this depletion?

(d) Sketch the equilibrium energy-band diagram along axis 1–1 (through the Pt-Si junction and into the substrate). Make the diagram qualitatively correct; show the vacuum level and assume that ideal Schottky theory *does* apply.

3.10   To treat avalanche from first principles, consider that an incident electron collides with the lattice and frees a hole-electron pair. Assume that after the interaction, the three particles each have equal kinetic energies. Also assume that all three have equal masses. Use conservation of energy and momentum principles to find that the threshold for avalanche occurs when the incident electron possesses $(3/2) E_g$ units of kinetic energy. (Despite the many approximations made in this problem the energy is a useful first order measure of what is found in practice.)

3.11   Is Zener breakdown more likely to occur in a reverse-biased silicon or germanium *pn*-junction diode if the peak electric field is the same in both diodes? Discuss. (Consider the size of the bandgap of each material.)

3.12[†]   Carry through the analysis following Eq. (3.4.20) and show that the tunneling distance and field derived are the proper values. The coefficient B given in Eq. (3.4.19) is 7.87×10^7 V cm^{-1} for a bandgap of 1.1 eV if m^* is taken to be the rest mass of the electron. Use the conductivity effective mass in the calculations.

✔ 3.13   For a JFET derive an expression for the major temperature variation of the conductance in the linear region at a fixed gate voltage. Assume that the mobility varies as T$^{-3/2}$.

3.14[†]   For a JFET derive an expression for the drain conductance $g = \partial I_D/\partial V_D$ at a given gate voltage in the saturation region. Assume that this conductance results from the widening of the depletion region near the drain and approximate the latter by a one-dimensional step junction with very heavy doping in the drain region.

3.15[†]   Figure P3.15 shows a JFET made in an annular geometry. The junctions shown can be approximated as being linearly graded with a doping gradient $dN/dx = a$. Series resistance at the source and drain is negligible.

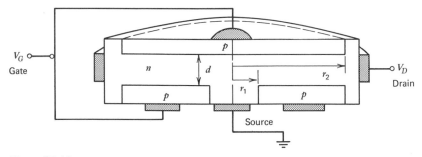

Figure P3.15

(a) What is the value of the gate turn-off voltage V_T in terms of the properties of the device?

(b) Write a differential equation that can be integrated to find the dependence of I_D on V_D, V_G, and device properties. Do not solve this differential equation, but set it in a form involving definite integrals.

(c) How does the transconductance $(\partial I_D/\partial V_G)$ for this structure depend upon the radii r_1 and r_2? Specifically, if a device is made with $r_1 = 10\,\mu m$ and $r_2 = 40\,\mu m$ and it has a g_m value of 10^{-2} mho, what value of g_m would be expected for a device made with all parameters the same except that r_2 were made $60\,\mu m$ in length?

1. Graded impurity distributions:

$$\phi = -\frac{1}{q}(E_f - E_i) \qquad - \text{Potential}$$

$$\mathcal{E}_x = -\frac{d\phi}{dx} = \frac{1}{q}dE_i \qquad - \text{Field is the change in band edge.}$$

at T.E. $\not\!\!{E}$ $\mathcal{E}_x = -\frac{kT}{q}\frac{1}{n}\frac{dn}{dx}$ or $\frac{kT}{q}\frac{1}{p}\frac{dp}{dx}$

$$\phi_3 - \phi_2 = \frac{kT}{q}\ln\frac{n_3}{n_2}$$

Poisson's Equation $\quad \frac{d^2\phi}{dx^2} = \frac{-\rho}{\epsilon_s} = -\frac{q}{\epsilon_s}(p - n + N_d - N_a)$

Quasi-neutrality: $\quad n \approx N_d \;,\; p \approx N_a$.

2. p-n Junctions:

potential barrier $q\phi_s = qX + (E_c - E_f)$

Depletion region in which $n \approx 0$, $p \approx 0$. $\;,\; \ell = N_d$ or N_a

$$N_d X_p = N_d X_n$$

$\not\!\!{E}$ $\mathcal{E}(x) = -\frac{q}{\epsilon_s}N_d(X_n - x)$ or $-\frac{q}{\epsilon_s}N_a(x + X_a)$

$$\phi(x) = \phi_n - \frac{qN_d}{2\epsilon_s}(X_n - x)^2 \;,\; \phi_n = \frac{kT}{q}\ln\frac{N_d}{n_i}$$

$$X_d = X_n + X_p = \left[\frac{2\epsilon_s}{q}\phi_i\left(\frac{1}{N_a} + \frac{1}{N_p}\right)\right]^{1/2}$$

3. Reverse - Biased.

X_d widens, low current flow. $\quad X_d = \left[\frac{2\epsilon_s}{q}\left(\frac{1}{N_a} + \frac{1}{N_p}\right)(\phi_i - V_a)\right]^{1/2}$

$$\mathcal{E}_{max} = 2(\phi_i - V_a)/X_d$$

capacitance C' (per unit area) $= \frac{dQ'}{dV_a} = \left[\frac{q\epsilon_s}{2\left(\frac{1}{N_a} + \frac{1}{N_d}\right)(\phi_i - V_a)}\right]^{1/2} = \frac{\epsilon_s}{X_d}$

CHAPTER 4

CURRENTS IN *pn* JUNCTIONS

Under reverse bias, little current flows across a *pn* junction unless the applied voltage exceeds a breakdown value. Thus far, our consideration has been of reverse-biased *pn* junctions; we have therefore neglected consideration of current flow at low bias values. Instead, we have focused attention on the barrier to majority-carrier transfers at the *pn* junction and on changes in the depletion-region width as the applied voltage was varied.

   The major topic in this chapter is current flow across a *pn* junction under both forward and reverse bias. A detailed understanding of this topic is important not only to gain insight into junction-diode behavior but, of even greater significance, to grasp the basis for junction-transistor operation. As a first step toward analyzing current flow, we derive a continuity equation for free carriers, that is, an equation that takes account of the various mechanisms affecting the population of carriers in an infinitesimal volume inside a semiconductor. To formulate some very significant terms in this equation, those which account for the generation and recombination process, it will be necessary to consider several basic physical processes in more detail than was done in Chapter 1. After deriving a continuity equation that treats generation and recombination, it will be possible to characterize the minority-carrier distributions in the quasi-neutral regions of a *pn* junction under bias. We shall consider in detail the solutions for two especially simple cases of *pn* junctions and carry out what is known as the *ideal-diode analysis*. Then, in order to relate this result to real silicon diodes it will be necessary to discuss generation and recombination in the space-charge region. The physical model developed for the steady-state current-voltage relationship will help in our consideration of charge storage and diode transients. Finally, we assess the role of *pn* junctions in integrated circuits and give a practical perspective to the earlier theory in a concluding section.

4.1 CONTINUITY EQUATION

To discuss current flow in a *pn* junction it is very useful to write an equation that is basically an accounting for the flux of free carriers into and out of an infinitesimal volume in space. A *continuity equation* such as this can be written for both majority and minority carriers in semiconductors. We shall find that solutions of the minority-carrier continuity equation in semiconductors have special importance in many device applications.

To derive a one-dimensional continuity equation for electrons, we consider an infinitesimal slice of thickness dx located at x (Fig. 4.1). The number of electrons in the slice may increase because of net flow into the volume and from net carrier generation in the slab. The overall rate of electron increase equals the algebraic sum of

(1) the number of electrons flowing into the slab, minus

(2) the number flowing out, plus

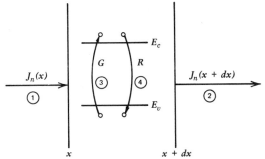

Figure 4.1 The increase in the electron density in an infinitesimal slice of thickness dx is related to the net flow of electrons into the slice and the excess of generation over recombination.

(3) the rate at which electrons are generated, minus
(4) the rate at which they recombine.

The first two components are found by dividing the currents at each side of the slice by the charge on an electron; for the present we symbolize the last two by G and R, respectively. The rate of change in the number of electrons in the slab is then

$$\frac{\partial n}{\partial t} A \, dx = \left(\frac{J_n(x)}{-q} - \frac{J_n(x + dx)}{-q} \right) A + (G_n - R_n) A \, dx \qquad (4.1.1)$$

where A is the cross-sectional area of the slice and G_n and R_n represent the generation and recombination rates per unit volume for electrons. Expanding the second term on the right-hand side in a Taylor series,

$$J_n(x + dx) = J_n(x) + \frac{\partial J_n}{\partial x} dx + \cdots \qquad (4.1.2)$$

we derive the basic continuity equation for electrons

$$\frac{\partial n}{\partial t} = \frac{1}{q} \frac{\partial J_n}{\partial x} + (G_n - R_n) \qquad (4.1.3a)$$

A similar continuity equation applies to holes except that the sign of the first term on the right-hand side of Eq. (4.1.3a) is changed because of the charge associated with a hole

$$\frac{\partial p}{\partial t} = -\frac{1}{q} \frac{\partial J_p}{\partial x} + (G_p - R_p) \qquad (4.1.3b)$$

To obtain equations that can be solved, we must relate the quantities on the right-hand side of Eqs. (4.1.3) to the carrier densities n and p. This is straightforward for the current terms, since J_n and J_p have been written in terms of carrier densities in Eqs. (1.2.19) and (1.2.20). When these equations are inserted, we obtain

$$\frac{\partial n}{\partial t} = \mu_n n(x) \frac{\partial \mathscr{E}(x)}{\partial x} + \mu_n \mathscr{E}(x) \frac{\partial n(x)}{\partial x} + D_n \frac{\partial^2 n(x)}{\partial x^2} + (G_n - R_n) \qquad (4.1.4a)$$

and

$$\frac{\partial p}{\partial t} = -\mu_p p(x) \frac{\partial \mathscr{E}(x)}{\partial x} - \mu_p \mathscr{E}(x) \frac{\partial p(x)}{\partial x} + D_p \frac{\partial^2 p(x)}{\partial x^2} + (G_p - R_p) \qquad (4.1.4b)$$

Note that we have assumed that mobility μ and diffusion constant D are not functions of x. Although this assumption is not valid in a number of important cases, the major physical effects are included in Eqs. (4.1.4) and the more exact formulations are seldom considered.

If the electric field is zero or negligible in the region under consideration, the first two terms on the right-hand side of Eqs. (4.1.4) can be neglected and the analysis is greatly simplified. Even if the field is not negligible, some of the terms in Eqs. (4.1.4) may be unimportant. For example, if the field is constant, the first term in each equation drops out. As we saw in Section 3.1, a constant field is present in an exponentially graded semiconductor. Rarely is it necessary to deal with the full complexity of Eqs. (4.1.4).

It is worthwhile to recall some basic facts of calculus before we seek specific solutions to the continuity equations. The continuity equations [Eqs. (4.1.4)] are partial differential equations because they are functions of time and position. Thus, they have an infinity of solutions, one of which applies to a given problem because it matches boundary and initial conditions. The continuity equations simplify to ordinary differential equations if, for example, one is interested in steady-state solutions. In that case, the time dependence on the left-hand side of the equations falls away and only position-dependent derivatives remain.

4.2 GENERATION AND RECOMBINATION

To formulate correct expressions for generation and recombination in terms of free-carrier densities, it is necessary to develop the physics of semiconductors beyond the discussion of Chapter 1. In Chapter 1, we noted that an electron could be excited by thermal energy from the valence to the conduction band, leaving behind a hole in the valence band. Both the hole and the electron could contribute to conduction. In thermal equilibrium this generation rate equals the rate at which the inverse process—the direct transfer of an electron into a valence-band site—occurs. These processes do describe one means for free-carrier generation and recombination. There are, however, other processes by which generation-recombination can take place.

Although direct transitions between the valence band and the conduction band occur in all semiconductors,* a complication of the crystal structures makes them very unlikely in silicon and germanium. In these materials, electrons at the lowest energy in the conduction band have a nonzero property that is equivalent to classical momentum. Since the holes at the valence-band edge do have a zero "momentum," a direct transition that conserves both energy and momentum is impossible without a lattice interaction occurring simultaneously. Thus, in silicon or germanium direct transitions across the forbidden energy gap correspond to a simultaneous interaction of three particles: the electron, the hole, and a phonon that represents the lattice interaction.

Three-particle interactions are far less likely than are two-particle interactions, such as those between a free carrier and a phonon, which can take place if there are

---

* Direct transitions are the most important generation-recombination process in gallium arsenide and gallium-arsenide-phosphide, two semiconductors that are used for luminescent diodes.

localized allowed energy states into which electrons or holes can make transitions. In practice localized states at energies between E_v and E_c are always present because of lattice imperfections caused by misplaced atoms in the crystal or, more usually, because of impurity atoms. Furthermore, they are always present in sufficient numbers to dominate the generation-recombination process in silicon and germanium. These localized states act as stepping stones. In a recombination event, for example, an electron falls from the conduction band to a state that we logically call a *recombination center*,* and then it falls further into a vacant state in the valence band, thus recombining with a hole.

Localized States: Capture and Emission

The four processes through which free carriers can interact with localized states are indicated in Fig. 4.2. The illustration shows a density N_t of states at an energy E_t within the forbidden gap. The states shown are acceptor type—that is, neutral when empty and negative when full—but the processes described apply also to donor-type states.

In the first process, *electron capture*, an electron falls from the conduction band into an empty localized state. The rate at which this process occurs is proportional to the density of electrons n in the conduction band, the density of empty localized states, and the probability that an electron passes near a state and is captured by it. The density of empty localized states is given by their total density N_t times one minus the probability $f(E_t)$ that they are occupied. When thermal equilibrium applies, f is just f_D, the Fermi function as given by Eq. (1.1.18). In the nonequilibrium case, f is not f_D, but we need not specify it further for this discussion.

The probability per unit time that an electron is captured by a localized state is given by the product of the electron thermal velocity v_{th} and a parameter σ_n called

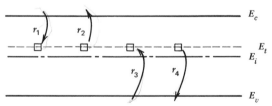

Figure 4.2 Free carriers can interact with localized states by four processes: r_1, electron capture; r_2, electron emission; r_3, hole capture, and r_4, hole emission. The localized state shown is acceptor type and at energy E_t within the forbidden-energy gap.

---

* Because the states behave symmetrically as interim sites either for generation or recombination of free carriers, they are properly termed generation-recombination centers. For brevity, however, this is usually shortened to "recombination centers."

the *capture cross section*. The capture cross section describes the effectiveness of the localized state in capturing an electron. This product $v_{th}\sigma_n$ may be visualized as the volume swept out per unit time by a particle with cross section σ_n. If the localized state lies within this volume, the electron is captured by it. The capture cross section is generally determined experimentally for a given type of localized state. A typical size for an effective recombination center is about 10^{-15} cm^2 for gold or iron.[1] An abnormally large cross section, 10^{-10} cm^2, is associated with beryllium.[2] Combining the factors discussed above, we may write the total rate of capture of electrons by the localized states as

$$r_1 = n\{N_t[1 - f(E_t)]\}[v_{th}\sigma_n] \tag{4.2.1}$$

The second process is the inverse of electron capture: that is, *electron emission*. The emission of an electron from the localized state into the conduction band occurs at a rate given by the product of the density of states occupied by electrons $N_t f(E_t)$ times the probability e_n that the electron makes this jump.

$$r_2 = [N_t f(E_t)]e_n \tag{4.2.2}$$

The emission probability can be expressed in terms of the quantities already defined in Eq. (4.2.1) by considering the capture and emission rates in the limiting case of thermal equilibrium. At thermal equilibrium, the rates of capture and emission of carriers must be equal and the probability function $f(E)$ is given by the Fermi function $f_D(E)$ [Eq. (1.1.18)]. Thus, we can write

$$r_1 = r_2 = nN_t[1 - f_D(E_t)]v_{th}\sigma_n = N_t f_D(E_t)e_n \tag{4.2.3}$$

and

$$e_n = v_{th}\sigma_n n_i \exp\left(\frac{E_t - E_i}{kT}\right) \tag{4.2.4}$$

The right-hand side of Eq. (4.2.4) can be used to replace e_n in the general case described by Eq. (4.2.2). From Eq. (4.2.4) we see that electron emission from the localized state becomes more probable when its energy is closer to the conduction band since $E_t - E_i$ is then greater.

Corresponding relationships describe the interactions between the localized states and the valence band. For example, the third process, *hole capture*, is proportional to the density of localized states occupied by electrons $N_t f(E_t)$, the density of holes and a transition probability. This probability may then be described by the product of the hole thermal velocity v_{th} and the capture cross section σ_p of a hole by the localized state. Thus,

$$r_3 = [N_t f(E_t)]p(v_{th}\sigma_p) \tag{4.2.5}$$

The fourth process, *hole emission*, describes the excitation of an electron from the valence band into the empty localized state. By arguments similar to those for electron emission, hole emission is given by

$$r_4 = \{N_t[1 - f(E_t)]\}e_p \tag{4.2.6}$$

The emission probability e_p of a hole may be written in terms of σ_p by considering the thermal-equilibrium case, for which $r_3 = r_4$, and is given by

$$e_p = v_{th}\sigma_p n_i \exp\left(\frac{E_i - E_t}{kT}\right) \tag{4.2.7}$$

Analogously to Eq. (4.2.4) the probability of emission of a hole from the localized state to the valence band becomes very much greater as the energy of the state approaches the valence-band edge.

Before making use of the equations (4.2.1, 4.2.2, 4.2.5, and 4.2.6) for the kinetics of interactions between the valence and conduction bands via the localized states at E_t, it is worthwhile to consider qualitatively the physics that they represent. First, we recognize that at thermal equilibrium $r_1 = r_2$ and $r_3 = r_4$ since thermal equilibrium requires every process to be balanced by its inverse. When we have a nonequilibrium situation, $r_1 \neq r_2$ and $r_3 \neq r_4$. To see what does apply about these rates, imagine specifically that holes in an *n*-type semiconductor are suddenly increased in number above their thermal-equilibrium value. This would cause r_3 to increase. The effect of this rate increase would be to increase r_4 and r_1 (both of which eliminate holes at E_t). If most of the holes disappear from E_t via r_1, they will remove electrons and the localized state will be an effective recombination center. If the holes are removed from the level at E_t predominantly by an increase in r_4, they will return to the valence band and the site will be effective as a *hole trap*. A given localized state will generally be effective in only one way: either as a *trap* or as a *recombination center*. Our interest for the present is in recombination centers.

Shockley, Hall, Read Recombination†

The equations describing generation and recombination through localized states or recombination centers were originally derived by Shockley and Read[3] and by Hall,[4] and the process is referred to frequently as Shockley, Hall, Read (or *SHR*) recombination. According to the *SHR* model, when nonequilibrium occurs in a semiconductor, the overall population of the recombination centers is not greatly affected. The reason for this is that they very quickly capture majority carriers (there are so many of them around) but have to wait for the arrival of a minority carrier. Thus, the states are nearly always full of majority carriers whether under thermal-equilibrium conditions or in nonequilibrium.

To clarify this behavior, consider a typical example: acceptorlike recombination centers in an *n*-type semiconductor. At thermal equilibrium, the Fermi level is near to E_c and, therefore, above the energy of the recombination centers. Hence, they are virtually all filled with electrons and r_1 and r_2 are both much greater than

r_3 or r_4. When equilibrium is disturbed by low-level excitation (in which the hole population is changed greatly while the electron population is hardly affected), r_1 will have to exceed r_2 by only a very slight amount in order to accommodate the increased rate of hole capture represented by r_3. Thus, the population of the localized states remains constant, and the net rate of electron capture $r_1 - r_2$ equals the net rate of hole capture by the states. These net rates are, in turn, just the net rate of recombination that we define by the symbol U.

$$U \equiv \frac{-dn}{dt} \equiv \frac{-dp}{dt} = r_1 - r_2 = r_3 - r_4 \qquad (4.2.8)$$

Inserting the expressions for r_1 through r_4 into Eq. (4.2.8), we can eliminate f and solve for U to obtain

$$U = \frac{N_t v_{th} \sigma_n \sigma_p (pn - n_i^2)}{\sigma_p \left[p + n_i \exp\left(\dfrac{E_i - E_t}{kT}\right) \right] + \sigma_n \left[n + n_i \exp\left(\dfrac{E_t - E_i}{kT}\right) \right]} \qquad (4.2.9a)$$

$$= \frac{(pn - n_i^2)}{\tau_{no} \left[p + n_i \exp\left(\dfrac{E_i - E_t}{kT}\right) \right] + \tau_{po} \left[n + n_i \exp\left(\dfrac{E_t - E_i}{kT}\right) \right]} \qquad (4.2.9b)$$

where $\tau_{no} = (N_t v_{th} \sigma_n)^{-1}$ and $\tau_{po} = (N_t v_{th} \sigma_p)^{-1}$.

Equation (4.2.9) shows that U is positive, and therefore there is net recombination if the pn product exceeds n_i^2. The sign changes and there is net generation if the pn product is less than n_i^2. The term $(pn - n_i^2)$ represents the restoring "force" for free-carrier populations in a nonequilibrium condition.

The dependence of U on the energy level of the recombination centers in Eq. (4.2.9) can be more easily grasped if we consider the case of equal electron and hole capture cross sections. For the case $\sigma_p = \sigma_n \equiv \sigma_o$, we can define $\tau_0 \equiv (N_t v_{th} \sigma_o)^{-1}$ and, therefore,

$$U = \frac{(pn - n_i^2)}{\left[p + n + 2n_i \cosh\left(\dfrac{E_t - E_i}{kT}\right) \right] \tau_0} \qquad (4.2.10)$$

The dependence on the energy level of the recombination center is contained in the hyperbolic cosine term in Eq. (4.2.10). This term is symmetric around $E_t = E_i$, which reflects a symmetry in the capture of holes and electrons by the center. The denominator is at its minimum value at $E_t = E_i$, so that U is a maximum value for recombination centers having energies near the middle of the gap. In Fig. 4.3 (solid curve) U normalized to its maximum value has been plotted versus $(E_t - E_i)/kT$ from Eq. (4.2.10) for a reasonable case of recombination in an n-type semiconductor. The conditions used for the figure are: $p \ll n$, $n = 10^{16}$ cm^{-3},

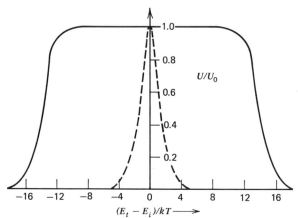

Figure 4.3   Recombination rate (solid curve) and generation rate (dotted curve) as functions of the difference between the energy of the recombination center E_t and the intrinsic Fermi energy E_i. The curves are normalized to the rates for $E_t = E_i$, and have been drawn using Eq. (4.2.10) with parameter values listed in the text.

$(pn - n_i^2) = 1.5 \times 10^{31}$ cm^{-6}, and $\tau_0 = 10^{-7}$ s. We shall find shortly that these values are appropriate in the quasi-neutral region near a forward-biased _pn_ junction. As a second illustration of Eq. (4.2.10), Fig. 4.3 (dotted curve) shows the dependence of U (normalized to its maximum value) on $(E_t - E_i)/kT$ for a case involving generation. Again, reasonable values have been used for the other terms in Eq. (4.2.10): $(pn - n_i^2) = -2.1 \times 10^{20}$ cm^{-6}, p and n much less than n_i, and $\tau_0 = 10^{-7}$ s. These conditions might fit the region near the center of a _pn_ junction depletion zone under reverse bias. The results sketched in Fig. 4.3 show that the dependence of the centers upon energy is more pronounced for the case of generation in a depleted semiconductor than for recombination in an undepleted region. The reason for this is that the energy level plays a greater role in equalizing the rates of transfer between the recombination center and the conduction and valence bands when all carrier densities are small. In either generation or recombination, however, the most effective recombination centers will have E_t close to E_i. As practical examples, gold and copper give rise to two very effective recombination centers. The values of $(E_t - E_i)$ in silicon for these elements are 0.03 and 0.01 eV, respectively.[5]

Equations (4.2.9) and (4.2.10) are the major results of the _SHR_ recombination analysis. They show that U, the net recombination rate through a recombination center, will be a function of the free-carrier densities as well as specific properties of the recombination site. The equations usually can be simplified for a particular problem. The predictions of material and device behavior based upon these equations have been generally corroborated by experiments.

Excess-Carrier Lifetime

To understand the physical significance of the net recombination rate U, we may consider a semiconductor with no current flow in which thermal equilibrium is disturbed by the sudden creation of excess electrons and holes. These excess carriers then decay spontaneously as the semiconductor returns to thermal equilibrium. Solutions of the continuity equations [Eqs. (4.1.3)] for this case give the excess electron density as a function of time. Let us consider this problem under an assumption that is very frequently satisfied in practice; we assume that the disturbance of equilibrium corresponds to *low-level injection*. In this condition the external disturbance does not appreciably change the total free-carrier density from its equilibrium value. For the case we are considering, if we call the extra injected electron density n' and the extra hole density p', then low level injection implies that n' and p' are both much less than $(n_o + p_o)$ where n_o and p_o represent the thermal-equilibrium densities of carriers in the semiconductor. From these definitions, $n' \equiv n - n_o$ and $p' \equiv p - p_o$.

If $\sigma_n = \sigma_p$, then the recombination rate U is given by Eq. (4.2.10) and hence the continuity equation (4.1.3a) can be written

$$\frac{dn'}{dt} = G - R = -U = \frac{-(n_o + p_o)n'}{\left(n_o + p_o + 2n_i \cosh\left[\dfrac{E_t - E_i}{kT}\right]\right)\tau_0} \qquad (4.2.11)$$

Solving for n', we find that the excess carrier density decays exponentially with time

$$n'(t) = n'(0)\exp(-t/\tau_n) \qquad (4.2.12)$$

where the *lifetime* τ_n is given by

$$\tau_n = \left[\frac{n_o + p_o + 2n_i\cosh\left(\dfrac{E_t - E_i}{kT}\right)}{(n_o + p_o)}\right]\tau_0 \qquad (4.2.13)$$

As we noted in the previous section, for recombination centers to be effective, the term $(E_t - E_i)$ is relatively small and therefore the third term in the numerator of Eq. (4.2.13) is negligible compared to the first two terms. Equation (4.2.13) then reduces to

$$\tau_n = \tau_0 = \frac{1}{N_t v_{th}\sigma_o} \qquad (4.2.14)$$

independent of carriers conc majority

and

$$U = \frac{n'}{\tau_n} \qquad (4.2.15)$$

Equation (4.2.14) shows that the excess-carrier lifetime is independent of the majority-carrier concentration for recombination through recombination centers under low-level injection. We may understand this behavior physically by considering the kinetics of the recombination process. For example, in a *p*-type semiconductor most of the recombination centers are empty of electrons since $E_f < E_t$ (assuming traps near midgap). The recombination process is therefore limited by the capture of electrons from the conduction band. Once an electron is captured by a recombination center, one of the many holes in the valence band is quickly captured. Thus, the rate-limiting step in the recombination process is the capture of a minority carrier by the recombination center; this is insensitive to the majority-carrier population.

Minority-carrier lifetimes can vary widely, dependent upon the density and type of recombination centers in the semiconductor. For devices such as radiation detectors that depend upon long lifetime for minority carriers, special care can yield silicon having millisecond lifetimes or greater. For integrated circuits, typical values range from a tenth of a microsecond to tens of microseconds.

Surface Recombination.[†] Thus far, we have considered generation-recombination centers that are uniformly distributed throughout the bulk of the semiconductor material. As discussed in Section 2.5, a semiconductor surface is the location of an abundance of extra localized states having energies within the forbidden gap. The presence of a passivating layer of silicon dioxide over the semiconductor surface, as is usual in devices made by the planar process, ties up many of the bonds that would otherwise contribute to surface states and protects the surface from foreign atoms. A passivating oxide can reduce the density of surface states from about 10^{15} to about 10^{11} cm^{-2}. Even with passivated surfaces, however, surface states provide additional generation-recombination centers over those present in the bulk. Since the properties of many practical semiconductor devices are affected by generation and recombination at the surface, a brief discussion is in order.

The kinetics of generation-recombination at the surface are similar to those considered for bulk centers with one significant exception. While we considered the volume density N_t (cm^{-3}) of bulk centers, we must discuss the area density N_{st} (cm^{-2}) of surface centers. Although the N_{st} surface centers are actually distributed over a thickness of many atoms, the poorly defined microscopic structure near the semiconductor surface makes useful a description in terms of an equivalent number of states located at the surface. We may write an expression for the recombination rate U per unit area at the surface analogous to Eq. (4.2.9):

$$U_s = \frac{N_{st} v_{th} \sigma_n \sigma_p (p_s n_s - n_i^2)}{\sigma_p \left[p_s + n_i \exp\left(\dfrac{E_i - E_{st}}{kT}\right) \right] + \sigma_n \left[n_s + n_i \exp\left(\dfrac{E_{st} - E_i}{kT}\right) \right]} \qquad (4.2.16)$$

where the subscript s denotes concentrations and conditions near the surface and E_{st} is the energy of the surface generation-recombination centers. To stress the physical significance of surface recombination, we again simplify the mathematics by considering the most efficient centers, which are located near midgap, and equal capture cross sections for electrons and holes. With these assumptions Eq. (4.2.16) reduces to

$$U_s = N_{st} v_{th} \sigma \; \frac{(p_s n_s - n_i^2)}{p_s + n_s + 2n_i \cosh\left(\dfrac{E_{st} - E_i}{kT}\right)} \qquad (4.2.17)$$

We saw in Chapter 2 that the surface of a semiconductor is often at a different potential than the bulk so that surface carrier concentrations may differ from their values in the neutral bulk region. Even in the case of oxide-passivated surfaces, there is generally a space-charge region near the surface of the semiconductor as shown in Fig. 4.4 for p-type silicon. As we shall make clear in Chapter 7, silicon surfaces tend to be less strongly p-type or more n-type than the bulk. If we assume that the pn product remains constant throughout the space-charge region, the product at the surface $p_s n_s$ may be expressed in terms of quantities at

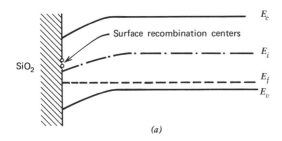

(a)

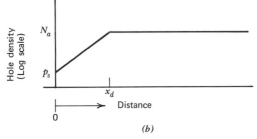

(b)

Figure 4.4 (*a*) Sketch of the energy-band diagram near the surface of p-type silicon that has been covered with a passivating oxide. (*b*) Hole density near the surface.

the neutral edge of the space-charge region:

$$p_s n_s = p_p(x_d)n_p(x_d) \approx N_a n_p(x_d) \qquad (4.2.18)$$

in a *p*-type semiconductor.* Equation (4.2.17) may then be written

$$U_s = N_{st}v_{th}\sigma \frac{N_a[n_p(x_d) - n_{po}]}{(p_s + n_s + 2n_i)} = N_{st}v_{th}\sigma \frac{N_a}{(p_s + n_s + 2n_i)} n_p'(x_d) \qquad (4.2.19)$$

where we have assumed $E_t \approx E_i$. In Eq. (4.2.19) we have expressed the surface recombination rate U_s in terms of the deviation n_p' of the minority-carrier concentration from its equilibrium value at the interior boundary of the surface space-charge region.

The coefficient of n_p' on the right side of Eq. (4.2.19) is usually defined as a parameter s, which describes the characteristics of the surface recombination process:

$$s = N_{st}v_{th}\sigma \frac{N_a}{(p_s + n_s + 2n_i)} \qquad (4.2.20)$$

The value of s depends on the physical nature and density of the surface generation-recombination centers as well as on the potential at the surface. If the surface region is depleted of mobile carriers, n_s and p_s are very small, and s is large. If the surface is neutral, $p_s \sim N_a$; s is small and given by

$$s = s_o = N_{st}v_{th}\sigma \qquad (4.2.21)$$

where the subscript o denotes that the surface and bulk are at the same potential; that is, the surface region is neutral. The dependence of s on surface potential is of great practical significance.

The dimensions of s are cm s^{-1}, and s is consequently called the surface recombination velocity, although it is not directly related to an actual velocity. A physical interpretation of s can be obtained by comparing Eq. (4.2.21) to Eq. (4.2.14) for the minority-carrier lifetime; s is related to the rate at which excess carriers recombine at the surface just as $1/\tau$ is related to the rate at which they recombine in the bulk.

As an example of the use of surface recombination velocity, consider an *n*-type semiconductor that is illuminated through a passivating oxide so that the hole and electron densities exceed their thermal-equilibrium values. At thermal equilibrium, assume that $n_s = 10^{16}$ cm^{-3} and $p_s = 2.1 \times 10^4$ cm^{-3}. If the excess carriers are created by the light at a rate of 10^{14} cm^{-2} s^{-1} and the surface hole density under illumination increases to 10^{12} cm^{-3}, what value of s would lead to 50% of the excess carriers recombining at the surface? *Answer:* $s \times 10^{12} = 0.5 \times 10^{14}$ or $s = 50$ cm s^{-1}.

---

* The subscripts p and n are used generally to denote carrier concentrations in *p*- and *n*-type material, respectively.

4.3 CURRENT-VOLTAGE CHARACTERISTICS OF *pn* JUNCTIONS

Through use of the continuity equations [Eqs. (4.1.4)] together with the concept of excess-carrier lifetime as derived from the *SHR* generation-recombination model, we are in a position to find expressions for current in a *pn* junction under bias. Solutions of the continuity equations in the quasi-neutral regions will give carrier densities in terms of position and time. Expressions for current are then obtained directly by using Eqs. (1.2.19) and (1.2.20), which define carrier flow in terms of carrier densities. The total current consists, in general, of the sum of four components: hole and electron drift currents and hole and electron diffusion currents.

We consider a *pn*-junction diode that is connected to a voltage source with the *n*-region grounded and the *p*-region at V_a volts relative to ground. The diode structure is of constant cross-sectional area A, and it has the longitudinal dimensions sketched in Fig. 4.5. The junction is not illuminated, and carrier densities within the diode are only influenced by the applied voltage. The applied voltage V_a is dropped partially across the quasi-neutral regions and partially across the junction itself. Since the voltage drops across the quasi-neutral regions are ohmic (current times resistance), they are very small at low currents. In fact, based on the structure of most *IC* junctions, the ohmic drops can be considered negligible up to relatively high values of bias and we shall assume these drops to be insignificant at this point. We shall reconsider this assumption, however, at the end of our analysis.

If we assume then that V_a is sustained entirely at the junction, the total junction voltage will be $\phi_i - V_a$ under bias where ϕ_i is the built-in voltage. If V_a is positive,

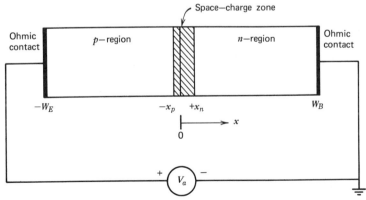

Figure 4.5 Sketch of a *pn*-junction diode structure used in the discussion of currents. The sketch shows the dimensions (cross-sectional area A is assumed to be uniform) and the bias convention.

that is, for *forward bias*, the applied voltage reduces the barrier to the diffusion flow of majority carriers at the junction. The reduced barrier, in turn, permits a net transfer rate of holes from the *p*-side into the *n*-side and of electrons from the *n*-side into the *p*-side. When these transferred carriers enter the quasi-neutral regions, they are minority carriers and they are very quickly neutralized by majority carriers that enter the quasi-neutral regions from the ohmic contacts at the ends. This neutralization of injected carriers is just the dielectric-relaxation process that we considered in Chapter 1. Once the minority carriers have been injected across the space-charge region, they tend to diffuse into the neutral interior.

If V_a is negative, that is, for *reverse bias*, the barrier height to diffusing majority carriers is increased. Since equilibrium is disturbed, minority carriers near the junction space-charge zone tend to be depleted. The majority-carrier concentration is likewise reduced by the process of dielectric relaxation.

From these few remarks, we can see that the minority-carrier densities deserve special attention because they really determine what currents flow in a *pn* junction. The majority carriers act only as suppliers of the injected minority-carrier current or as charge neutralizers in the quasi-neutral regions. It is helpful to consider the majority carriers as willing "slaves" of the minority carriers. Accordingly, we shall seek solutions of the continuity equations for the minority-carrier densities in each of the quasi-neutral regions.

Boundary Values of Minority-Carrier Densities

To write these solutions in a useful fashion, however, it will be necessary to relate the boundary values of the minority-carrier densities to the applied voltage V_a. The most straightforward way of doing this is to make two additional assumptions: first, that the applied bias leads to low-level injection and, second, that the applied bias is small enough so that the detailed balance between majority and minority populations across the junction regions is not appreciably disturbed. The first of these assumptions, low-level injection, has been considered in Section 4.2. Briefly, it implies that the majority-carrier populations are negligibly changed at the edge of the quasi-neutral regions by the applied bias. The assumption that detailed balance nearly applies allows the use of Eq. (3.1.9) across the junction where the potential difference is known to be $\phi_i - V_a$.

Both of these assumptions are certainly valid when V_a is very small ($|V_a| \ll \phi_i$). Their validity at higher biases deserves more careful consideration. As with our previous assumptions we shall defer this consideration until the analysis is completed.

By the assumption of low-level injection, the electron density at the quasi-neutral boundary in the *n*-region next to the *pn* junction is equal to the dopant density whether at equilibrium or under bias. As in Chapter 3 we call this position x_n (Fig. 4.5), and we denote the thermal equilibrium density with an extra subscript *o*.

Likewise, the boundary of the quasi-neutral *p*-region is at $-x_p$ and the hole density there is equal to the acceptor dopant density at equilibrium as well as under bias. Summarizing these statements with equations, we have

$$n_{po}(-x_p) = n_{no}(x_n) \exp\left(\frac{-q\phi_i}{kT}\right)$$

$$= N_d(x_n) \exp\left(\frac{-q\phi_i}{kT}\right) \tag{4.3.1}$$

$$p_{no}(x_n) = p_{po}(-x_p) \exp\left(\frac{-q\phi_i}{kT}\right)$$

$$= N_a(-x_p) \exp\left(\frac{-q\phi_i}{kT}\right) \tag{4.3.2}$$

$$n_p(-x_p) = N_d(x_n) \exp\left[\frac{-q(\phi_i - V_a)}{kT}\right] \tag{4.3.3}$$

and

$$p_n(x_n) = N_a(-x_p) \exp\left[\frac{-q(\phi_i - V_a)}{kT}\right] \tag{4.3.4}$$

These four equations can be combined to express the excess minority-carrier densities at the boundaries in terms of their thermal equilibrium values. We define the excess densities by

$$n' \equiv n - n_o \tag{4.3.5}$$

and

$$p' \equiv p - p_o \tag{4.3.6}$$

Then

$$n'_p(-x_p) = n_{po}(-x_p) \left[\exp\left(\frac{qV_a}{kT}\right) - 1 \right] \tag{4.3.7}$$

and

$$p'_n(x_n) = p_{no}(x_n) \left[\exp\left(\frac{qV_a}{kT}\right) - 1 \right] \tag{4.3.8}$$

Equations (4.3.7) and (4.3.8) are extremely important results that we shall use to define specific solutions for the continuity equations for minority carriers in the quasi-neutral regions near a *pn* junction. The equations show that the minority-carrier density is an exponential function of applied bias while the majority-carrier density has been assumed to be insensitive (to first order) to it. Since the minority-carrier density at thermal equilibrium is typically 11 or 12 orders of magnitude below the majority-carrier density, Eqs. (4.3.7) and (4.3.8) will not be in conflict with the low-level injection assumption until the exponential factor is typically of the order 10^{11} or 10^{12}. We shall look into the second assumption, that of quasi-equilibrium of carrier fluxes, after we have obtained the voltage dependence of currents.

Ideal-Diode Analysis

Having reviewed the reasons for our focus on the minority carriers and having obtained the bias dependence of excess minority carriers, we are ready to consider solutions of the continuity equations (4.1.4) in the quasi-neutral regions. Let us first do this under a series of idealizations that comprise what has been called the *ideal-diode analysis*.

First, we consider excess holes injected into the *n*-regions, where bulk recombination through generation-recombination centers is dominant. Thus, the term $(G_p - R_p)$ in Eq. (4.1.4) can be expressed through Eq. (4.2.9). Because of our assumption of low-level injection, the arguments used in the discussion of excess-carrier lifetime that led to Eq. (4.2.15) apply also to this case. Therefore, the continuity equation in the quasi-neutral approximation discussed in Section 4.1 becomes

$$\frac{\partial p_n}{\partial t} = D_p \frac{\partial^2 p_n}{\partial x^2} - \frac{p_n - p_{no}}{\tau_p} \tag{4.3.9}$$

where the subscripts *n* emphasize that the holes are in the *n*-region.

Fortunately, the simplest case of impurity doping, that of a constant donor density along *x*, is frequently encountered in practice. Taking this case and considering steady state ($\partial p/\partial t = 0$), we can rewrite Eq. (4.3.9) as a total differential equation in terms of the excess density p', which was defined in Eq. (4.3.6). This equation

$$0 = D_p \frac{d^2 p'_n}{dx^2} - \frac{p'_n}{\tau_p} \tag{4.3.10}$$

has the simple exponential solution

$$p'_n(x) = A \exp\left(-\frac{x - x_n}{\sqrt{D_p \tau_p}}\right) + B \exp\left(\frac{x - x_n}{\sqrt{D_p \tau_p}}\right) \tag{4.3.11}$$

where A and B are constants determined by the boundary conditions of the problem. The characteristic length $\sqrt{D_p \tau_p}$ in Eq. (4.3.11) is called the diffusion length and denoted by L_p. (The diffusion length of an electron in a *p*-type region is denoted by L_n.) For specific applications of the solution to the continuity equation given in Eq. (4.3.11), let us consider two limiting cases based upon the length W_B of the *n*-region from the junction to the ohmic contact (Fig. 4.5).

Long-Base Diode. First, if W_B is very long compared to the diffusion length L_p, essentially all of the injected holes recombine before traveling completely across it. This case is popularly known as the *long-base diode* for reasons that will become clear later. For the long-base diode, L_p is the average distance traveled in the neutral zone before an injected hole recombines (Problem 4.8). Since p'_n must decrease with increasing x, the constant B in Eq. (4.3.11) must be zero. The constant A in the solution is determined by applying Eq. (4.3.8), which specifies $p'_n(x_n)$ as a function of applied voltage. The complete solution is therefore

$$p'_n(x) = p_{no}(e^{qV_a/kT} - 1)e^{-(x-x_n)/L_p} \tag{4.3.12}$$

as shown in Fig. 4.6. Having derived in Eq. (4.3.12) an expression for the excess hole density, we obtain in a straightforward manner an expression for hole current. The hole current flows only by diffusion, since we have assumed that the field is negligible in the neutral regions; therefore, from Eq. (1.2.20)

$$J_p(x) = -qD_p \frac{dp_n}{dx} = qD_p \frac{p_{no}}{L_p}(e^{qV_a/kT} - 1)e^{-(x-x_n)/L_p}$$

$$= qD_p \frac{n_i^2}{N_d L_p}(e^{qV_a/kT} - 1)e^{-(x-x_n)/L_p} \tag{4.3.13}$$

The hole current is therefore a maximum at $x = x_n$ and decreases away from the junction (Fig. 4.7) because the hole gradient decreases as carriers are lost by recombination. Since the total current must remain constant with distance from the junction in the steady-state case, the electron current must increase as we move away from the junction. This electron current supplies the electrons with which the holes are recombining.

The total current is entirely carried by electrons at the ohmic contact at W_B. Moving toward the junction, the electron current decreases as recombination

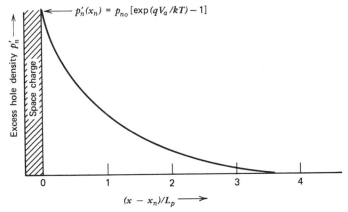

Figure 4.6 Spatial variation of holes in the quasi-neutral *n*-region of a long-base diode under forward bias V_a. The excess density p'_n has been calculated from Eq. (4.3.12).

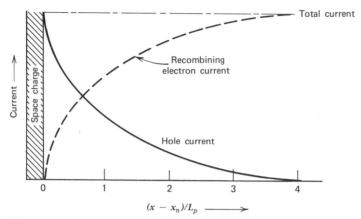

Figure 4.7 Hole current (solid line) and recombining electron current (dotted line) in the quasi-neutral *n*-region of the long-base diode of Fig. 4.5. The sum of the two currents *J* (dot-dash line) is constant. The hole current is calculated from Eq. (4.3.13).

occurs with the injected holes. At the junction, the only electron current flowing is the one injected into the *p*-region. The electrons injected into the *p*-region constitute the minority-carrier current there. Thus, we see that the total current is obtained by summing the two minority-carrier injection currents: holes into the *n*-side plus electrons into the *p*-side.

The minority-carrier electron current that is injected into the *p*-region can be found by an analogous treatment to the anlaysis used to obtain Eq. (4.3.13). If the ohmic contact is at $-W_E$ where $W_E \gg L_n \equiv \sqrt{D_n \tau_n}$, then

$$J_n = qD_n \frac{n_i^2}{N_a L_n} (e^{qV_a/kT} - 1) e^{(x+x_p)/L_n} \qquad (4.3.14)$$

Because we have chosen the origin for *x* at the physical junction (Fig. 4.5), *x* is a negative number throughout the *p*-region. Hence, J_n decreases away from the junction as did J_p in the *n*-region. To obtain an expression for the total current J_t, we sum the minority-carrier components at $-x_p$ and $+x_n$, respectively, as expressed in Eqs. (4.3.13) and (4.3.14):

$$J_t = J_p(x_n) + J_n(-x_p) = qn_i^2 \left(\frac{D_p}{N_d L_p} + \frac{D_n}{N_a L_n} \right)(e^{qV_a/kT} - 1)$$

$$= J_0(e^{qV_a/kT} - 1) \qquad (4.3.15)$$

where J_0 is the magnitude of a saturation current density predicted by this theory when a negative bias of a few kT/q volts is applied. Equation (4.3.15) is identical in form to Eq. (2.3.6), which was derived for a metal-semiconductor Schottky-barrier diode. The similar dependence on voltage arises from the application in both cases of quasi-equilibrium assumptions, which led to Eqs. (4.3.7) and (4.3.8).

Short-Base Diode. A second limiting case occurs when the lengths W_B and W_E of the *n*- and *p*-type regions are much shorter than the diffusion lengths L_p and L_n. In this case, there is little recombination in the bulk of the quasi-neutral regions. In the limit, all injected minority carriers recombine at the ohmic contacts at either end of the diode structure. We can obtain solutions most easily in this case by approximating the exponentials in Eq. (4.3.11) by the first two terms of a Taylor series expansion so that

$$p_n'(x) = A' + B' \frac{(x - x_n)}{L_p} \tag{4.3.16}$$

Because of the ohmic contact at $x = W_B$, $p_n'(W_B) = 0$. The boundary condition at $x = x_n$ is given by Eq. (4.3.8) as before, so that the solution for the excess hole density in the *n*-region becomes

$$p_n'(x) = p_{no}(e^{qV_a/kT} - 1)\left(1 - \frac{x - x_n}{W_B'}\right) \tag{4.3.17}$$

where $W_B' = W_B - x_n$ is the length of the quasi-neutral *n*-type region (Fig. 4.5). From Eq. (4.3.17) we see that the excess hole density decreases linearly with distance across the *n*-type region (Fig. 4.8). The approximation $L_p \gg W_B$ is equivalent to stating that all the holes diffuse across the *n*-type region before recombining. The assumption that no recombination occurs in the *n*-type region could also have been made by letting the lifetime τ_p approach infinity in Eq. (4.3.10). The differential equation that results has a linear solution.

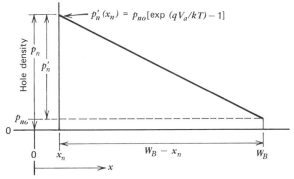

Figure 4.8 Hole density in the quasi-neutral *n*-region of an ideal short-base diode under forward bias of V_a volts. The excess hole density p_n' is calculated from Eq. (4.3.17).

A linearly varying concentration indicates that the hole current remains constant throughout the *n*-type region, and that no electron current is needed to compensate for recombining holes. Therefore,

$$J_p = -qD_p\frac{dp}{dx} = qD_p\frac{p_{no}}{W'_B}(e^{qV_a/kT} - 1) = qD_p\frac{n_i^2}{N_dW'_B}(e^{qV_a/kT} - 1) \quad (4.3.18)$$

Comparing Eqs. (4.3.18) and (4.3.13) for the hole currents in the short- and long-base diodes, we see that the solutions obtained are similar except for the characteristic length appropriate to each geometry. In the long-base diode, the characteristic length is the minority-carrier diffusion length. In the short-base diode, it is the length of the quasi-neutral region. As in the long-base diode, the total current in the short-base diode is made up of electrons injected into the *p*-region and of holes injected into the *n*-region. Hence, for the short-base diode with $W_E \ll L_n = \sqrt{D_n\tau_n}$ and $W_B \ll L_p = \sqrt{D_p\tau_p}$,

$$J_t = qn_i^2\left(\frac{D_p}{N_dW'_B} + \frac{D_n}{N_aW'_E}\right)(e^{qV_a/kT} - 1) \quad (4.3.19)$$

Of course, a given diode may be approximated by a combination of these two limiting cases; that is, it may be short-base in the *p*-region and long-base in the *n*-region or vice versa. Handling these cases is straightforward.

Both diode cases, Eqs. (4.3.15) and (4.3.19) predict a large current flow under forward bias and a small saturation current in reverse bias. Physically, this marked nonuniformity results because forward bias aids the injection of majority carriers from each region across the junction. These carriers are replaced at the end ohmic contacts. Under reverse bias, the net flow across the junction is composed of minority carriers from each region. These are very few in number and not replaced at the end contacts. Hence, only a very small saturation current flows under reverse bias (Problem 4.7).

We see from Eqs. (4.3.15) and (4.3.19) that the most lightly doped side of the junction determines the reverse saturation current. For example, if the dopant concentration on the *n*-side of the junction is much less than that on the *p*-side, the hole current injected across the junction into the *n*-side is much greater than the electron current injected into the *p*-region. Such a situation would characterize a heavy *p*-type diffusion into a lightly doped *n*-type wafer.

Validity of Approximations.   Before taking a more searching look at the physics of *pn*-junction diodes, we return to consider two of the initial assumptions used to carry out the analysis. These assumptions are first, that ohmic drops in the quasi-neutral regions are small so that V_a is sustained entirely across the junction space-charge region and, second, that applied bias does not greatly alter the detailed balance between drift and diffusion tendencies that exists at thermal equilibrium.

To consider these assumptions, let us take the fairly typical case of $N_d = 5 \times 10^{15}$ and $N_a = 5 \times 10^{18}$. If we take first a short-base diode with $W_B = W_E = 3 \ \mu m$, then from Eq. (4.3.19) we calculate a saturation current of approximately 10^{-9} A cm^{-2}. For a typical cross-sectional area of 10^{-5} cm^2 and a typical forward-bias voltage of ~ 0.65 V, this would lead to a current of ~ 1 mA or 100 A cm^{-2}.

This current would consist almost entirely of holes injected into the *n*-side since the first term in the saturation current of Eq. (4.3.19) far exceeds the second. Thus, the ohmic drop would only exist in the highly conducting *p*-region that has a resistivity of 0.03 Ω-cm. The field in this region is just 3 V cm^{-1} and, hence, the potential dropped in the 3 μm length is approximately 1 mV, which is certainly negligible compared to the applied 0.65 volts. Note that there is essentially no ohmic drop in the *n*-region for this case because there is only a negligible flow of electron current. If we take a long-base diode with lower doping on either side, the assumption is on shakier grounds. To see this, assume that $N_a = N_d = 10^{16}$ and that we are considering a long-base diode of cross-sectional area 10^{-5} cm^2 with 0.65 V applied bias. If $L_n \simeq L_p = 30 \ \mu m$, using Eq. (4.3.15), we calculate that such a diode would pass approximately 32×10^{-6} A. If the *p* and *n* regions were each 100 μm long there would be ohmic voltage drops of about 0.03 V in the *p*-region and 0.02 V in the *n*-region. Again, this can be considered negligible to first order as compared to the 0.65 V applied. These results are typical of many practical cases, and we conclude that it is reasonable to assume that the entire applied voltage changes the height of the potential barrier at the *pn* junction for low-to-moderate currents.

If the applied forward voltage is nearly as large as the built-in potential, however, the barrier to majority-carrier flow is substantially reduced and large currents may flow. In that case a significant portion of the applied voltage is dropped across the neutral regions, and the series resistance of the neutral regions must be considered. The effect of series resistance can easily be included using circuit analysis, and we shall not consider it further here.

To determine the validity of the quasi-equilibrium assumption we must compare the magnitude of typical currents to the balanced drift and diffusion tendencies at thermal equilibrium. For a typical integrated-circuit *pn* junction in which the hole concentration changes from 10^{18} to 10^4 cm^{-3} across a depletion region about 10^{-5} cm wide, the hole diffusion current for the average gradient is of the order of 10^5 A cm^{-2}. As we saw in Chapter 3, at thermal equilibrium this tendency of holes to diffuse is exactly balanced by an opposite tendency to drift under the influence of the electric field in the depletion region. We have already argued that typical forward-bias diode currents are roughly 10^2 A cm^{-2}—only 0.1% of the two current tendencies in balance at thermal equilibrium. Thus, it is reasonable to treat the case of small and moderate biases by considering only slight deviations from thermal equilibrium. This means that we may relate the carrier concentrations on either side of the junction space-charge region by considering the effective barrier height to be $(\phi_i - V_a)$ and by using Eqs. (4.3.7) and (4.3.8). These two equations depend also on the validity of the low-level injection assumption. The current tendencies at *pn* junctions that are in detailed balance at thermal equilibrium are so large relative to currents that flow in practice that a more general relationship than Eqs. (4.3.7) and (4.3.8) is valid. This relationship, which we will not prove here, is that the carrier densities at either side of a biased *pn* junction* are dependent upon applied voltage according to the equation

$$pn = n_i^2 \exp\left(\frac{qV_a}{kT}\right) \tag{4.3.20}$$

In the low-level injection case, Eq. (4.3.20) reduces to the boundary conditions for minority carriers, which were obtained previously. Equation (4.3.20) has special utility in cases where injected carrier densities approach the thermal-equilibrium densities of majority carriers.

Space-Charge-Region Currents[†]

The analysis of *pn* junctions that led to the diode equations [Eqs. (4.3.15) and (4.3.19)] was based upon events in the quasi-neutral regions. The space-charge region was treated solely as a barrier to the diffusion of majority carriers, and it played a role only in the establishment of minority-carrier densities at its boundaries [Eqs. (4.3.7) and (4.3.8)]. This is a reasonable first order description of events and the equations derived from it [Eqs. (4.3.15) and (4.3.19)] are referred to as the *ideal-diode* equations. Over a significant range of useful biases, however, the ideal-diode equations are quite inaccurate, especially for silicon *pn* junctions. It is necessary to consider corrections to these equations that arise from a more complete treatment of events in the space-charge region at the junction.

As we noted in Chapter 3, the space-charge region typically has dimensions of the order of 10^{-4} cm. Like the quasi-neutral zones of the diode, it contains generation-recombination centers; unlike the quasi-neutral zones, it is a region of steep impurity gradients and very rapidly changing populations of holes and electrons. Since the injected carriers, under forward bias, must pass through this region, some carriers may be lost by recombination. Under reverse bias, generation of carriers in the region leads to excess current above the saturation value predicted by the ideal-diode equations.

To find expressions for generation-recombination in the space-charge zone we make use of the Shockley, Hall, Read theory. For simplicity, we take the case of equal hole and electron capture cross sections [Eq. (4.2.10)] and evaluate it in the space-charge region with an applied bias V_a. The *pn* product there is given by

---

* Equation (4.3.20) is also true within the space-charge region.[6]

Eq. (4.3.20) and thus the overall recombination-generation rate $U = -dn/dt = -dp/dt$ is

$$U = \frac{n_i^2 (e^{qV_a/kT} - 1)}{\left(p + n + 2n_i \cosh \dfrac{(E_t - E_i)}{kT} \right) \tau_0} \tag{4.3.21}$$

The recombination rate is thus positive for forward bias and negative under reverse bias. The total current arising from generation and recombination in the space-charge region is given by the integral of the recombination rate across the space-charge region:

$$J_r = q \int_{-x_p}^{x_n} U \, dx \tag{4.3.22}$$

Although this integral is not easily evaluated, a qualitative discussion permits us to deduce the dependence of current on voltage. Consider first forward bias and, as in Section 4.2, note that recombination centers located near the middle of the forbidden gap are the most effective (Fig. 4.3). Thus, we consider $E_t \approx E_i$ in Eq. (4.3.21). From this equation, it is evident that the recombination rate is maximized when the sum $p + n$ is minimized. If we consider this sum as a variable function in p and n that is subject to the product of p and n being given by Eq. (4.3.20), we can easily show (Problem 4.15) that U is maximized when

$$p = n = n_i \exp \left(\frac{qV_a}{2kT} \right) \tag{4.3.23}$$

For typical forward-bias conditions, the sum $p + n$ is much greater than n_i within the space-charge region. If the dominant contribution to the integral in Eq. (4.3.22) is given by this maximum value of U extending across a portion x' of the space-charge region, the recombination current may be expressed as

$$J_r = \frac{qx'n_i^2 (e^{qV_a/kT} - 1)}{2n_i (e^{qV_a/2kT} + 1) \tau_0}$$

$$\approx \frac{qx'n_i}{2\tau_0} \exp \left(\frac{qV_a}{2kT} \right) \tag{4.3.24}$$

where $\tau_0 = 1/N_t \sigma v_{th}$ is again the lifetime associated with the recombination of excess carriers in a region with a density N_t of recombination centers. Thus, unlike current arising from carrier diffusion and recombination in the quasi-neutral regions, the current resulting from recombination in the space-charge region varies with applied voltage as $\exp (qV_a/2kT)$ (assuming x' is only a weak function of voltage). This different exponential behavior can be observed in real diodes, especially at low currents. Because terms like τ_0 are not known with high precision, space-charge-zone recombination is usually not given more elaborate analysis than our treatment and x', in particular, is often approximated by the entire space-charge-region width x_d. If we make this approximation and express the ratio between the ideal-diode current J_t (Eq. 4.3.15) and the space-charge-zone recombination-current J_r under foward bias, we find

$$\frac{J_t}{J_r} = \frac{2n_i}{x_d}\left[\frac{L_n}{N_a} + \frac{L_p}{N_d}\right]\exp\left(\frac{qV_a}{2kT}\right) \tag{4.3.25}$$

Thus, space-charge recombination current J_r becomes less significant relative to ideal-diode current as bias increases. Also, the more defect-free is the material, the longer are the diffusion lengths and the more dominant is J_t over J_r. Taking typical silicon diode values of $L_n = 30$ μm, $x_d = 0.25$ μm, and $N_a = 10^{16}$, it is easily seen that J_t exceeds J_r for V_a greater than about 0.41 V.

Under reverse bias, the numerator of Eq. (4.3.21) approaches $-n_i^2$. Therefore U is negative, indicating net generation in the space-charge zone. From Eq. (4.3.20) the *pn* product becomes vanishingly small. As before, the maximum value of U occurs when both p and n are equal; in this case both are much less than n_i. It would take more elaborate analysis than we carry out here to show that p and n are both substantially less than n_i over a portion of the space-charge region x_i bounded by the points at which the intrinsic Fermi level E_i crosses the quasi-Fermi levels.[7]* This region may be considerably smaller than the total space-charge region width x_d. Outside x_i either p or n is greater than n_i, and the generation rate falls precipitously. The net generation rate in the space-charge region can be approximated by the product of the maximum generation rate and the width x_i:

$$J_g = \frac{qn_i x_i}{2\tau_0} \tag{4.3.26}$$

where we have again assumed that the most effective centers are located at E_i. For a one-sided *pn* junction with a very heavily doped *p*-type side, virtually all of the space-charge region extends into the lightly doped *n*-type semiconductor, and we may solve for the width of the total space-charge region x_d and for the region x_i with maximum generation rate:

$$x_d = \left[\frac{2\epsilon_s}{qN_d}(\phi_i - V_a)\right]^{1/2} \tag{4.3.27}$$

$$= \left[\frac{2\epsilon_s kT}{q^2 N_d}\left(\ell n \frac{N_d N_a}{n_i^2} - \frac{qV_a}{kT}\right)\right]^{1/2}$$

and

$$x_i = \left(\frac{2\epsilon_s kT}{q^2 N_d}\right)^{1/2}\left[\left(\ell n \frac{N_d}{n_i} - \frac{qV_a}{kT}\right)^{1/2} - \left(\ell n \frac{N_d}{n_i}\right)^{1/2}\right] \tag{4.3.28}$$

---

* A discussion of quasi-Fermi levels, which are useful concepts to treat nonequilibrium conditions in semiconductors, is provided in reference 8.

where V_a is the applied voltage ($V_a < 0$ for reverse bias). Both x_d and x_i depend on the square root of the applied voltage for large reverse bias, and the difference between the two quantities becomes very small at high reverse bias. Figure 4.9 shows the ratio x_i/x_d as a function of voltage for several donor concentrations in a one-sided step junction. Because the density of recombination centers in practical diodes may vary with position and is generally not well known, it is often not worthwhile to distinguish between x_i and x_d and the latter is frequently used in Eq. (4.3.26).

From Eqs. (4.3.26) and (4.3.28) we see that the generation current in the space-charge region is only a slight function of the reverse bias, varying roughly as the square root of applied voltage. We may estimate the relative importance of the contributions to the reverse current from the quasi-neutral region and from the space-charge region for a long-base diode by taking the ratio of Eq. (4.3.15) under reverse bias to Eq. (4.3.26):

$$\frac{J_t}{J_g} = \frac{2n_i}{x_i}\left[\frac{L_n}{N_a} + \frac{L_p}{N_d}\right] \tag{4.3.29}$$

Values of the parameters for practical diodes are such that J_t/J_g is much less than unity. Hence, the current in reverse-biased silicon diodes is generated dominantly in the space-charge region. Generation-recombination centers in the space-charge

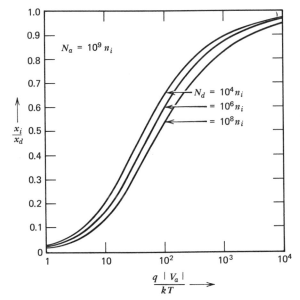

Figure 4.9 The ratio of generation-layer width x_i to space-charge-region width x_d as a function of reverse voltage for several donor concentrations in a one-sided step junction.[7]

region are also responsible for excess currents in reverse-biased Schottky diodes (Chapter 2). In that case, however, the dominant currents are associated with recombination centers at the metal-semiconductor interface.

Summary. Our analysis of current in _pn_ junctions has touched on several points and has become rather lengthy. It is worthwhile, therefore, to summarize the results at this point, and to make several comments about them.

We have considered solutions of the continuity equation for two cases of uniformly doped step-junction diodes. Both of these led to a dependence of current on voltage [Eqs. (4.3.15) and (4.3.19)], which is the same as was found earlier for Schottky-barrier diodes, and which is generally known as the ideal-diode law [i.e., $J = J_0(\exp(qV_a/kT) - 1)$]. In practical diodes, currents that result from events in the space-charge region must be added to this ideal law to obtain the overall current-voltage relationship. These added currents dominate in silicon diodes under reverse bias and under low forward bias. A sketch of the currents flowing in a long-base diode at moderate forward bias is shown in Fig. 4.10. The total current close to the junction in each of the quasi-neutral regions consists of (1) injected minority carriers diffusing away from the junction, (2) majority carriers drifting toward the junction to recombine with minority carriers injected into the quasi-neutral regions, (3) majority carriers drifting toward the junction to be injected into the opposite quasi-neutral region, (4) majority carriers drifting toward the junction space-charge region where they recombine with injected carriers

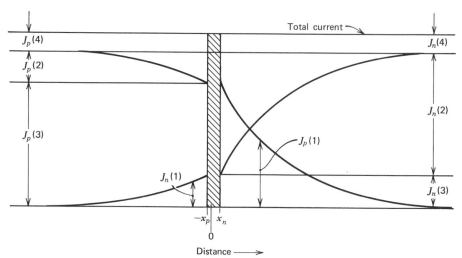

Figure 4.10 The current components in the quasi-neutral regions of a long-base diode under moderate forward bias: $J(1)$ injected minority-carrier current, $J(2)$ majority-carrier current recombining with $J(1)$, $J(3)$ majority-carrier current injected across the junction, and $J(4)$ space-charge-zone recombination current.

entering the junction region from the opposite side. Far from the junction in either region, the entire current is carried by drifting majority carriers.

A next logical step is to analyze *pn* junctions in which the dopant densities are nonuniform in the quasi-neutral regions. We can expect nonuniform dopant distributions in many diffused junctions. Nonuniform doping is, however, most gainfully discussed by using approximation techniques—in particular, by methods that we shall develop when we discuss transistors. We therefore defer direct considerations of this situation to Chapter 5. We shall see that the ideal-diode law is again obtained in practical cases of nonuniform doping, but the saturation current J_0 has a different formulation than our earlier results. As with the ideal diode, generation-recombination in the space-charge region leads to added current components. The observed steady-state dependence of current on voltage is made up of the superposition of the ideal-diode current and these components.

4.4 CHARGE STORAGE AND DIODE TRANSIENTS

In the previous section we saw that forward bias across a *pn* junction causes the injection of electrons from the *n*-type region into the *p*-type region and of holes in the opposite direction. After injection across the junction, these minority carriers move into the quasi-neutral regions. The resulting distribution of minority carriers leads both to current flow and to charge storage in the junction diode. In this section we consider the stored charge, its relation to the current, and its effect on the transient response of the *pn* junction to a change in diode bias conditions.

On a fundamental basis, the time-dependent behavior of minority carriers is included in the continuity equations [Eqs. (4.1.4)]. Since these partial differential equations are functions of time and position, they can be solved for various forcing functions and initial conditions to yield the transient behavior of minority carriers. This is generally not done in practice for several reasons. First, explicit solutions can be obtained only in special cases and for idealized forcing functions that can only approximate the real circuit conditions. It therefore makes little sense to carry out the exact mathematics of a solution when that solution will only apply approximately to the real problem. A second reason for not carrying out the solution of the partial differential equation is that the diode transient is not governed only by the minority charges stored in the quasi-neutral regions, but is also a function of charge storage in the depletion region. Thus, one must assess the changes in both charge stores simultaneously. Overall, the best way to look at the transient problem is to consider the physical behavior of these charges directly.

Minority-Carrier Storage

The total injected minority-carrier charge per unit area stored in the quasi-neutral *n*-type region can be found by integrating the excess hole distribution across the quasi-neutral region.

$$Q'_p = q \int_{x_n}^{W_B} p'_n(x)\, dx \qquad (4.4.1)$$

Since the region is quasi-neutral, the *n*-region contains Q'_p/q extra electrons above the thermal-equilibrium value to balance the $+Q'_p$ charge on the holes.

We first consider the long-base diode, in which all of the injected minority carriers recombine before they reach the end of the *n*-type region. For simplicity, we also assume the ideal-diode structure: that is, the semiconductor is uniformly doped. From Eq. (4.3.12) we know that the minority-carrier distribution decays exponentially as the holes diffuse further into the *n*-type region. Inserting Eq. (4.3.12) into Eq. (4.4.1) and integrating from the edge of the space-charge region across the quasi-neutral region, we find the charge Q'_p resulting from holes stored in the *n*-type region to be

$$Q'_p = qL_p p_{no}(e^{qV_a/kT} - 1) \qquad (4.4.2)$$

The stored charge, like the current, depends exponentially on applied bias. This behavior could be expected since the exponential term arises from the magnitude of the excess minority-carrier density at the edge of the space-charge region for both the current and the stored charge. We can gain further insight into the physical behavior by expressing the stored charge in terms of the corresponding hole current. By using the expression for the hole current [Eq. (4.3.13)] at $x = x_n$ in Eq. (4.4.2), we find the simple equation

$$Q'_p = \frac{L_p^{\,2}}{D_p} J_p(x_n) = \tau_p J_p(x_n) \qquad \text{long base} \qquad (4.4.3)$$

Thus, the amount of stored charge is the product of the current and the lifetime for this case of the ideal long-base diode. This relationship is reasonable, since the injected carriers diffuse farther into the *n*-type region before recombining if their lifetime is greater; more holes are then stored.

The situation in the ideal short-base diode is slightly different. To treat it we use the expression for the hole distribution [Eq. (4.3.17)] in Eq. (4.4.1) to obtain the stored hole charge as

$$Q'_p = \frac{q(W_B - x_n)}{2} p_{no}(e^{qV_a/kT} - 1) \qquad (4.4.4)$$

where W_B is the length of the *n*-type region (Fig. 4.5). Again, we can express the stored hole charge in terms of the hole current [Eq. (4.3.18)]:

$$Q'_p = \frac{(W_B - x_n)^2}{2D_p} J_p \qquad \text{short base} \qquad (4.4.5)$$

Although the group of constants in front of J_p in Eq. (4.4.5) has the dimensions of time, it is not the lifetime, as was found in Eq. (4.4.3) for the long-base diode. To see the physical significance of this group of constants, we rewrite Eq. (4.4.5) as

$$\frac{Q'_p}{J_p} = \frac{(W_B - x_n)^2}{2D_p} \quad \text{Average time} \tag{4.4.6}$$

The amount of stored charge divided by the rate at which the charge enters or leaves the n-type region must be equal to the time an average carrier spends in this region. Therefore, the right-hand side of Eq. (4.4.6) is equal to the average *transit time* τ_{tr} of a hole moving through the n-type region of the short-base diode.

Transient-Behavior of Minority-Carrier Storage.[†] We can extend the physical picture of minority-carrier storage to discuss qualitatively the nature of the transient buildup and decay of Q_p. Consider the behavior of holes in the n-region of an initially uncharged, ideal, long-base diode to which a positive constant current source is suddenly applied. Before steady state can be reached, holes must be transported into the quasi-neutral n-region to establish a particular hole distribution. Theoretically, at $t = 0^+$ a near-infinite gradient of holes is possible because the n-region is nearly free of holes at that time. The actual gradient is set, however, by the size of the current source. The hole store Q_p builds with time as the holes are supplied, and the voltage across the junction rises to reflect the boundary value of the hole density. A sketch of the transient buildup for this situation is shown in Fig. 4.11. The total time required to reach the steady-state hole distribution is given by the ratio of the steady-state stored hole charge to the size of the current source.

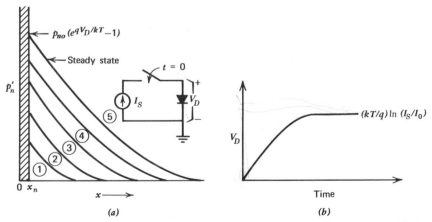

(a)   (b)

Figure 4.11 (a) Qualitative sketches of the transient buildup of excess stored holes in a long-base ideal diode. The case shown is one in which a constant current drive is supplied starting at time zero with the diode initially unbiased. Note the constant gradient at $x = x_n$ as time increases from (1) through (5), which indicates a constant injected hole current. (Circuit is shown in inset.) (b) Sketch of diode voltage V_D versus time.

The turn-off time of the diode is limited by the speed with which stored holes can be removed from the quasi-neutral region. When a reverse bias is suddenly applied across the forward-biased junction, the current can reverse direction quickly since the gradient near the edge of the space-charge region can change with only a small change in the number of stored holes (curve 1 of Fig. 4.12a). As long as minority carriers are present at the edge of the space-charge region, the diode will be able to conduct a large amount of current in the reverse direction. The junction will, in fact, remain forward biased until the injected minority carriers near the edge of the depletion region are removed. This means that a plot of current versus time (Fig. 4.12b) will initially be nearly constant (and determined by the external circuit), and will remain in this condition until t_4, at which time $p_n'(x_n)$ goes to zero. Thereafter, (curves 5 and 6 in Fig. 4.12a) the rate at which holes can be delivered across the space-charge region becomes more and more limited. The current then decays on a time scale that is determined largely by carrier lifetime in the quasi-neutral region.

Since the switching time of a diode depends on the amount of stored charge that must be injected or removed, we may decrease the switching time by reducing the stored charge. From Eq. (4.4.3) we see that this reduction in stored charge may be achieved either by restricting the forward current or by decreasing the lifetime of minority carriers. Consequently, optimal design for a given diode may not maximize the minority-carrier lifetime by forming the purest material possible. In fact, in many applications where devices must switch rapidly, recombination centers

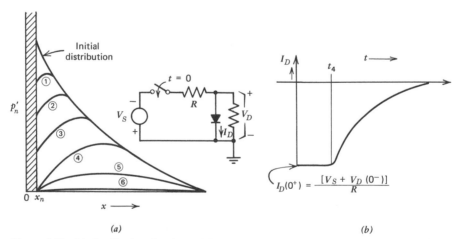

(a)   (b)

Figure 4.12   (a) Qualitative sketches of the transient decay of excess stored holes in a long-base ideal diode. In the case shown, a diode with an initial forward bias is subjected at time zero to an abrupt reverse bias supplied through a series resistor. (Circuit shown in inset.) (b) Sketch of diode current I_D versus time.

are purposely introduced into the semiconductor to reduce the lifetime so that little charge is stored. The critical concern is, of course, to introduce these "lifetime-killing" impurities in a controlled manner so that the amount of charge storage is reduced without increasing the reverse leakage current $[-J_0$ in Eq. (4.3.15)] to unacceptable levels. For example, gold may be diffused into a semiconductor wafer at a precisely controlled temperature.

The forward current through a *pn* junction can be restricted by the Schottky-diode clamping technique discussed in Section 2.6. Since a Schottky diode has a lower turn-on voltage than a *pn* junction, the voltage across a *pn* junction and the current through it may be limited by placing a Schottky diode in parallel with the *pn* junction. Charge storage in the *pn* junction is then decreased, and the turn-off time is reduced. Additional current flows through the Schottky diode, of course, but to first order no minority charge is stored there.

Thus far, we have discussed hole storage in the quasi-neutral *n*-type side of the *pn* junction. There is, of course, a similar storage of electrons in the *p*-region. Expressions analogous to Eqs. (4.4.2) and (4.4.4) can be derived to express this stored electron charge. The total minority-carrier charge stored in the diode is the sum of the two components.

We shall find these concepts of stored minority charge, transit time, and the interrelationship between them to be useful in discussing transients in *pn* junctions. To treat the transient problem, we can neglect changes in the majority-carrier concentrations since majority carriers respond to changes in fields within a dielectric relaxation time $\tau_r = \epsilon_s/\sigma$, which is of the order of 10^{-12} s (one picosecond) for 1 Ω-cm silicon. This time is much shorter than the lifetime [Eq. (4.3.3)] or the transit time [Eq. (4.4.6)] of minority carriers. For integrated-circuit devices these times are typically between 10^{-10} and 10^{-5} s.

Total Junction Storage. In Section 3.3 we examined one mechanism of charge storage in a *pn* junction. We saw that majority carriers near the edges of the depletion region move as the depletion region expands or contracts in response to a changing reverse bias. We saw that this charge storage in the depletion region could be modeled by a small-signal capacitance. Capacitance owing to charge storage in the depletion zone is usually symbolized by C_j. An example is the abrupt-junction depletion capacitance derived in Eq. (3.3.8).

Similarly, the variation of stored minority-carrier charge in the quasi-neutral regions under forward bias can be modeled by another small-signal capacitance. This capacitance is usually called the diffusion capacitance (denoted C_d), since in the ideal-diode case the minority carriers move across the quasi-neutral region by diffusion. We can find the contribution to C_d from the stored holes in the *n*-region by applying the definition $C_d' = dQ_p'/dV_a$ to Eq. (4.4.2) for the long-base ideal diode or to Eq. (4.4.4) for the short-base diode. (The prime denotes capacitance per unit area.)

Since both diodes have the same voltage dependence, we find that in general

$$C'_d = \frac{qQ'_{po}e^{qV_a/kT}}{kT} \qquad (4.4.7)$$

where

$$Q'_{po} = J_0\tau_p = qp_{no}L_p \qquad (4.4.8)$$

for the long-base diode and

$$Q'_{po} = J_0\tau_{tr} = qp_{no}(W_B - x_n) \qquad (4.4.9)$$

for the short-base diode.

Again, it is straightforward to add components to C_d, which represent electron storage in the *p*-region in cases where that storage is significant. As should be expected, Eq. (4.4.7) shows that C_d is inconsequential under reverse bias because the minority-carrier storage is negligible. Under forward bias, both Q_p and C_d rise exponentially with voltage.

From the foregoing we can see that the relative significance of charge storage in the space-charge zone (as represented by C_j) and charge storage in the quasi-neutral regions depends strongly on the actual junction voltage. Under reverse bias, storage in the quasi-neutral regions is negligible and the storage represented by the junction-capacitance term will dominate. Under forward bias, although C_j increases (because x_d decreases), the exponential factor in the formula for C_d generally makes diffusion capacitance and its associated charge storage dominant. For accurate solution of practical problems, however, it is frequently necessary to account for both types of charge storage in the forward-bias region.

Circuit Modeling

The nature of charge storage at the *pn* junction strongly complicates the use of equivalent circuits in hand calculations of transient problems. For example, the most frequent use of diodes in integrated circuits is as switches, which are successively forward and reverse-biased during circuit operation. In this application, the dominant charge storage shifts during the transient itself between the diffusion and depletion-layer charge. To make hand calculations in such situations, the most successful technique is, first, to determine the total charge storage at the beginning and end of the transient. Then, the time required for switching is calculated by dividing the increment in stored charge ΔQ by the driving current I that causes the diode to switch (Problem 4.18).

This technique does not, of course, give an accurate picture of the current and voltage of the diode as functions of time. For many applications, such detail is not of interest, but if it has importance, the diode can be *piecewise-linear* approximated. This phrase means that the nonlinear charge-storage effects can be approximated to first order by linear elements over a small voltage range. If the voltage increment is made small enough, this approximation will be accurate, and an arbitrary precision can be maintained by joining together a sufficient number of piecewise-linear approximations to represent the entire voltage variation in a given transient problem.

Although we did not label them as such, the small-signal capacitances C_j and C_d in Eqs. (3.3.8) and (4.4.7) are piecewise-linear approximations because they represent an overall nonlinear charge-storage effect in terms of linear circuit elements (capacitors). A complete piecewise-linear model for a *pn* junction must include a conductance as well as a capacitance to represent the real current through the device. The conductance (per unit area) in the ideal diode is [from Eq. (4.3.15)]

$$g_d' = \frac{dJ}{dV} = \frac{q}{kT} J_0(e^{qV_a/kT}) = \frac{q}{kT}(J + J_0) \qquad (4.4.10)$$

The total piecewise-linear circuit for the diode consists of g_d in shunt with C_j and C_d (Fig. 4.13). Since we have discussed only basic junction processes, the circuit in Fig. 4.13 does not include elements that are inherent to the planar fabrication. We shall refer to these in Section 4.5.

For specialized applications such as the low-voltage sinusoidal excitation of a diode that is biased quiescently (i.e., at dc) to some operating point, analysis with the circuit shown in Fig. 4.13 provides adequate accuracy. It is, therefore, referred to as the *diode small-signal equivalent circuit*. The diode small-signal equivalent circuit is especially useful for analysis of a wide range of communication and linear amplification circuits.

Hand calculations using the small-signal equivalent circuit in piecewise-linear analysis become overly cumbersome very quickly. The complexity is such that the problem is gratefully and rapidly given over to a computer in all cases except exercises for university students. To carry out the analysis on a computer, an initial

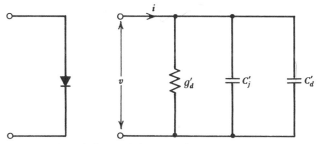

Figure 4.13 Diode small-signal equivalent circuit. A piecewise-linear representation for the nonlinear current-voltage relationship of a junction diode. The element values for a long-base ideal diode are given (per unit area) by Eqs. (3.3.8) for C_j', (4.4.7) for C_d', and (4.4.10) for g_d'.

state is specified in which starting values of the currents, voltages, and capacitances are found.

As an example to clarify these comments, we consider the circuit in Fig. 4.14. The initial state might correspond to an initially unbiased circuit ($V_S = 0$ in Fig. 4.14). At time $t = 0$, the source voltage V_S in Fig. 4.14 is changed. Since the voltage across a capacitor cannot change instantaneously, the size of the capacitor and the value of the current source that represents the diode can be calculated from the initial state (at $t = 0^-$). The elements in the circuit are assumed to hold these values for a time increment Δt. At the end of Δt, the voltage across the diode will have changed to a new value V_D. This new voltage V_D can then be used to compute new values for the components of the equivalent circuit, and the procedure is repeated in an iterative manner until the diode reaches its final state.

If the circuit of Fig. 4.14 were analyzed by a computer program in a piecewise-linear manner, the sequence of operations might be the following:

Computer Operation	*Parameter Values*
Initialize I_D, C, and V_D, for steady state with $V_S = 0$.	$V_D = 0,\ I_D = 0,\ C = C_0$
Set time $t = 0$.	$t = 0$
Set V_S to new value.	$V_S = V_1$
→ Increment time,	$t = t + \Delta t$
Compute currents I_R and I_C.	$I_R = \dfrac{V_S - V_D}{R_S}$
	$I_C = I_R - I_D$
Compute change in voltage V_D during Δt.	$\Delta V_D = \dfrac{I_C\,\Delta t}{C} = \dfrac{(I_R - I_D)}{C}\,\Delta t$
Calculate new voltage.	$V_D = V_D + \Delta V_D = V_D + \dfrac{I_R - I_D}{C}\,\Delta t$
Calculate new diode current.	$I_D = I_0(e^{qV_D/kT} - 1)$
Test whether I_D has reached its steady-state value.	
No. Continue.	
Yes. Exit from loop.	
Calculate new capacitance. (An abrupt junction is assumed).	$C = \dfrac{C_0}{\sqrt{1 - V_D/\phi_i}} + \dfrac{q}{kT}\,\tau I_D$
└ Go to beginning of loop for next time increment.	
└→ STOP	

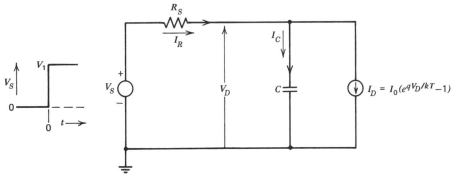

Figure 4.14 Circuit illustrating a typical diode switching problem. The source voltage V_S changes from zero to V_1 at $t = 0$. The analysis of this problem is discussed in the text.

The analysis continues until the diode current approaches to within some predetermined increment of the steady-state value. The computer instructions may also include output commands so that any variable, such as voltage or current, can be plotted on a recorder to give the user an indication of the circuit behavior.

This piecewise-linear numerical approach can be utilized in any nonlinear circuit where computer simulation is the only reasonable means for studying the behavior of the total circuit. It is one of several alternative computational approaches (or algorithms) that enable the computer solution of nonlinear problems. Models for diodes and other nonlinear circuit elements based either on piecewise-linear approximations or on other techniques are typically included as subroutines in more elaborate computer-analysis programs for integrated circuits. The use of these computer-simulation programs is an invaluable aid to the design and analysis of integrated circuits.

4.5 DEVICES: INTEGRATED-CIRCUIT DIODES

In order to form a *pn*-junction diode using integrated-circuit technology, it is only necessary to diffuse a *p*-type region into an *n*-type wafer and make contact to the front and back of the wafer (Fig. 4.15). If the diode is part of an integrated circuit, however, many *pn* junctions are typically formed on the same wafer, and simple diffusion as described above results in all of the *n*-regions being electrically connected. This, of course, is generally unacceptable. Interconnections between diodes, or between other devices in an integrated circuit, can be avoided by surrounding them by *pn* junctions that are kept under reverse bias at all times. This technique of *junction isolation* for integrated-circuit devices is the largest application of junction diodes to integrated circuits. It is not the only isolation method, but it is the one

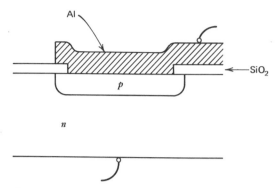

Figure 4.15 Structure of a diffused *pn* junction. A *p*-region is diffused into an *n*-type wafer. The surface is passivated with silicon oxide and ohmic contacts are made to the front and the back of the wafer.

most frequently used. To illustrate some aspects of diode structures in integrated circuits, we consider that an array of diodes is desired in an integrated circuit.

To achieve a junction-isolated array of diodes, the most standard procedure would be to start with a *p*-type wafer on which a thin *n*-type epitaxial silicon layer is grown (Fig. 4.16*a*). A typical "epi layer" may be from 2 to 5 *μ*m thick for a digital logic switching circuit and as thick as 10 to 20 *μ*m for a linear circuit. The *pn* junction at the epi-substrate interface provides electrical isolation in the vertical direction as long as it is reverse biased. The substrate is usually relatively lightly doped *p*-type material ($N_a \approx 2 \times 10^{15}$ cm^{-3}) in order to reduce undesired or "parasitic" junction capacitance between the active region and the substrate and to maintain a high breakdown voltage.

The following sequence of steps includes oxide growth, photomasking, and etching to define the web of isolating *pn* junctions. The acceptor dopant diffused into the *n*-epitaxial region to form these junctions must penetrate completely through the epitaxial layer in order to surround the *n*-regions with *p*-type material (Fig. 4.16*b*). The resulting isolated *n*-regions are commonly and picturesquely called "tubs" in the integrated-circuit industry. The *p*-isolation diffusion, because of its depth, typically takes the longest of any diffusion time used in the *IC* process. Because of the lateral diffusion under the oxide mask that takes place at roughly the same rate as the vertical diffusion toward the substrate, the designer must allow for a considerable surface area to be consumed by the isolation junctions. This is a major disadvantage of junction isolation. The greatest advantage of the isolation technique is its compatibility with other *IC* processing steps.

After isolation, the next step in producing a junction diode is to introduce a second diffused *p*-region—this time keeping the diffusion time relatively short so that a *pn* junction is formed within the *n*-type tub (Fig. 4.16*c*). It is important at this point to keep the two space-charge regions, one from the diode junction and one

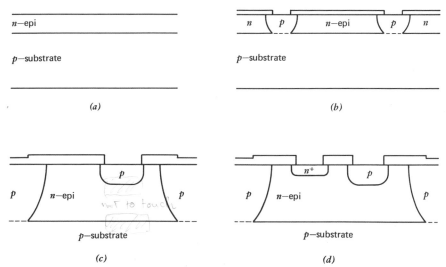

Figure 4.16 Planar-process steps in the production of junction diodes for integrated circuits: (a) substrate junction formed by n-type epitaxial layer on p-type substrate, (b) p-type isolation diffusions through n-type epitaxial layer, (c) shallower p-type diode diffusion into epitaxial layer, (d) highly doped, n-type diffusion to provide ohmic contact to n-type region. [Scale expanded in parts (c) and (d).]

from the substrate isolation junction, from coming into contact. Proper design must insure that they do not touch, regardless of the voltage conditions that may be imposed on the diode and substrate junctions. Since higher reverse bias widens space-charge regions, the design focus is on the highest reverse-bias conditions. If the two space-charge regions do meet, the condition is described by the term *punch-through* since the quasi-neutral n-region separating the two p-regions has been "punched through." Under this condition the p-region for the diode and the p-type substrate are not isolated and large currents can flow between them (Problem 4.20). To avoid punch-through to the sidewalls, the edge of the diode diffusion must also be located sufficiently far from the edge of the isolation diffusion.

The next step is to make electrical connection to both sides of the diode from the top of the wafer. First, a highly doped n^+ contact diffusion is added to the n-type region (Fig. 4.16d) in order to achieve good ohmic contact. As we saw in Chapter 2, aluminum does not make ohmic contact to lightly doped n-type silicon. After contact areas are opened, the junction-isolated diode is completed by covering the wafer with aluminum, defining the aluminum pattern by selective etching, and alloying it. In Fig. 4.17, the diode structure is drawn with a single scale for both horizontal and vertical dimensions. In Fig. 4.16, as in most sketches of *IC* cross sections, the vertical dimension has been exaggerated for clarity.

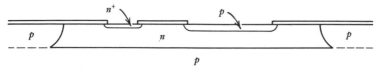

Figure 4.17   The diode structure of Fig. 4.16 sketched with the same scale for both *x* and *y* axes to give a proper perspective. The relative dimensions shown are typical for switching diodes.

An integrated-circuit diode made by the process we have outlined above suffers from a very serious disadvantage, however, and in practice some modification of the structure is almost always necessary. To obtain a relatively large current-carrying area it is desirable to have the current flow vertically through the diode. However, this requires that majority-carrier electrons must flow to the junction region through a path that includes the long, thin *n*-region bordered by the two *pn* junctions. This is a path very like the pinch resistor described at the end of Chapter 1. For a typical epitaxial-layer resistivity of 0.5 Ω-cm and typical integrated-circuit dimensions, the series resistance introduced by this path would be several kilohms. Such a high series resistance would produce sufficient ohmic drop, even at relatively low forward currents, to cause the *pn* junction bias to vary along its length. Because current is exponentially related to the junction voltage under forward bias, there would be *current crowding* toward the regions of the junction that are nearest the n^+ contact. The effective area of the diode would be reduced from its geometric area, leading to many undesirable effects.

To eliminate this problem, a heavy doping of *n*-type impurity is made on the substrate before growing the epitaxial layer. This diffused region is usually included beneath each junction region (Fig. 4.18*a*). After the epitaxial layer has been grown and the diffusions discussed above have been carried out, the structure appears as shown in Fig. 4.18*b*. By using this *buried layer* of *n*-type material, the series resistance is reduced to a few ohms, which is generally quite acceptable. An added

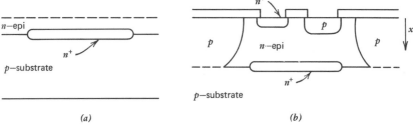

(*a*)   (*b*)

Figure 4.18   (*a*) A heavily doped *buried layer* may be diffused into the *p*-type substrate before the epitaxial layer is grown; this reduces resistance in series with the diode. (*b*) An integrated-circuit diode incorporating a buried layer.

advantage of the buried layer is that it reduces the possibility of punch-through between the p-region at the surface and the substrate. This is true because the added donors in the buried layer inhibit the extension of the space-charge region into that zone. Although punch-through is inhibited, the breakdown voltage of the junction diode may be reduced by the presence of the buried layer because it acts to confine the space-charge region and thus to increase the maximum field.

The incorporation of a buried layer does add some fabrication complications to integrated-circuit production. First, since the diffusion of the buried-layer impurities is carried out before the epitaxial-growth step, these impurities go through the temperature cycling needed for the growth and also for all the subsequent processing steps. This means that a slow-diffusing, relatively nonvolatile impurity must be used. The slow diffusion is required to maintain the desired dimensions for the devices; nonvolatility is necessary to avoid auto-doping the epitaxial layer during the growth stage. In practice, relatively good results are obtained by using arsenic or antimony for the buried-layer dopant. The p-type diffusions are then carried out with boron, which diffuses very much faster. Phosphorus is used for subsequent n-type diffusions that must consist of any appreciable fraction of the epitaxial-layer width. The lack of a p-type dopant with the desirable slow-diffusing properties of arsenic or antimony has made it difficult to fabricate practical p-type buried layers.

The n-type buried layer, by narrowing the space-charge region between the n-type region and the p-type substrate, does add to the parasitic capacitance between the epitaxial layer and the substrate. This is one reason that the buried-layer region is usually introduced only beneath the active regions. One unavoidable disadvantage of the incorporation of a buried layer is that the strain associated with its high impurity density generally reduces the lifetime and mobility in the epitaxial region above it. Since this is precisely the critical region with respect to device performance, the drawback can be severe. Optimum design requires choosing a buried-layer doping that balances these disadvantages with the advantages that motivated the inclusion of a buried layer.

As a final topic concerned with the practical aspects of junction diodes for integrated-circuits, we give some perspective to the widespread use of tables and nomographs for design and analysis. Properties of pn junctions formed by (1) a Gaussian diffusion profile into a constant background doping and (2) a complementary-error function profile into a constant background doping have been published by Lawrence and Warner.[9] These authors obtained their results by finding exact solutions of the junction equations using computer routines. Their calculations permit the determination of junction capacitance, total depletion-layer thickness, and the extent of the depletion layer on either side of the junction. Reference 9 presents the data in especially useful fashion. To make use of the Lawrence and Warner results, it is necessary to know the background dopant density of the region of constant doping as well as the density of the dopant at the

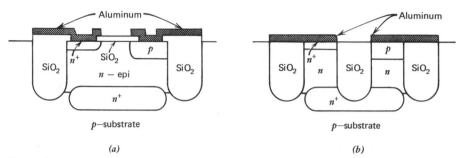

Figure 4.19 Two possible designs of oxide-isolated *pn* junctions. The structures can be smaller than junction-isolated devices because lateral-diffusion problems can be avoided.

surface from which diffusion is taking place. These data specify a family of curves that are plots of the output data as functions of total junction voltage. Because it is a closely related calculation, the Lawrence-Warner analysis also provides curves of the maximum junction field versus total voltage across the junctions. Reference to these results can provide a major, time-saving aid to the designer of an integrated circuit.

Tables of other parameters for typical *IC* processing conditions are also consulted frequently in the design of circuits. These tables relate to specific processes that are employed by a manufacturer to produce integrated circuits. An example is included at the end of Chapter 5. In addition to referring to tables and graphs from published references, designers are increasingly carrying out their own exact analyses on computers. These analyses are based upon solution by numerical methods of many of the equations that we have presented.

Finally, we note that although the *pn*-junction isolation technique that we have just described is widely used for integrated circuits today, several newer techniques are being investigated and adopted for demanding applications. Several processes have been worked out to produce regions containing silicon oxide between devices. This technology greatly reduces the area needed for the isolation regions and decreases some parasitic capacitances. These advantages are gained at a considerable increase in processing complexity, however. Two possible realizations of an oxide-isolated diode are shown in Fig. 4.19. Note the area reduction possible because the diode and contact diffusions can extend to the oxide isolation regions.

SUMMARY

The continuity equation for free carriers, a partial differential equation that accounts for the mechanisms causing an increase or a decrease in carrier densities, is a powerful tool for semiconductor device analysis. By solving the continuity equation for minority carriers in the quasi-neutral regions near a *pn* junction, we can obtain the carrier densities. Boundary values for minority-carrier densities at the edges of the quasi-neutral regions may be written as functions of applied junction bias. Using these boundary values and the defining relationships for currents as a function of free-carrier density, it is possible to obtain current-voltage characteristics in the steady state for several simple cases with important practical applications. These solutions lead to the *ideal-diode equation* $J = J_0[\exp(qV_a/kT) - 1]$. Solutions to the time-dependent continuity equation can be carried out in simple cases to give meaning to the concept of the lifetime of excess carriers. The recombination rate, which determines the carrier *lifetime*, is a function of the properties of recombination centers in silicon and germanium. These recombination centers consist of localized electronic states having energies within the forbidden gap, typically close to the intrinsic Fermi level. Generation and recombination through recombination centers located inside the space-charge region are responsible for significant currents, especially under reverse and low forward bias, in silicon junction diodes. Because there are extra localized states at a silicon surface and also because a space-charge region is usually present in addition, surface recombination can be an added important effect. Surface recombination is usually described by a parameter *s* known as the *surface-recombination velocity*. The transient behavior of junction diodes is governed by the charge and discharge of stored minority carriers as well as by the variation of stored depletion-layer charge. These components of charge are nonlinear functions of voltage, and linear circuit representation is only approximately valid for small bias variations. A *small-signal equivalent circuit* can be developed with capacitors representing the charge storage. This circuit can be used in piecewise fashion to make accurate calculations for large-signal transients. Any practical problems are usually carried out by computer analysis. The *pn*-junction diode is a basic building block for integrated circuits. There is widespread use of *pn* junctions as isolating elements to avoid unintentional interactions between the devices making up an integrated circuit. Series resistance between surface contacts and forward-biased *pn* junctions can be reduced by adding buried layers of donor dopant atoms below diffused junctions. Many of the properties of *pn* junctions for integrated circuits can be obtained by reference to curves of published results obtained from numerical solutions of the junction equations. Alternatives to junction isolation, such as oxide isolation, have advantages in area and parasitic element reduction, but they require a more complicated technology.

REFERENCES

1. K. Thiessen and G. Zech, *Phys. Stat. Sol.*, (a) *10*, K133 (1972).
2. W. Fahrner and A. Goetzberger, *J. Appl. Phys.*, *44*, 725 (1973).
3. W. Shockley and W. T. Read, *Phys. Rev.*, *87*, 835 (1952).

4. R. N. Hall, *Phys. Rev.*, *87*, 387 (1952).
5. E. M. Conwell, *Proc. IRE*, *46*, 1281 (1958).
6. J. L. Moll, *Physics of Semiconductors*, McGraw-Hill, New York, 1964, page 117.
7. P. U. Calzolari and S. Graffi, *Solid-State Electron*, *15*, 1003 (1972).
8. W. Shockley, *Electrons and Holes in Semiconductors*, Van Nostrand, Princeton, 1950, page 302.
9. H. Lawrence and R. M. Warner, Jr., *Bell Syst. Tec. J.*, *34*, 105 (1955).
10. P. E. Gray, D. DeWitt, A. R. Boothroyd, and J. F. Gibbons, *Physical Electronics and Circuit Models of Transistors*, Wiley, New York, 1964, page 75.

PROBLEMS

4.1 A 0.6 Ω-cm, n-type silicon sample contains 10^{15} cm^{-3} generation-recombination centers located at the intrinsic Fermi level with $\sigma_n = \sigma_p = 10^{-15}$ cm^2. Assume $v_{th} = 10^7$ cm s^{-1}
  (a) Calculate the generation rate if the region is depleted of mobile carriers.
  (b) Calculate the generation rate in a region where only the minority-carrier concentration has been reduced appreciably below its equilibrium value.

4.2 Light is incident on a silicon bar doped with 10^{16} cm^{-3} donors, creating 10^{21} cm^{-3} s^{-1} electron-hole pairs uniformly throughout the sample. There are 10^{15} cm^{-3} bulk recombination centers at E_i with electron and hole capture cross sections of 10^{-14} cm^2.
  (a) Calculate the steady-state hole and electron concentrations with the light turned on.
  (b) At time $t = 0$ the light is turned off. Calculate the time response of the total hole density and find the lifetime. The thermal velocity is 10^7 cm s^{-1}, and there is no current flowing.

4.3 (This problem is intended to provide perspective on the concept of charge neutrality in semiconductors.) Assume that the distribution of free electrons in Fig. P4.3a could be maintained in a block of silicon without any neutralizing charges. Consider that the electric field associated with the charge vanishes at $x = W$ and take: $n_1 = 10^{18}$ cm^{-3}, $W = 1$ μm, $D_n = 7$ cm^2 s^{-1}, and $T = 300°$K.
  (a) Calculate the field and current at $x = 0$ that is required to sustain this distribution of free charge. (Note that both a drift and diffusion component will be present.)
  (b) In light of the results of (a), is the charge configuration reasonable?
  (c) If a current density of approximately 10^5 A cm^{-2} can be sustained before power dissipation in the silicon causes irreversible damage, what is the maximum field that can exist in silicon having a uniform electron density equal to n_1? (The current

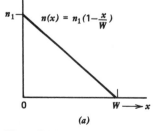

(a)

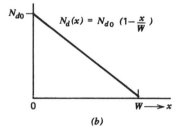

(b)

Figure P4.3

densities in most operating IC devices are two or more orders of magnitude lower than this value.)

(d) If the compensating positive charge density shown in Fig. P4.3b exists in the silicon, calculate the value of N_{d0} such that the current and field of part (c) will be present. Express this as a fraction of the free-electron density n_0, and note thereby how close to electrical neutrality the block of silicon will be. (Use values at $x = 0$).

4.4[†] If recombination of electrons takes place through a donorlike recombination center, the capture cross section σ_n for the site can be crudely estimated by considering that the free electron enters a region where its thermal energy $\frac{3}{2}kT$ is less than the energy associated with Coulombic attraction; this means that the carrier is drawn to the center. The radius at which this occurs describes an area that may be taken to be equal to σ_n.

(a) Obtain an expression for σ_n based upon this model and evaluate it for silicon at $300°K$.

(b) Considering the dependence on temperature of σ_n from this model together with other temperature-dependent terms, how would the electron capture rate for the center depend on temperature? (Assume that the center exists in extrinsic silicon.)

4.5[†] The analysis of recombination through recombination centers in Section 4.2 assumed low level injection with no current flowing so that we could find the characteristic time in which excess carriers decayed. Following a similar analysis, find the "effective lifetime" in the opposite case of *very* high level injection in an n-type semiconductor, assuming that charge neutrality holds. Let $n' = p' \gg n_0, p_0$, and compare the decay time or effective lifetime to that found in the case of low level injection for both $\sigma_n \neq \sigma_p$ and $\sigma_n = \sigma_p$.

4.6 Consider the continuity equation for minority-carrier holes that are injected across a pn junction into an n-type region that is very short so that recombination of the holes takes place essentially only at the contact at $x = W_B$. Show by the direct use of Eq. (4.3.10) that the holes are distributed linearly along x as was found in Eq. (4.3.17).

4.7 An ideal pn-junction diode has long p- and n-regions and negligible generation or recombination in the space-charge zone, so that the current-voltage characteristics are described by Eq. (4.3.15). Show that the limiting reverse-bias current predicted by Eq. (4.3.15) corresponds to the integrated generation rate of minority carriers on either side of the pn junction. *Hint. Use Eq. (4.3.12) and its equivalent for electrons and consider that* $G_p - R_p = -p_n'/\tau_p$.

4.8 Holes are injected across a forward-biased pn junction into an n-region that is very much longer than L_p, the hole diffusion length. Show, by using Eq. (4.3.12) that L_p is the average length that a hole diffuses before it recombines in the n-region.

4.9 For an ideal pn-junction diode show that the ratio of hole current to total current can be controlled by varying the relative doping on the two sides of the junction. If we call the ratio of hole current to total current γ, express γ as a function of N_a/N_d. Calculate γ for a silicon pn junction for which resistivity of the n-side is 0.001 Ω-cm and that of the p-side is 1 Ω-cm. Assume that $\tau_p = 0.1 \tau_n$ and that both neutral regions are much longer than the minority-carrier diffusion lengths in them.

4.10[†] A pn-junction diode has the configuration shown in Fig. P4.10. Assume that
1. $N_a = N_d = N_o \gg n_i$.
2. $W \ll L$, the minority-carrier diffusion length; $L_1 \gg L$.

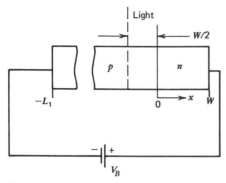

Figure P4.10

3. All the diffusion constants $= D$, all lifetimes $= \tau$.
4. The space-charge-region width $\ll W$.
5. The external applied voltage is V_B, where $V_B \gg \phi_i$, the built-in voltage.
6. The recombination velocity is infinite at $x = +W$.

If a light beam that produces G_0 hole-electron pairs per unit area per unit time in a slab of negligible width along the x direction is incident on the diode at the plane $x = -W/2$ on the p-type side:

(a) Assuming low level injection, find and sketch the minority-carrier concentrations in the neutral regions of the diode.

(b) Calculate the current that flows in the illuminated diode in terms of G_0 and the diode properties.

(c) In terms of the constants associated with the diode, what current flows when the light beam is removed (steady-state)?[10]

4.11 Consider an ideal, long-base, silicon abrupt *pn*-junction diode with uniform cross section and constant doping on either side of the junction. The diode is made from 1 Ω-cm p-type and 0.2 Ω-cm n-type materials in which the minority-carrier lifetimes are $\tau_n = 10^{-6}$ s and $\tau_p = 10^{-8}$ s, respectively. ("Ideal" implies that effects within the space-charge zone are negligible and that minority carriers flow only by diffusion mechanisms in the charge neutral regions.)

(a) What is the value of the built-in voltage?

(b) Calculate the density of the minority carriers at the edge of the space-charge zone when the applied voltage is 0.589 V (which is $23 \times kT/q$).

(c) Sketch the majority- and minority-carrier currents as functions of distance from the junction on both sides of the junction, under the applied bias voltage of part (b).

(d) Calculate the location(s) of the plane or planes at which the minority-carrier and majority-carrier currents are equal in magnitude for the applied bias voltage of part (b).

4.12 Construct a sketch showing the distribution of hole and electron currents in the neutral regions of a *pn*-junction diode (with forward bias) having *significant* recombination in the space-charge layer. Assume that:

(a) The injected hole current is twice the injected electron current.

(b) The net rate at which pairs recombine in the space-charge layer is equal to half of the net rate at which electrons recombine in the p-type region.

Demonstrate that the total diode current is given by adding a term that describes the space-charge-region recombination current to the sum of the diffusion currents at the edges of space-charge region.

4.13 Consider the integrated-circuit structures shown in cross section in Problem 3.9.

(a) Ignoring series resistance but considering all junctions, sketch the two sets of $I-V$ characteristics that you would expect to see for measurements of current when (i) voltage V_C (both polarities) is applied to contact C and contact A is grounded while contact B is left open. (ii) voltage V_B (both polarities) is applied to contact B and contact A is grounded while contact C is left open.

(b)† If it were possible to deplete the n^- layer fully under the conditions of part (a,i) above, what would you expect to happen to the current flowing through contact A? Explain the reasons for your conclusions.

4.14† A silicon diode of junction area 10^{-5} cm^2 is made with a uniformly doped p-type region having 5×10^{18} acceptors cm^{-3} and a step junction to an n-type region doped with 10^{15} donors cm^{-3}. The diode is found to be adequately described by the simplest theory; that is, one-dimensional flow with dominance of bulk recombination and transport of minority carriers under forward bias by pure diffusion into the neutral regions from the edge of the space-charge zone. Recombination-generation in the space-charge zone is negligible, and the minority-carrier lifetime in each region is 100 ns.

This junction diode is to be used in a circuit that requires a rectification ratio between forward current I_f and reverse current I_r ($|I_f/I_r|$) of 10^4 at 0.5 V and a maximum reverse saturation current of 100 nA. What is the maximum temperature at which the diode will perform adequately? Consider only the major temperature dependences (i.e., neglect changes in D, μ, τ_p, N_c, and N_v with temperature).

4.15 Prove that the recombination rate in the space-charge region of a pn junction in which $\sigma_n = \sigma_p$ is maximized when $p = n$, as given in Eq. (4.3.23).

4.16† Use the values for a silicon junction diode given in the paragraph following Eq. (4.3.25) together with the equation itself to make a semilogarithmic plot of J_t/J_r versus applied voltage V_a under forward bias. The diode consists of a step junction to a highly doped n-region (i.e., $N_a \ll N_d$).

4.17 Consider an abrupt diode biased so that $(\phi_i - V_a) = 5$ V. The diode has the following properties: $N_a = 10^{17}$ cm^{-3}, $\tau_n = 10^{-6}$ s, $N_d = 10^{18}$ cm^{-3}, and $\tau_p = 10^{-8}$ s.

(a) Use Eq. (4.3.29) to determine the ratio J_t/J_g (take $x_i = x_d$).

(b) Discuss the expected temperature dependence of this ratio.

4.18 An ideal short-base diode has been built with an abrupt junction in which $N_d \gg N_a$ and $N_a = 10^{17}$ cm^{-3}. (Assume that the n-region is degenerately doped.) The width of the p-region between the junction edge and the contact at which all recombination takes place is 3 μm. The area of the diode is 10^{-5} cm^2.

(a) Calculate the charge stored in the neutral p-region if 0.5 mA flows through the diode.

(b) Determine the charge stored in the narrowed space-charge layer under forward bias. (c) How much time would it take a current source of 0.5 mA to cause the diode to switch from an "off" condition (with $V_a = 0$) to the steady state of 0.5 mA of current flow?

4.19 Consider a small-signal equivalent circuit for a Schottky diode under forward bias. Compare the circuit to that for a *pn*-junction diode as sketched in Fig. 4.13. Discuss the differences and similarities in the two circuits.

4.20† Consider qualitatively the effects of punch-through on the structure shown in Fig. 4.16*d* as follows: (a) sketch a band diagram at thermal equilibrium along an axis that passes through the middle of the diffused *p*-region and runs through to the substrate; (b) with the substrate grounded, consider the imposition of sufficient reverse bias at the upper *pn* junction so that the junction between the epitaxial *n*-region and the *p*-substrate becomes forward biased. When this occurs, the holes in the substrate can be injected into the upper *p*-region. (In steady-state, a *space-charge-limited current* of holes will flow.) Sketch a band diagram for this condition.

4.21 Calculate the small-signal, incremental resistance and capacitance for an ideal, long-base silicon diode using the following parameters:

$$N_d = 10^{18} \text{ cm}^{-3}, \qquad N_a = 10^{16} \text{ cm}^{-3}, \qquad \tau_n = \tau_p = 10^{-8} \text{ s},$$
$$A = 10^{-4} \text{ cm}^2, \qquad T = 300°\text{K}$$

(a) For 0.1, 0.5, and 0.7 V forward bias.
(b) For 0, 5, and 20 V reverse bias.
(c) What is the series resistance of the *p*-type quasi-neutral region if this region is 0.1 cm long. This resistance must be added to the elements modeling the ideal diode when considering the response of a real diode.

4.22 A *pn* junction is initially off. A step of current is applied to the device with a polarity that turns the device on. Explain physically why it takes much less time for the width of the space-charge region to change than for the minority-carrier distributions to reach steady state in the quasi-neutral regions. (That is, compare the physical processes in the two situations.)

4.23 Consider the silicon integrated-circuit diode shown in Fig. P4.23. Can injected holes flowing from the *p*-type diffusion to the n^+ buried layer best be described by the long-base or short-base approximation? What about carriers flowing sideways from the edge of the *p*-type diffusion to the n^+ contact region? The hole lifetime is 1 *μ*sec in the *n*-type epitaxial layer, and the n^+ buried layer is an effective sink of excess minority carriers. Justify your answers. (Note that the figure is not to scale.)

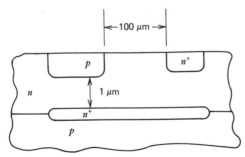

Figure P4.23

4.24 At high current densities, a significant fraction of the voltage applied across a diode can be dropped across the neutral regions of the device. Consider the current-voltage relation for a one-sided step junction with N_d donors on the lightly doped side of the junction. Find the current and applied voltage V_a at which 10% of V_a appears across the n-type neutral region for a typical integrated-circuit diode. Assume that the cross-sectional area is 10^{-5} cm^2, the length of the neutral region in the n-type silicon is 10 μm, $N_d = 5 \times 10^{15}$ cm^{-3}, and $\tau_p = 1$ ns.

Continuity equation: $\dfrac{\partial n}{\partial t} = \dfrac{1}{q} \dfrac{\partial J_n}{\partial x} + (G_n - R_n)$

$\dfrac{dn'}{dt} = G - R = -U = \dfrac{-(n_0 + P_0)n'}{(n_0 + P_0 + 2n_i \cosh[(E_t - E_i)/kT])\tau_0}$ low level injection

$n'(t) = n'(0) \exp(-t/\tau_n)$; $\tau_n \sim \tau_0 = (1/N_t v_{th} \sigma_0)$

$U = n'/\tau_n$

Long base diode: $P_n'(x) = P_{n_0}(e^{qV_a/kT} - 1)e^{-(x-x_n)/L_p}$

CHAPTER 5

BIPOLAR TRANSISTORS I : BASIC OPERATION

In Chapter 4, we saw that a *pn* junction, under bias such that the *p*-region is positive with respect to the *n*-region, conducts current because holes are injected from the *p*-region and electrons are injected from the *n*-region. These are majority carriers in each region, and their supply is not limited near the junction. Consequently, current rises rapidly as voltage is increased. Currents are very much smaller under the reverse-bias condition because they are carried only by minority carriers that are generated either in the junction space-charge zone or in other nearby regions. The current passed by a reverse-biased junction will, however, be increased if the supply of minority carriers in the vicinity of the junction is enhanced. This enhancement can, for example, be the result of the incident radiation of energetic particles in diode photodetectors or radiation sensors.

Another means of enhancing the minority-carrier population in the vicinity of a reverse-biased *pn* junction is simply to locate a forward-biased *pn* junction very close to it. This method is especially advantageous because it brings the minority-carrier population under electrical control, that is, under the control of the bias applied to the nearby forward-biased junction.

The modulation of the current-flow properties of one *pn* junction by changing the bias on a nearby junction is called *bipolar transistor action*. It is one of the most

200

significant ideas in the history of device electronics, and the investigations that led to it were honored with a Nobel Prize in Physics for the inventors of the bipolar junction transistor (or *BJT*), William Shockley, John Bardeen, and Walter Brattain.

In this chapter we formulate the basic physical description for bipolar transistor action. From the formulation we discover that transistor operation may be described more simply when various specific bias ranges are considered separately. We concentrate first on the so-called *active region* in which transistors act as amplifiers, and we consider the other bias regions in terms of transistor switching. We introduce the Ebers-Moll model, which provides an extremely useful and informative means of describing basic transistor operation in each of the bias ranges. The relationship of this model to our discussion of transistor action at the beginning of the chapter is then made clear. In a final section, we consider the design of planar diffused bipolar amplifying and switching transistors.

5.1 TRANSISTOR ACTION

A simple realization of a structure showing transistor action is sketched in Fig. 5.1. Two *pn* junctions are shown spaced a distance W units apart in a single bar of semiconductor material. The bar has a uniform cross section of area A. The junctions are located "close" to one another so that electrons that are injected across

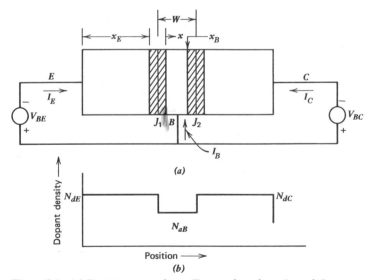

Figure 5.1   (*a*) Prototype transistor. Two *pn* junctions J_1 and J_2 are spaced W units apart. (*b*) Each region has a constant doping density. The doping changes abruptly at J_1 and J_2. The quasi-neutral portion of the middle *p*-region is bounded by the edges of space-charge zones at $x = 0$ and $x = x_B$, respectively.

Junction 1 when V_{BE} is positive are in the vicinity of Junction 2. From our discussion in Chapter 4, we recognize that the qualitative term "close" means that the junction spacing W is small enough so that the loss of electrons by recombination in the middle p-region is minimal. The middle region is commonly called the *base* of the transistor. For our initial discussion, we assume that electron recombination or generation in the base region is not significant; this assumption is reconsidered at a later time.

An important observation about current in this structure is that there is a negligible flow of holes (base majority carriers) between junctions 1 and 2 or in the reverse direction. This is true under all bias conditions because there can be only a vanishingly small flow of holes from either n-region into the p-type base. If we call the longitudinal dimension x and write the expression for hole current in the x direction (assuming no recombination) that was derived in Eq. (1.2.20), we have

$$J_p = 0 = q\mu_p p \mathscr{E}_x - qD_p \frac{dp}{dx} \tag{5.1.1}$$

and

$$\mathscr{E}_x = \frac{D_p}{\mu_p} \frac{1}{p} \frac{dp}{dx}$$

$$= \frac{kT}{q} \frac{1}{p} \frac{dp}{dx} \tag{5.1.2}$$

Thus, the condition of zero hole current in the base has led to an expression (Eq. 5.1.2) for the longitudinal electric field.* The field is seen to be dependent both on the magnitude of the majority-carrier (hole) density and on its gradient. If, for example, the base is homogeneously doped, then dp/dx is zero and there is no electric field. In general, however, there will be a field in the base that is proportional to the ratio between the majority-carrier (hole) density gradient and the majority-carrier (hole) density itself.

In contrast to a current of holes, however, a current of *electrons* flowing between the two junctions is possible since either junction can freely supply electrons from the n-regions to the middle p-region if that junction is forward biased. The electron current (from Eq. 1.2.19) is

$$J_n = q\mu_n n \mathscr{E} + qD_n \frac{dn}{dx} \tag{5.1.3}$$

$$= kT\mu_n \frac{n}{p} \frac{dp}{dx} + qD_n \frac{dn}{dx}$$

and

$$J_n = \frac{qD_n}{p} \left(n \frac{dp}{dx} + p \frac{dn}{dx} \right) = \frac{qD_n}{p} \frac{d(pn)}{dx} \tag{5.1.4}$$

---

* Strictly, Eqs. (5.1.1) and (5.1.2) must be considered as approximations because second order effects (to be described later) can account for small quantities of x-directed hole current.

It is useful to write an integrated form of Eq. (5.1.4) with arbitrary limits x and x' and to treat recombination as negligible so that J_n can be removed from the integral:

$$J_n \int_x^{x'} \frac{p \, dx}{q \, D_n} = \int_x^{x'} \frac{d(pn)}{dx} dx \tag{5.1.5}$$

$$= p(x')n(x') - p(x)n(x)$$

where we have used the fact that the right-hand side of Eq. (5.1.4) is a perfect differential. Equation (5.1.5) shows the minority-carrier (electron) current in the base to be dependent on the difference in the hole-electron products across a region divided by the integrated majority-carrier density in the same region. If we use the junctions themselves as the two boundaries of the region, then $x = 0$ becomes the lower limit and $x' = x_B$, the upper limit for the integrals in Eq. (5.1.5). From Eq. (4.3.20), the pn products at the boundaries can then be related to the junction voltages.

$$p(0)n(0) = n_i^2 \exp\left(\frac{qV_{BE}}{kT}\right)$$

$$p(x_B)n(x_B) = n_i^2 \exp\left(\frac{qV_{BC}}{kT}\right) \tag{5.1.6}$$

Thus, the longitudinal electron current in the base can be expressed as a function of the junction voltages V_{BE} and V_{BC}.

$$J_n = \frac{qn_i^2 \left[\exp\left(\frac{qV_{BC}}{kT}\right) - \exp\left(\frac{qV_{BE}}{kT}\right) \right]}{\int_0^{x_B} \frac{p \, dx}{D_n}} \tag{5.1.7}$$

Before we discuss the many implications of Eq. (5.1.7), we shall modify it slightly. First, we note that D_n is frequently a weak function of position in the base and can therefore be expressed by an average value $\tilde{D}_n$ and removed from the integral in the denominator of the right-hand side of Eq. (5.1.7).* With D_n removed, the integral expresses the total majority-carrier density per unit area in the base. We shall find it useful to convert this density into the total majority-carrier charge in the base by multiplying by qA (assuming area to be constant across the base) to define the base majority charge Q_B.

$$Q_B = qA \int_0^{x_B} p \, dx \tag{5.1.8}$$

* Strictly, $\tilde{D}_n$ is not a direct spatial average, but rather

$$\tilde{D}_n \equiv \int_0^{x_B} p \, dx \bigg/ \int_0^{x_B} (p/D_n) \, dx$$

With these changes, we can express the electron current flowing from the first junction to the second junction.

$$I_n = I_S \left[\exp\left(\frac{qV_{BC}}{kT}\right) - \exp\left(\frac{qV_{BE}}{kT}\right) \right]$$ (5.1.9)

where

$$I_S = \frac{q^2 A^2 n_i^2 \tilde{D}_n}{Q_B}$$ (5.1.10)

From Eq. (5.1.9) we see that the current I_n can be switched on and off through the action of the junction voltages. If both V_{BC} and V_{BE} are negative and significantly greater than kT/q, I_n will be insignificant. If, on the other hand, either V_{BE} or V_{BC} is positive and greater than kT/q, I_n will be a sensitive function of the most positive voltage.

Equation (5.1.9) is an elegant representation of the physics of transistor action, and it is capable of describing more phenomena than just switching action in the device. Before we explore these phenomena more fully, however, let us consider the physical basis for transistor action.

Prototype Transistor

The equation that we have derived to represent transistor action [Eq. (5.1.7)] is not a function of the base doping profile; rather it depends only on the integrated base majority charge. We shall see that this feature makes the equation extremely useful. For further insight into transistor action, however, it is worthwhile to consider currents in a very simple transistor from a different point of view. The simple transistor consists of the structure in Fig. 5.1 with abrupt junctions and constant base doping. We call this the *prototype transistor* because it resembles closely the device described by W. Shockley in his original discussion of the bipolar junction transistor.[1]

In Fig. 5.2 we sketch the energy-band diagrams and electron densities (base minority carriers) for the prototype transistor at equilibrium and under various bias conditions. Figure 5.2a refers to the transistor under equilibrium (no bias) conditions. There are only a very few electrons in the base region and transfers of electrons from either end region are inhibited by energy barriers. Negative bias applied to both junctions tends to increase these barriers and to deplete the base region of the few electrons present under equilibrium conditions. The band diagram and a sketch of the electron population for this case are shown in Fig. 5.2b. Positive bias applied to both junctions injects electrons into the base region by lowering the built-in barrier height, and the resultant vast increase in the base electron density allows the ready flow of current between the two junctions (Fig. 5.2c).

Active Bias. There is another bias condition in which one junction is forward biased and the other is reverse biased. This arrangement holds the greatest interest

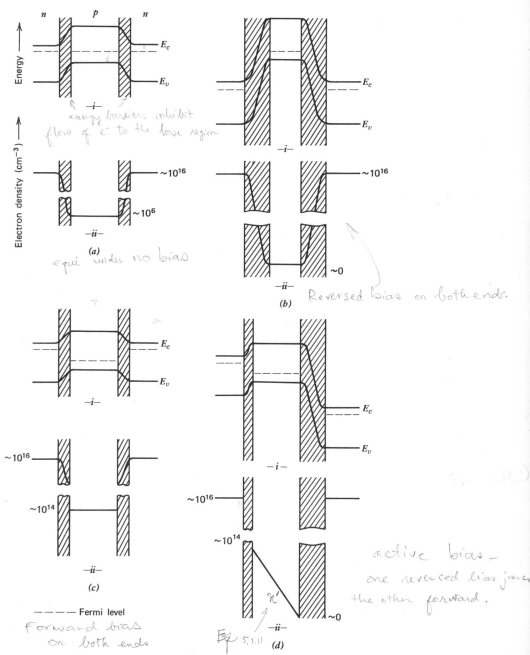

The handwritten annotations on the figure read:

energy barriers inhibit flow of e^- to the base region

equi. under no bias

Reversed bias on both ends.

Forward bias on both ends

active bias — one reversed bias junc. the other forward.

Eq 5.1.11

Figure 5.2 Energy-band diagrams (*i*) and electron-density sketches (*ii*) for the transistor sketched in Fig. 5.1. (*a*) Equilibrium condition, (*b*) both junctions reverse-biased, (*c*) both junctions forward-biased, (*d*) one junction reverse-biased, and one junction forward-biased. The cross-hatched areas are the space-charge zones.

because, as we shall see, it makes signal amplification possible. Diagrams for this bias condition, called *active bias*, are sketched in Fig. 5.2d. As we see in the figure, the forward-biased junction injects electrons into the base because the barrier is reduced from its equilibrium value. The reverse-biased junction, on the other hand, sweeps any nearby electrons from the base into the *n*-region at the end. There is thus a flow of electrons (base minority carriers) from the forward-biased junction, which we call the *emitter* junction, to the reverse-biased junction, known as the *collector* junction. Electrons in the vicinity of the collector junction are swept rapidly through the space-charge zone and into the *n*-type collector region.

As an introduction to transistors under active bias, and as clarification of some of the qualitative discussion that we have presented, we consider briefly the amplifying or active-bias operation of the prototype transistor. In Fig. 5.2d we have sketched as a straight line the electron density (minority-carrier density) in the transistor base. This result is justified by the analysis of the short-base diode that was carried out in Chapter 4 because the electrons in the base of a uniformly doped transistor are spatially distributed in the same way as they would be in the ideal short-base diode (Section 4.3). That is, the continuity equation that applies to this case has a negligible recombination term, and the doping density is constant. Hence, the solution for the electron density is linear in x. The boundary value at the base-emitter edge is exponentially dependent on the voltage V_{BE}; at the collector edge of the base, the electron density is negligible. The distribution for n' is thus linear, as found in Eq. (4.3.17) (for holes in that case). In terms of the geometry of Fig. 5.1, the excess electron density $n' = n - n_{po}$ is

$$n' = n_{po}\left[e^{qV_{BE}/kT}\left(1 - \frac{x}{x_B}\right) - 1 \right] \qquad 0 < x < x_B \qquad (5.1.11)$$

The current of electrons through the base for V_{BE} greater than kT/q, as is usual for active bias, is easily found since it is a purely diffusive flow. In terms of base doping N_a, since $I_n = qD_nA(dn/dx)$, we find

$$I_n = \frac{-qD_nAn_i{}^2 \exp\left(qV_{BE}/kT\right)}{N_ax_B} \qquad (5.1.12)$$

To derive this result from Eq. (5.1.9), we note that Q_B for the prototype transistor as given by Eq. (5.1.8) is

$$Q_B = qAN_ax_B \qquad (5.1.13)$$

When this value for Q_B is used in Eq. (5.1.10) and the negligible $\exp\left(qV_{BC}/kT\right)$ term in Eq. (5.1.9) is dropped, the expression for current that we derived in Eq. (5.1.12) is obtained. The significance of the integrated equations for transistor action [Eqs. (5.1.7) to (5.1.10)] lies in their ability to deal with transistors that are not uniformly doped. The original development of this approach[2] to transistor analysis had as its purpose the optimization of various alternative transistor doping profiles. We shall return to consider further aspects of this point of view in Chapter 6.

Transistors for Integrated Circuits

Before deriving further equations for transistors under active bias, we devote a few paragraphs to integrated-circuit transistors. This permits us to contrast *IC* devices with the simple prototype structure of Fig. 5.1 and allows us to develop theory and to discuss practical applications at the same time. The discussion here will be amplified in Section 5.5 after we have developed a perspective that will allow a more searching examination of the *IC* transistor.

Transistors for integrated circuits are made universally by the planar process described in Section 1.3. They therefore differ considerably in structure and in dopant configuration from the prototype device that was sketched in Fig. 5.1. The prototype transistor is a fairly good model for a grown-junction transistor, a device of little commercial interest at present. We have used it in our discussion because, regardless of the details of its structure, it permits a clear focus on many phenomena that are important to transistor operation.

A top view and a cross-sectional view of a typical integrated-circuit transistor are shown in Figs. 5.3*a* and 5.3*b*, respectively. The structure is produced by a sequence of steps identical to the one outlined in Section 4.5 where the fabrication of an array of junction isolated diodes was described. To make transistors instead of diodes, a heavily doped *n*-region is diffused into the *p*-region. This diffusion step produces the emitter-base junction. As with the diode array, a lightly doped *p*-type wafer is used for the substrate. A heavily doped *n*-type buried layer added to reduce series resistance underlies the collector *pn* junction. More specific discussion about dopant densities, geometries, and trade-offs in the processing of transistors is taken up in Section 5.5. Our focus for the present is on the basic device geometry.

One of the most critical parameters in transistor production, the width of the quasi-neutral base region (x_B in Fig. 5.3*c*), is typically a few tenths of a micron ($\sim 10^{-5}$ cm). Most of the other dimensions in the transistor are substantially larger—10^{-4} cm and greater. Transistor action, as we have defined it, is quite critically dependent upon the proximity of the two interactive junctions. Because of this, under most transistor bias conditions, transistor action in the device is confined to the shaded region of the cross-sectional view in Fig. 5.3*b*. In Fig. 5.3*c*, the region that includes the base diffusion has been expanded to facilitate discussion. This sketch indicates by cross-hatching the emitter-base and base-collector space-charge zones. It has been rotated 90° to maintain an *x*-axis orientation consistent with Fig. 5.1.

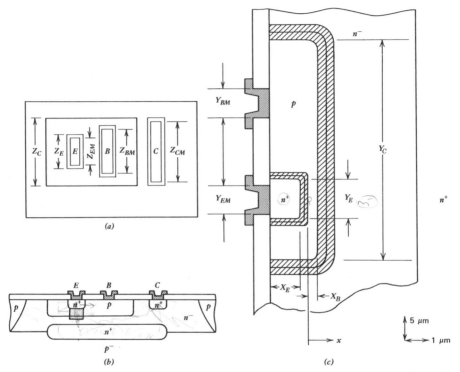

Figure 5.3 Top view (*a*) and cross sections (*b*) and (*c*) of a typical *npn IC* transistor. The region dominating transistor action is shaded in (*b*) and the area bounded by the base diffusion is rotated 90° and expanded in (*c*); *x* and *y* scales are indicated for (*c*) only.

Figure 5.3*c* reveals that transistor action is essentially confined to an area defined by Y_E, the width of the emitter quasi-neutral region. This is nearly equal to the emitter-stripe width because Y_E is typically at least five or ten times as wide as the penetration of space charge into the emitter. Thus, although *IC* processing unavoidably makes the collector width Y_C very much larger than the emitter width Y_E, the transistor is nonetheless well approximated for many purposes as a one-dimensional device. We have already made use of this fact in specifying a single area *A* in Eq. (5.1.8); for the transistor sketched in Fig. 5.3, that area *A* is the product of Y_E and Z_E. As the theory of junction transistors is developed further, we shall refer to other dimensions on Fig. 5.3, and examine the constraints imposed by the *IC* device on transistor equations.

In Section 5.2 we consider transistor action under *active bias* in more detail. As we shall see, the concepts involved are important for understanding both amplifying and switching uses of the transistor.

5.2 ACTIVE BIAS

We have described the condition V_{BE} positive and V_{BC} zero or negative as active bias* for an *npn* transistor. This bias results in electron injection at one junction (the *emitter-base junction*) and electron collection at the other (*base-collector junction*). When V_{BC} is zero or negative and V_{BE} is substantially larger than kT/q, we see from Eq. (5.1.9) that an electron current of value

$$I_n \approx -I_S \exp\left(\frac{qV_{BE}}{kT}\right) \tag{5.2.1}$$

will flow from left to right across the collecting junction (J_2 in Fig. 5.1). Using the standard convention that currents into a transistor are positive, I_n in Eq. (5.2.1) is equal to $+I_C$, a positive collector current. Equation (5.2.1) shows that under active bias, collector current is exponentially related to emitter-base voltage. Experimentally, this relationship is found to hold over many decades of collector current (Fig. 5.4).

Figure 5.4 shows experimental measurements of collector current I_C plotted on a logarithmic scale as a function of the base-emitter bias voltage V_{BE}. Over nearly the entire range of voltage, these measurements confirm an exponential dependence. Furthermore, after accounting for the ratio of natural to common logarithms, the slope of the I_C plot is just q/kT, as expected from Eq. (5.2.1). The experimental precision of this dependence is sufficiently high to make possible the construction of accurate thermometers that take advantage of collector-current changes with temperature in transistors under fixed emitter bias. The intercept of an extrapolated line drawn through the I_C measurements with the current axis at $V_{BE} = 0$ provides a value for I_S in Eq. (5.2.1). From Eq. (5.1.10) we find that the built-in base charge in the quasi-neutral region can be obtained once I_S is known.

$$Q_{B0} = \frac{q^2 A^2 n_i^2 \tilde{D}_n}{I_S} \tag{5.2.2}$$

All of the other parameters in Eq. (5.2.2) are well known.[†] We called Q_{B0} in Eq. (5.2.2) the built-in base charge because, as shown in Eq. (5.1.8), it represents the total hole charge in the quasi-neutral base as the base-emitter bias tends to zero. This charge is built-in during processing of the transistor. As we see above, a value for Q_{B0} can be obtained through current-voltage measurements on the transistor. Because this was first described by H. K. Gummel,[3] the number of base dopant

---

* Strictly, these polarities correspond to *forward-active bias*; if the polarities of V_{BE} and V_{BC} are interchanged, the transistor is under *reverse-active bias*.

[†] A possible exception is $\tilde{D}_n$, but diffusivity is not a very strong function of concentration (Fig. 1.12) and thus can be approximated.

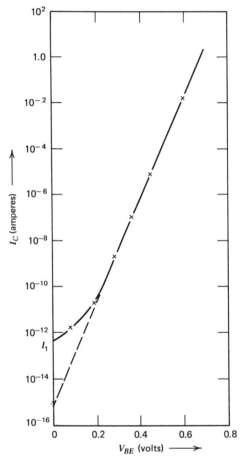

Figure 5.4   Semilogarithmic plot of collected current versus base-emitter voltage for an *IC npn* transistor under active bias.

atoms (per cm^2) in the quasi-neutral region given by

$$\int_0^{x_B} N_a(x)\, dx = \frac{Q_{B0}}{qA} = \frac{qAn_i^2\tilde{D}_n}{I_S} \tag{5.2.3}$$

is often called the *Gummel number*.

Equation (5.2.3) emphasizes that I_S, the multiplying factor for transistor current at a given bias, is inversely proportional to the Gummel number, that is, to the total base doping. The lower the built-in base charge Q_{B0}, the higher is the current at a given bias. Therefore, one might propose to design a prototype transistor with a very low constant base doping. A major disadvantage of this design is that even a small forward bias might make the low-injection approximation invalid near the base-emitter junction in the transistor, [i.e., $n_p(0)$ would approach N_a]. As we shall see in Chapter 6, high injection in a transistor causes a loss in performance. It can be avoided, and Q_{B0} can be kept small at the same time by making the base dopant profile to taper from a maximum near the base-emitter junction. The dopant density is then large in the region where the injected minority density is large and small where the minority density becomes negligible. Fortunately, diffusion technology automatically provides transistors with graded base dopings and the advantage that we have described is obtained in *IC* transistors. Other advantages of graded-base transistors were first pointed out by Kroemer[4], and are described in Chapter 6.

Control of Q_{B0} during transistor processing is the key step in the production of integrated circuits. For high-gain transistors, in which the Gummel number is below 10^{12} cm^{-2}, extraordinary care is taken to control this parameter.

Current Gain

Our discussion of transistor action under active bias has thus far considered only current flowing between the collector and the emitter. This represents the output current in an active-biased transistor. Collector current is an exponential function of base-emitter voltage [Eq. (5.2.1)], because forward bias on the base-emitter junction causes electron injection into the base to vary exponentially. Under active bias, these electrons are collected efficiently by the field at the base-collector, space-charge region. The base-emitter terminals are thus the control electrodes for the collector current under active bias. The smaller the current that flows through these terminals for any given positive V_{BE}, the more effective will the transistor be as an amplifier since the lower will be the input power (the product of V_{BE} and the base-emitter current).

Several mechanisms give rise to base-emitter currents in an active-biased transistor. Based on our previous considerations, the most straightforward of these is recombination of injected electrons with majority-carrier holes in the base. Another component of base-emitter current is due to recombination in the base-emitter space-charge region. A third component of current flows because forward bias on the junction not only injects electrons into the base, but it also causes holes to be injected into the emitter. These various components are indicated pictorially in Fig. 5.5. If a transistor is meant to amplify effectively, all of these current components should be kept appreciably smaller than the collector current.

Most transistors are designed so that recombination of injected electrons in the base-emitter, space-charge zone is of less consequence than the other components mentioned except under very low current conditions. Accordingly, we shall consider recombination in the space-charge zone separately in Chapter 6 when we discuss limitations on transistor performance.

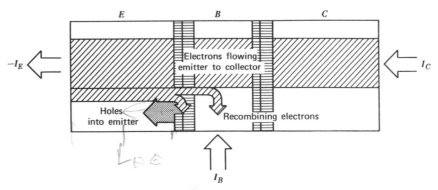

Figure 5.5 Terminal currents and major current components in an active-biased transistor. Not shown is the collector leakage current (I_1 in Fig. 5.4).

Recombination within the base itself can be expressed readily using the theory that we have developed in Section 4.3. There we found that recombination of excess minority carriers is directly proportional to their density (in this case n' the excess base electron density) so that the total base-region recombination current is

$$I_{rB} = qA_E \int_0^{x_B} \frac{[n - (n_i^2/N_a)]\, dx}{\tau_n} \tag{5.2.4}$$

where $A_E = Y_E \times Z_E$ in Fig. 5.3 is the area of significant minority-carrier injection. To make I_{rB} as small as possible at a given bias, we see from Eq. (5.2.4) that base lifetime τ_n should be maximized and base width x_B should be minimized.

Under active bias, the injected excess electron density n' is much greater than the equilibrium electron density (n_i^2/N_a) over most of the base. Also, lifetime is not strongly x dependent so that Eq. (5.2.4) can be simplified to

$$I_{rB} = \frac{qA_E}{\tau_n} \int_0^{x_B} n\, dx \tag{5.2.5}$$

For the special case of a transistor with a uniformly doped base such as the prototype transistor sketched in Fig. 5.1., the dependence of n' on x is linear. Thus, the integration in Eq. (5.2.5) is easily carried through to obtain

$$I_{rB} = \frac{qA_E n_i^2 x_B}{2N_a \tau_n} \left[\exp\left(\frac{qV_{BE}}{kT}\right) - 1 \right] \tag{5.2.6}$$

Although Eq. (5.2.6) was derived for a special case, the proportionality of base recombination current to $[\exp(qV_{BE}/kT) - 1]$ is obtained in general.

The loss of carriers to recombination in the base region is measured by the base transport factor, which is usually given the symbol α_T and defined by

$$\alpha_T = \frac{I_{nE} - I_{rB}}{I_{nE}} = 1 - \frac{I_{rB}}{I_{nE}} \qquad (5.2.7)$$

where I_{nE} is the electron current injected from the emitter. Using Eqs. (5.1.12) and (5.2.6) for a transistor with a uniformly doped base, we obtain

$$\alpha_T = 1 - \frac{x_B{}^2}{2D_n\tau_n} = 1 - \frac{x_B{}^2}{2L_n{}^2} \qquad (5.2.8)$$

Although Eq. (5.2.8) is not directly applicable to IC transistors, it is sometimes used for these devices. The equation does not apply because base doping in the IC transistor is not constant. Since a graded base, as obtained by diffusion technology, increases I_S, less minority-carrier injection is needed for a given output current. Thus, a graded base improves the transport factor and α_T as calculated by Eq. (5.2.8) can be regarded as a "worst-case" parameter. If we take a typical diffusion length of 30 μm and consider x_B to be 0.5 μm, we find $\alpha_T = 0.99986$. Indeed, the loss of minority carriers to recombination in the quasi-neutral base is very small in IC transistors.

The injection of base majority carriers (holes) into the emitter is the predominant cause for base current in most integrated-circuit transistors. Expressions for this current have already been derived in Section 4.3, since the base-emitter junction is just a forward-biased diode. For a given device, however, it is necessary to determine where recombination of the holes injected into the emitter takes place in order to write a correct expression for the hole current. Consider first the prototype transistor shown in Fig. 5.1; it has an emitter contact spaced a distance x_E from the edge of the base-emitter space-charge zone. If x_E is much greater than a diffusion length for holes, virtually all injected holes will recombine before reaching the ohmic contact. Hence, by the theory developed in Section 4.3, the excess holes are distributed in the emitter according to an exponentially decaying function [Eq. (4.3.12)] as sketched in Fig. 5.6a. The hole current is proportional to the hole-density gradient at the edge of the emitter quasi-neutral region.

$$I_{pE} = \frac{-qA_E n_i{}^2 D_{pE}}{N_{dE}L_{pE}} (e^{qV_{BE}/kT} - 1) \qquad (5.2.9)$$

long _short_ (5.2.9)

If the emitter contact is close to the base $(x_E \ll L_{pE})$, the hole concentration is a linear function of x (Fig. 5.6b) and the hole current is

$$I'_{pE} = \frac{-qA_E n_i{}^2 D_{pE}}{N_{dE}x_E} (e^{qV_{BE}/kT} - 1) \qquad (5.2.10)$$

long base-
short

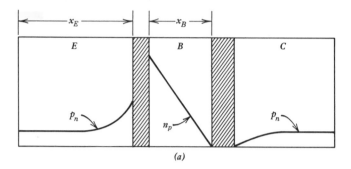

$$x_E \gg \sqrt{D_p \tau_p}$$

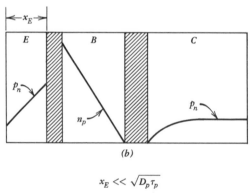

$$x_E \ll \sqrt{D_p \tau_p}$$

Figure 5.6 Minority-carrier distributions in the prototype transistor of Fig. 5.1 under active bias. (*a*) $x_E \gg$ hole diffusion length in emitter. (*b*) $x_E \ll$ hole diffusion length.

For many integrated-circuit transistors, the emitter contact is closer to the base than a diffusion length. In these devices, however, emitter doping is not constant so that injected holes in the emitter are influenced by a built-in electric field. Equations for this condition could be derived using a nearly analogous set of arguments to those employed in Section 5.1 for the transport of electrons in the base. The difference for this case is the presence of a nonzero majority-carrier (electron) current.

Emitter Injection: Nonuniform Doping. To discuss nonuniform doping, consider the *IC* transistor of Fig. 5.3. The impurity profile causes a built-in electric field in the emitter at thermal equilibrium. Aside from the junction space-charge region, however, we noted in Section 3.1 that most nonuniformly doped regions can be treated as nearly neutral or quasi-neutral. This assumption means that $n_0(x) \approx$

$N_d(x)$ and the built-in field at thermal equilibrium has the value obtained in Eq. (3.1.13), which we repeat here for reference.

$$\mathscr{E}_0(x) = -\frac{kT}{q}\frac{1}{N_{dE}(x)}\frac{dN_{dE}(x)}{dx} \tag{5.2.11}$$

If we consider the emitter-base junction under forward bias, we can expect that the field in the quasi-neutral emitter will be altered somewhat from the thermal-equilibrium value in Eq. (5.2.11). For first order analysis we express the field under bias as the sum of $\mathscr{E}_0$ plus an added term $\mathscr{E}_a$ that results from $V_{BE} \neq 0$. This is a reasonable assumption consistent with our consideration of low-level injection (i.e., $p_n' = n_n' \ll n_o$). With these considerations, the expressions for electron and hole currents in the emitter become

$$J_n = q\mu_n(n_{no} + p_n')(\mathscr{E}_0 + \mathscr{E}_a) + qD_n\left(\frac{dn_{no}}{dx} + \frac{dp_n'}{dx}\right)$$

and

$$J_p = q\mu_p(p_{no} + p_n')(\mathscr{E}_0 + \mathscr{E}_a) - qD_p\left(\frac{dp_{no}}{dx} + \frac{dp_n'}{dx}\right) \tag{5.2.12}$$

Since the drift and diffusion components balance at equilibrium (when $\mathscr{E} = \mathscr{E}_0$), Eqs. (5.2.12) can be written more simply as

$$J_n = q\mu_n n_{no}\mathscr{E}_a + qD_n\frac{dp_n'}{dx} \tag{5.2.13}$$

and

$$J_p = q\mu_p(p_{no} + p_n')\mathscr{E}_a + q\mu_p p_n'\mathscr{E}_0 - qD_p\frac{dp_n'}{dx} \tag{5.2.14}$$

The first term in Eq. (5.2.13) represents the ohmic flow of majority-carrier electrons. As with the ideal diode, we expect that only a small field $\mathscr{E}_a$ is necessary to provide this ohmic current. Typically $\mathscr{E}_a$ will be less than $\mathscr{E}_0$. Making this assumption in Eq. (5.2.14) for the minority-carrier hole current, we conclude that the second term is appreciably larger than the first. Thus,

$$J_p = q\mu_p p_n'\mathscr{E}_0 - qD_p\frac{dp_n'}{dx} \tag{5.2.15}$$

By using Eq. (5.2.11) for the thermal-equilibrium field and combining terms, we get

$$J_p = -\frac{qD_p}{N_d(x)}\frac{d}{dx}[p_n'(x)N_d(x)] \tag{5.2.16}$$

an equation similar to Eq. (5.1.4) for electron current in the base.

Now let us consider the important case of a short emitter region in which $x_E \ll L_{pE}$ so that negligible recombination of holes occurs in the bulk of the n-type region. Then the hole current is not a function of position, and Eq. (5.2.16) can be integrated in the negative x direction from an arbitrary point x within the emitter to the ohmic contact where $p'_n = 0$

$$J_p \int_x^{x_E} \frac{N_d(x')\, dx'}{D_p} = qp'_n(x)N_d(x) \tag{5.2.17}$$

The hole current can be found by evaluating Eq. (5.2.17) at the edge of the emitter-base, space-charge region $-x_n$, where the minority-carrier density is related to its thermal equilibrium value by Eq. (4.3.8). The hole current in the emitter is then found to be

$$I''_{pE} = -\frac{q\tilde{D}_p n_i^2 A_E (e^{qV_{BE}/kT} - 1)}{\int_{-x_n}^{x_E} N_d\, dx} \tag{5.2.18}$$

For a uniformly doped emitter Eq. (5.2.18) reduces to Eq. (5.2.10).

If recombination in the quasi-neutral region is not negligible and some of the carriers recombine before traversing the emitter, Eq. (5.2.16) is still valid, but it is not possible to remove J_p from the integral for an explicit solution as in Eq. (5.2.18). The hole current is, however, increased over the value predicted by Eq. (5.2.18). Further approximating techniques may be used to extract a solution, but we will not consider them here.

For the two cases in which surface recombination dominates in determining I_{pE} [Eqs. (5.2.10) and (5.2.18)], the proper value to use for A_E may not be straightforward. This is because surface recombination will almost certainly be far more effective at an ohmic contact than at an oxide surface. Hence, I_{pE} may be specified more accurately by using the contact area ($Y_{EM} \times Z_{EM}$ in Fig. 5.3) instead of the junction area for A_E in either Eqs. (5.2.10) or (5.2.18). We comment further on this point in Section 5.5 when we consider diffused planar transistors in more detail.

The effectiveness of an emitter junction in injecting electrons into the base is measured by the emitter efficiency, usually denoted by the symbol γ.

$$\gamma = \frac{I_{nE}}{I_{nE} + I_{pE}} = \frac{1}{1 + I_{pE}/I_{nE}} \tag{5.2.19}$$

Since $(I_{nE} + I_{pE})$ is the total emitter current I_E, the electron current crossing the emitter-base junction I_{nE} is just γI_E. Let us make an approximation for γ for the IC transistor using the approach taken previously for α_T. We apply the simple theory for the prototype transistor (Fig. 5.1) and take dimensions and average doping appropriate to the IC transistor. Consider an emitter-region width of $x_E = 1.5$ μm, a base quasi-neutral region width of $x_B = 0.5$ μm, an average emitter doping $N_{dE} = 5 \times 10^{17}$ cm^{-3}, and an average base doping $N_{aB} = 2 \times 10^{16}$ cm^{-3}. Then using Eqs. (5.1.12) and (5.2.10) in Eq. (5.2.19) together with diffusion constants from Fig. 1.12, we can write

$$\gamma = \cfrac{1}{1 + \cfrac{x_B N_{aB} D_{pE}}{x_E N_{dE} D_{nB}}} \tag{5.2.20}$$

$$= 0.9973$$

Values for γ of this general order are readily obtained in *IC* transistors.

The magnitude of the ratio of the collector current I_C to the emitter current I_E under active bias is frequently given the symbol α_F. For our analysis, α_F is the product of γ and α_T.

$$\boxed{\alpha_F = \gamma \alpha_T} \qquad = \frac{I_e}{I_E} \quad \text{Forward current gain.} \tag{5.2.21}$$

Since all currents into the transistor sum to zero by Kirchhoff's current law, we have

$$I_B + I_E + I_C = 0$$

$$I_B - \frac{I_C}{\alpha_F} + I_C = 0$$

or

$$I_C = \frac{\alpha_F I_B}{(1 - \alpha_F)} = \beta_F I_B \tag{5.2.22}$$

where $\beta_F \equiv I_C / I_B$ is the current gain for the case in which input current flows between the base and emitter and output current flows into the collector. Because α_F is nearly unity, β_F is very large (typically of order 100). For the case that we took as an example, the prototype transistor with *IC* dimensions and average doping values, $\alpha_F = 0.997$ and $\beta_F = 351$. Small changes in α_F caused, for example, by process variations in fabricating the transistor are likewise magnified to large changes in $\beta_F [d\beta_F = d\alpha_F/(1 - \alpha_F)^2]$. This means that β_F is not a precisely controlled transistor parameter.

The theory that we have developed in this section applies to quiescent bias conditions and the equations have been derived for total currents. For applications to amplifiers, equations that express the results of incremental excursions in bias around a quiescent bias point are needed. It is straightforward to obtain them, but we defer taking this step until Chapter 6 when we derive various equivalent circuits that are useful for design with transistors. We shall also consider frequency effects in transistors at that time. Our analyses thus far have considered only the dc (and low-frequency) case.

Before resuming our discussion of transistor action, we should point out that the analysis of hole injection into the emitter that we carried out for the case of nonuniform doping applies in general to *pn* junctions in integrated circuits. For example, the injected hole current given by Eq. (5.2.18) is analogous to the result we would obtain for electron injection into the *p*-type region in the diffused *pn*-junction diode considered in Section 4.5 (Fig. 4.18*b*).

5.3 TRANSISTOR SWITCHING

Our discussion of transistor action in Section 5.1 has emphasized that injection of electrons (base-minority carriers) into the base region is necessary for current to flow between the collector and the emitter or vice versa. Thus, transistor switching can be understood in terms of storage, extraction, and transport of electrons in the base. Transistor models having widespread utility are derived from this approach to understanding transistor switching.

Let us consider first the switching transistor in the "cut-off state." This condition is assured by depleting the base region of minority carriers, that is, by making both V_{BE} and V_{BC} zero or negative. With these bias voltages applied, the base has at most its equilibrium population of electrons and, thus, only very small currents can flow between collector and emitter. When negative bias is simultaneously applied to both junctions, the base becomes depleted of even the few "built-in" minority carriers and the dc behavior of the transistor is very close to that of an open circuit. The minority-carrier populations in all three regions of a homogeneously doped transistor biased into cutoff are shown in Fig. 5.7.

Although the cut-off transistor will not pass dc current, its behavior is not identical to that of an open circuit. As described in Section 3.3, a reverse bias on a *pn* junction exposes fixed donor and acceptor charges by extracting compensating free charges. For calculations, it is useful to make a graph that shows the variation in stored junction charge, which we denote q_V, as a function of voltage. This is readily accomplished by using the theory of Chapter 3 for step or linearly graded junctions or else using the numerical data of Lawrence and Warner (reference 9 in Chapter 4). A useful plot for estimating q_V is given in Fig. 5.8. Derivation of the plot is considered in Problems 5.8 and 5.9. The inset in Fig. 5.8 identifies q_V, the stored charge, and makes clear that only the excess above the space charge at thermal equilibrium is plotted.

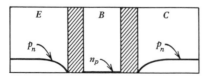

Figure 5.7 Minority-carrier densities in the prototype transistor of Fig. 5.1 under cut-off conditions.

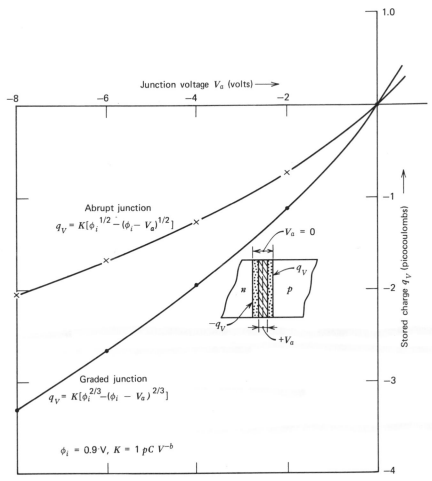

Figure 5.8 Stored charge versus applied junction bias in the space-charge regions of an abrupt junction and a linearly graded junction.

To place a transistor in cut-off, the switching source must supply the junction with the incremental charge stored in the space-charge zone; thus, the cut-off device has a capacitorlike behavior although the capacitance (dq_V/dV) is not constant. For example, consider the graded-junction plot in Fig. 5.8 to represent q_{VC}. If a transistor were switched from $V_{BC} = +0.6$ V (where $q_{VC} \approx +0.5$ pC) to $V_{BC} = -8$ V (where $q_{VC} \approx -3.2$ pC) the switching circuit would have to withdraw $[0.5 - (-3.2) = 3.7]$ picocoulombs of charge from the base lead to charge the base-collector junction. A similar calculation must be carried out for the emitter-base junction to determine the total charge withdrawn. Such calculations are typical of the analysis needed for the design of switching circuits. We shall discuss this topic in greater detail in Chapter 6 when we introduce charge-control modeling of transistors.

A measure of the very small currents that flow in a cut-off transistor is available in the active-bias data plotted in Fig. 5.4. We noted there that if the collector-base junction is reverse biased and V_{BE} is reduced nearly to zero, the collector current will begin to deviate from the exponential relationship predicted by Eq. (5.2.1). Instead, I_C approaches a low value I_1 that is independent of V_{BE}. The fraction of a picoamp indicated in Fig. 5.4 is typical. At very low base-emitter bias, the base-emitter junction is injecting such a small number of electrons that transistor action between the junctions is negligible. The collector-base junction behaves electrically like an isolated pn-junction diode under reverse bias, just as it would in cut-off, and I_1 flows from collector to base.

As was discussed in Chapter 4, three separate mechanisms can be responsible for the small current in a pn junction under reverse bias: generation of holes and electrons in the space-charge region, generation of electrons in the p-type base, and generation of holes in the n-type collector. The first of these three components, generation in the space-charge zone, dominates in silicon pn junctions as we discovered in evaluating Eq. (4.3.29).

The space-charge-zone generation component I_g is approximately proportional to the width of the space-charge layer as was shown in Section 4.3. It depends relatively weakly on collector-base bias [Eqs. (4.3.26-28)], and is most often written

$$I_1 = I_g = \frac{1}{2} \frac{qn_i x_d A_C}{\tau_0} \tag{5.3.1}$$

where x_d is the space-charge-zone thickness, A_C is the collector-base junction area and τ_0 is the effective lifetime in the zone itself. Because of its small size I_1 is of no consequence to the operation of silicon transistors biased in the active mode. It may have relevance for transistor switches, however. In terms of the IC transistor structure in Fig. 5.3, the area A_C is roughly $Y_C \times Z_C$ plus the vertical portions of the collector-base junction that meet the surface. There is also a contribution to I_g from recombination centers at the surface (Section 4.3). With careful processing this contribution can be made very small.

Regions of Operation

Figure 5.9 is a useful aid for understanding transistor switching because it delineates the regions of device operation in terms of the applied junction voltages. For example, the cut-off region that we have been discussing consists of the third quadrant in this map of a bias "space" having coordinates V_{BE} and V_{BC}. The fourth quadrant (V_{BE} positive, V_{BC} negative) corresponds to the forward-active region that was considered in Section 5.2.

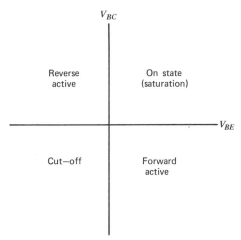

Figure 5.9 Regions of operation for an *npn* transistor as defined by base-emitter and base-collector bias polarities.

In the second quadrant the polarities of both V_{BC} and V_{BE} are inverted from the forward-active condition and the transistor is said to be biased in the reverse-active region. In this region an *npn* transistor injects electrons at the collector and collects them at the emitter. When operated in the reverse-active region, a one-to-one correspondence can be made to the parameters defined in this section for forward-active bias. These parameters are subscripted with an R to denote a reverse measurement. For example, output current is delivered to the emitter lead and the ratio of output current to input (base) current in a reverse-active condition will define $\beta_R = I_E/I_B$. The distribution of minority carriers in the prototype transistor of Fig. 5.1 under reverse-active bias is sketched in Fig. 5.10. Its similarity to Fig. 5.6 should be noted.

In contrast to the prototype transistor, the integrated-circuit transistor (Fig. 5.3) is asymmetric both in geometry and doping and a sketch of minority-carrier distributions under reverse-active bias differs markedly from one made in the forward-active region. First, electrons injected from the normal collector travel against the built-in base field as they move toward the emitter. Second, injection

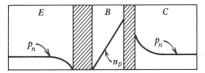

Figure 5.10 Minority-carrier densities in the prototype transistor of Fig. 5.1 under bias in the reverse-active region.

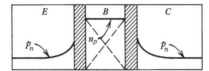

Figure 5.11   Minority-carrier densities in the prototype transistor of Fig. 5.1 under bias in the saturation region.

from the collector results in losses at the base contact and at the passivating oxide that are not significant in the forward-active region. Third, the injection efficiency in the reverse-active mode is very much lower than in the forward-active region. These asymmetries have important practical consequences, which we discuss in Chapter 6.

The first quadrant in Fig. 5.9 is defined by positive bias on both the base-emitter and base-collector junctions. This bias condition is referred to as *saturation*.* A transistor switch in the closed or "on" position will be biased in this region. In saturation, both junctions inject electrons, and the minority carriers are distributed throughout the prototype transistor as shown in Fig. 5.11. Note that in saturation the electron concentration is markedly increased throughout the entire base region. If the minority-carrier populations in Fig. 5.11 are compared to the diagrams of Figs. 5.6 and 5.10, we see that the saturated condition corresponds to the superposition of forward-active and reverse-active operation. The physical basis for this fact is that both junctions are injecting and collecting electrons *at the same time.* They inject because the built-in potential is reduced from its equilibrium value; they collect because the junction field is still of the proper polarity to sweep electrons from the base. This property of the saturated transistor being made up of a superposition of forward-active and reverse-active bias conditions does not depend on transistor geometry; it is valid for the *IC* transistor as well as for the prototype device.

We began this section with a general discussion of transistor switching and of the conditions in a cut-off transistor. This led us to a review of the regions of transistor operation, which are summarized in Fig. 5.9. A transistor switch is biased alternately in regions 1 and 3 (saturation and cut-off) and moves through regions 2 and 4 only during switching transients. To move from saturation to cut-off, the charges stored in and near the base of a transistor must be altered. Much of the design of transistor switching circuits consists of assuring that these charging and discharging requirements are adequately met. Models that account for the physical mechanisms we have discussed and that are useful for transistor switching calculations are presented in Chapter 6. In the next section we introduce a tran-

---

* The term "saturation" refers to the collector current that is determined by conditions in the circuit external to the transistor in this bias mode.

sistor model that has useful dc applications and that quantifies some of the physical pictures we have developed.

5.4 EBERS-MOLL MODEL

A simple and very useful model of the carrier injection and extraction phenomena occurring in transistors was developed in 1954 by J. J. Ebers and J. L. Moll of the Bell Telephone Laboratories.[5] The model is represented by the equivalent circuit shown in Fig. 5.12 and by the following equations (for *npn* transistors)

$$I_E = -I_F + \alpha_R I_R$$
$$I_C = \alpha_F I_F - I_R \tag{5.4.1}$$

where

$$I_F = I_{ES}\left[\exp\left(\frac{qV_{BE}}{kT}\right) - 1\right]$$
$$I_R = I_{CS}\left[\exp\left(\frac{qV_{BC}}{kT}\right) - 1\right] \tag{5.4.2}$$

To complete the definition of the model, at this point we postulate* a subsidiary reciprocity relationship between its parameters:

$$\alpha_F I_{ES} = \alpha_R I_{CS} \tag{5.4.3}$$

The Ebers-Moll model thus represents the transistor by two diodes that are connected in parallel with current sources. Each current source supplies current

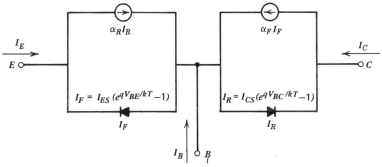

Figure 5.12 Equivalent circuit for the Ebers-Moll model of an *npn* transistor.

---

* A proof of this relationship is provided in an appendix to the original Ebers-Moll paper.[5] We shall consider Eq. (5.4.3) further at the conclusion of this section.

that is linearly related to the diode current of the opposite junction. To see the correspondence between the Ebers-Moll (*E-M*) model and our earlier discussion of transistor action, we perform the following analysis. First, we note that through the use of Eqs. (5.4.1), (5.4.2) and the Kirchhoff current law we can express the base current as follows:

$$I_B = I_{BC} + I_{BE} \tag{5.4.4}$$

where

$$I_{BC} = I_{CS}(1 - \alpha_R)\left[\exp\left(\frac{qV_{BC}}{kT}\right) - 1\right]$$

$$I_{BE} = I_{ES}(1 - \alpha_F)\left[\exp\left(\frac{qV_{BE}}{kT}\right) - 1\right] \tag{5.4.5}$$

The two components of base current in Eqs. (5.4.5) have been separated according to their voltage dependences. In terms of these base-current components, the emitter and collector currents in Eq. (5.4.1) can be written

$$I_E = \alpha_R I_{CS}\left[\exp\left(\frac{qV_{BC}}{kT}\right) - \exp\left(\frac{qV_{BE}}{kT}\right)\right] - I_{BE}$$

$$I_C = \alpha_F I_{ES}\left[\exp\left(\frac{qV_{BE}}{kT}\right) - \exp\left(\frac{qV_{BC}}{kT}\right)\right] - I_{BC} \tag{5.4.6}$$

In deriving Eq. (5.4.6) we have made use of the reciprocity relationship of Eq. (5.4.3). Equations (5.4.6) divide both the collector and emitter currents into two components, one linking the two junctions and the other flowing into the base. The expression for the linking current in Eqs. (5.4.6) is identical to that derived in Eq. (5.1.9) provided that we take

$$\alpha_F I_{ES} = \alpha_R I_{CS} = I_S \tag{5.4.7}$$

where I_S has been defined in Eq. (5.1.10). The base-current components I_{BE} and I_{BC} represent both recombination-generation processes in the base and space-charge zones and hole injection from the base into the emitter and collector, respectively.

To see how the *E-M* model represents the transistor in the various regions of operation, consider first the cut-off region in which V_{BE} and V_{BC} are both negative. From Eqs. (5.4.5) and (5.4.6), we obtain the equivalent circuit shown in Fig. 5.13. The model reduces to two current sources that represent the reverse saturation currents of the two junctions. In the forward-active region, the base-emitter junction is forward biased and the base-collector junction is reverse biased. We can rearrange Eqs. (5.4.1) to express collector current in terms of emitter current:

$$I_C = -\alpha_F I_E - I_R(1 - \alpha_F\alpha_R) \tag{5.4.8}$$

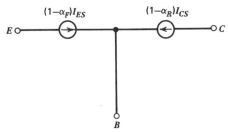

Figure 5.13   The Ebers-Moll representation for a transistor biased in the cut-off region.

which, under active-bias conditions, becomes

$$I_C = -\alpha_F I_E + I_{CS}(1 - \alpha_F \alpha_R) \qquad (5.4.9)$$

Similarly, reverse-active bias results in

$$I_E = -\alpha_R I_C + I_{ES}(1 - \alpha_F \alpha_R) \qquad (5.4.10)$$

Inspection of Eqs. (5.4.9) and (5.4.10) shows that the parameters of the E-M model, α_F, α_R, I_{ES}, and I_{CS}, can be obtained by measuring I_C versus I_E under forward-active bias or I_E versus I_C under reverse-active bias. If the forward-active data are used, the measured data should plot linearly with slope equal to $-\alpha_F$. The intercept of the plot with $I_E = 0$ corresponds to a measurement with open-circuited emitter, and the current flowing in this case is usually denoted I_{CO} (sometimes I_{CBO}) where

$$I_{CO} = I_C\big|_{I_E = 0} = I_{CS}(1 - \alpha_F \alpha_R) \qquad (5.4.11)$$

This current can be compared to the current I_{CEO} in the collector when the base is open circuited. From Eqs. (5.4.1) and (5.4.2),

$$I_{CEO} = I_C\big|_{I_B = 0} = \frac{I_{CS}(1 - \alpha_F \alpha_R)}{(1 - \alpha_F)} = \frac{I_{CO}}{(1 - \alpha_F)} \qquad (5.4.12)$$

The difference in magnitudes between I_{CO} and I_{CEO} can be traced to the boundary conditions enforced by the bias arrangements on the emitter-base junction. When I_{CO} flows, the emitter-base junction assumes a built-in reverse bias because of the extraction of some electrons from the emitter without their replacement from the external circuit. In this case the collector current is carried only by electrons generated in the base and by holes generated in the collector. In the second case, I_{CEO}, the base is open circuited instead of the emitter. Then, the base-emitter junction becomes forward biased, and the greater part of the collector current results from electrons carried across the base from the emitter. The leakage I_{CO} is therefore effectively multiplied by the transistor gain.

It is often very useful to obtain models in which the active elements (generators) are actuated by terminal currents. This is easily done for active bias with emitter current as the terminal variable using the equations we have developed. The equivalent circuit shown in Fig. 5.14a is an emitter-actuated model that is consistent with Eq. (5.4.9) if the leakage term at the collector is represented by I_{CO} (Eq. 5.4.11). It is straightforward to derive expressions for the active-biased transistor when driven from the base and to show the validity of the circuit sketched in Fig. 5.14b.

In saturation, the quantity of greatest interest is $V_{CE\,sat}$, the voltage drop across the "on-state" switch. In this case the E-M model allows one to derive

$$V_{CE\,sat} = \frac{kT}{q}\ln\left\{\frac{\left[1 + \frac{I_C}{I_B}(1 - \alpha_R)\right]}{\alpha_R\left[1 - \frac{I_C}{I_B}\left(\frac{1 - \alpha_F}{\alpha_F}\right)\right]}\right\} \qquad (5.4.13)$$

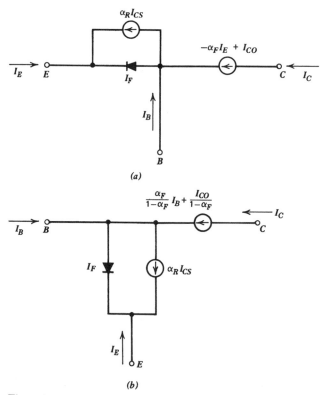

(a)

(b)

Figure 5.14 Equivalent circuits for *npn* transistors actuated by terminal currents. (a) Emitter-current actuated circuit, (b) base-current actuated circuit.

Since $V_{CE\,sat}$ is very small, a first approximation that is often used for design purposes is to consider it negligible. If not considered negligible, Eq. (5.4.13) shows that $V_{CE\,sat}$ changes only very slowly with collector current. Hence, an equivalent circuit consisting of a voltage source from collector to emitter or, preferably of two sources connected from base to emitter and from base to collector, respectively, would be appropriate. The *E-M* equations do not take account of any resistance in series with the junctions. The drops across series resistances, particularly in the collector regions of *IC* transistors, often exceed $V_{CE\,sat}$ predicted by Eq. (5.4.13); thus the "on-state" transistor switch is often modeled by a series voltage source and resistor denoted by $R_{C\,sat}$.

In this section we have been discussing the static *E-M* model, the model for quiescent conditions. Although this model can be modified to allow dynamic calculations (i.e., to solve for transient conditions), we shall find it advantageous to make use of another technique known as charge-control modeling for these calculations. The charge-control model is discussed in Chapter 6, where we shall also discuss a modification of the Ebers-Moll model that accounts for some important second-order effects.

Reciprocity[†]

In introducing the Ebers-Moll model, we postulated a reciprocity relationship between the model parameters [Eq. (5.4.3)]. Later, we discovered that this reciprocity equation was necessary to find a correspondence between the *E-M* model and our earlier theory of transistor action. The key relating the two models is in the reciprocity of transistor action between the emitter and collector junction as succinctly expressed in Eq. (5.4.7).

We have proven that reciprocity applies to a transistor with arbitrary base doping, constant base cross section and with negligible recombination through our discussion in Sec. 5.1. The validity of reciprocity is, however, considerably wider in scope. It extends, as we shall argue, to *IC* transistors under their usual operating conditions. Even when not strictly valid, deviations can usually be ignored without having theory stray too far from experiment. This is a very important point because the Ebers-Moll model is the cornerstone for virtually all large-signal, transistor-analysis techniques.

To explore this concept further, let us build upon the limited conditions for which reciprocity is proven: namely the case cited in Section 5.1. As a first complication, one might object to the condition $J_n =$ constant that was used to remove current from the integral [Eq. (5.1.5)]. An *IC* transistor in the reverse-active mode should, for example, inject a great deal more charge at a given $V_{BC} = V_a$ than would the same transistor in the forward-active mode with $V_{BE} = V_a$. Thus, a greater component of base-recombination current would flow in the reverse-active mode than in the forward-active mode. Consider then the currents in an *IC* transistor under reverse-active bias. The magnitude of the linking current to the emitter

(I_n) would be given by Eq. (5.1.9) $|I_n| \approx I_S \exp{(qV_a/kT)}$ because the linking current is, in fact, not a function of x; thus the derivation of the equation is still valid. By the arguments of Section 5.3, the recombination component would be proportional to the exponential factor; we shall call it $I_{rB} = I_{RB} \exp{(qV_a/kT)}$. The total current crossing the collector will, by Eqs. (5.4.1) and (5.4.2), be $I_{CS} \exp{(qV_a/kT)}$ neglecting small saturation components. Since we are, for the present, considering only extra current resulting from recombination, the total current crossing the collector junction is just the sum of the linking component and the recombination component:

$$I_{CS} \exp{\left(\frac{qV_a}{kT}\right)} = I_S \exp{\left(\frac{qV_a}{kT}\right)} + I_{RB} \exp{\left(\frac{qV_a}{kT}\right)} \qquad (5.4.14)$$

Because the magnitude of the ratio between emitter current and collector current is α_R and the magnitude of the current at the emitter is just $|I_n|$, we have

$$\alpha_R = \left|\frac{I_E}{I_C}\right| = \left|\frac{I_S \exp{(qV_a/kT)}}{(I_S + I_{RB}) \exp{(qV_a/kT)}}\right| \qquad (5.4.15)$$

which, using Eq. (5.4.14) reduces to

$$\alpha_R = \frac{I_S}{I_{CS}} \qquad (5.4.16)$$

As we can easily see from the derivation, a component of junction current resulting from any parasitic (or unwanted) loss can be added to the linking component $|I_n|$ and we will still be able to obtain Eq. (5.4.16) provided that the parasitic loss is proportional to $\exp{(qV_a/kT)}$ so that the voltage dependence can be factored and cancelled. Furthermore, if parasitic losses are considered for the IC transistor in the forward-active mode, we obtain $\alpha_F = I_S/I_{ES}$. This equation for α_F and Eq. (5.4.16) for α_R are identical with Eq. (5.4.7). Thus, reciprocity is established for the transistor with parasitic losses as well as for the simpler structure of Section 5.1. In particular, the asymmetry of the IC transistor does not invalidate reciprocity because the alphas for forward- and reverse-active operation vary inversely with the respective junction saturation currents. The constraint on this statement is that all junction currents must vary with voltage in the same way as the linking component. This constraint is violated if recombination in the space-charge zone becomes significant [its voltage variation is $\exp{(qV_a/2kT)}$ as shown in Eq. (4.3.24)] or if high-level effects occur. We consider the latter in Chapter 6.

5.5 DEVICES: PLANAR BIPOLAR AMPLIFYING AND SWITCHING TRANSISTORS

In discussing bipolar transistors for integrated circuits, it is helpful to divide them into two broad categories defined by their ultimate use: either amplification or switching. Transistors of either type are nearly always fabricated in an epitaxial layer of relatively high resistivity silicon. Typical values for resistivity of the epitaxial layer and for other design parameters of *IC* transistors are given in Table 5.1. The purpose of this epitaxial layer is to obtain a controlled, lightly doped region of collector material adjacent to the base junction. A buried layer of heavily doped silicon is added to maintain a high conductivity collector region below this.

The lightly doped region allows the collector-base junction to sustain relatively high voltages without breaking down (Section 3.4), while the higher doped buried-layer region reduces series resistance between the junction and the metallic collector contact. Inclusion of the buried layer reduces series ohmic resistance from the kilohm range to a few hundred ohms in typical devices. For some applications even this resistance is too much, and an extra processing step is added in which the heavily doped n^+ region under the collector contact is diffused until it reaches the buried layer. This extended diffused region is called a *collector plug*, and its inclusion reduces series resistance to the order of 10 ohms.

The base, emitter, collector-contact, and junction isolation regions are usually established by successive diffusions. Top and cross-sectional views of a typical

Table 5.1 **Typical Design Parameters for *IC* Transistors**

	Amplifying	Switching
Epitaxial Film		
Thickness	10 μm	3.5 μm
Resistivity	1 Ω-cm	0.3 $-$ 0.8 Ω-cm
Sheet resistance	1000 $\Omega/\square$	1500 $\Omega/\square$
Buried Layer		
Sheet resistance	$\sim 20\Omega/\square$	
Up diffusion	2.5 μm	1.4 μm
Emitter		
Diffusion depth in base	2.5 μm	0.8 μm
Sheet resistance	5 $\Omega/\square$	12 $\Omega/\square$
Base		
Diffusion depth	3.25 μm	1.3 μm
Sheet resistance	100 $\Omega/\square$	200 $\Omega/\square$
Oxide Thickness		
1. Background	0.8 μm	0.5 μm
2. Base	0.4 μm	0.33 μm
3. Emitter	0.3 μm	0.3 μm
Substrate		
Resistivity	~ 10 Ω-cm	
Orientation	$\langle 111 \rangle$	

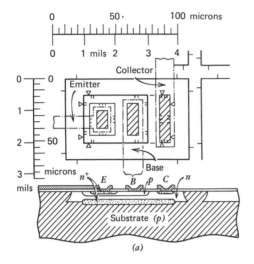

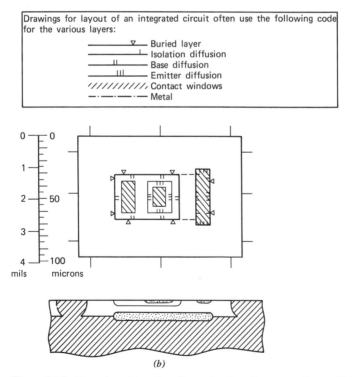

Figure 5.15 Top views (showing dimensions) and cross-sections of (*a*) a standard digital *IC* transistor, (*b*) an amplifying (35 V breakdown) transistor.[8]

junction-isolated *IC* transistor are shown in Fig. 5.15. Plots of the impurity concentrations measured along a coordinate perpendicular to the surface and passing through the emitter, base, and collector are given in Figs. 5.16 and 5.17. These figures show that the major differences in the design of switching and amplifying transistors lie in the thickness and resistivity of the epitaxial layer. Both quantities are greater in amplifying devices than in switching transistors; this means increased breakdown voltage and reduces a parasitic effect (the Early effect) that will be described in Chapter 6. For switching devices, saturation ("on-state") resistance should be minimized, which necessitates epitaxial thicknesses of only a few microns at resistivities in the order of some tenths of an Ω-cm. If Schottky clamping is to be used on the switching transistor (as described in Chapter 2), the resistivity must be greater than 0.1 Ω-cm in order to obtain a good metal-semiconductor barrier. Schottky-clamped transistors differ from unclamped switching devices only in the extended finger of metal from the base contact to the undoped epitaxial region forming the collector as shown in Fig. 5.18. The base doping level is ideally very low in order to maximize emitter injection efficiency γ, but it cannot be too low (not less than about 5×10^{16} cm^{-3}) without the possibility of a poor metal-semiconductor contact or even surface inversion at the metallic contact to the base.

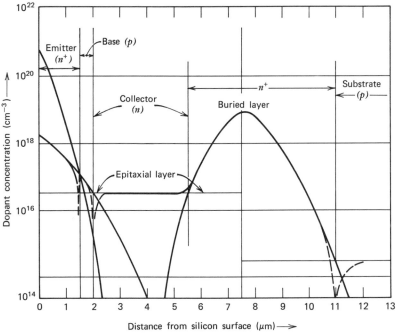

Figure 5.16 Cross section of the diffusion profile of the transistor in Fig. 5.15a.[8]

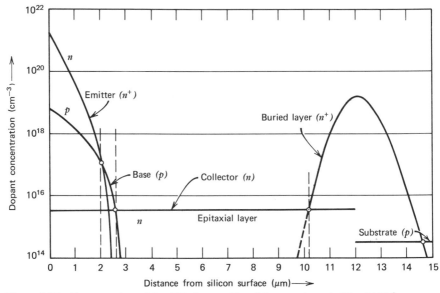

Figure 5.17 Cross section of the diffusion profile of the transistor in Fig. 5.15b.[8]

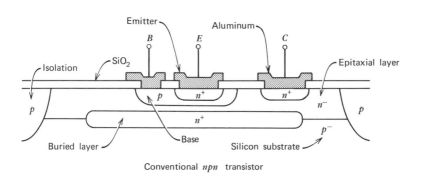

Conventional *npn* transistor

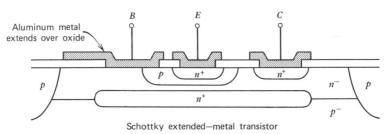

Schottky extended—metal transistor

Figure 5.18 Comparison of cross sections of (*a*) a conventional *npn* transistor and (*b*) a Schottky-clamped, extended metal *npn* transistor.

If the doping is too low, series resistance in the base also becomes problematic. The emitter is usually doped very heavily to increase emitter efficiency. When doped above about 10^{20} cm^{-3}, however, efficiency falls because of decreased hole lifetime in the emitter (which enhances hole injection from the base) and because of effects associated with the narrowing band gap of degenerately doped silicon.[6]

Reducing the base width and decreasing the base conductivity are two means of increasing gain, as we have noted. As the base width is reduced to submicron dimensions, however, it becomes increasingly likely that the collector-base space-charge region may reach through to the emitter-base space-charge region, completely depleting the base region. This condition, called *punch-through* results in a low resistance path from emitter to collector and can lead to ruinous currents in a transistor. This is one mode of failure that is influenced by the collector-base bias. The other failure mode that is sensitive to V_{CB} is avalanching of the collector junction. Avalanche breakdown has been described in connection with diode effects in Section 3.4.

The symbol adopted to represent a bipolar transistor is so widely used that it is doubtless familiar to the reader. As sketched in Fig. 5.19a and 5.19b, it consists of two angled lines that identify the collector and emitter leads; these are in contact with a straight line that stands for the base. An arrow in the direction of the *pn* junction of the emitter-base diode differentiates an *npn* transistor (Fig. 5.19a) from a *pnp* transistor (5.19b). In reality, *npn* transistors in integrated circuits are automatically coupled to a parasitic *pnp* structure with the substrate acting as a collector (Fig. 5.19c). For forward-active bias on the *npn* transistor, the substrate *pnp* is cut-off. In other modes of *npn* operation, the *pnp* can become active. It is important to be certain that the *pnp* transistor does not have any appreciable gain to avoid

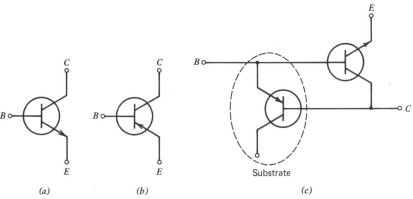

(a)   (b)   (c)

Figure 5.19 Standard symbols for (*a*) an *npn* transistor and (*b*) a *pnp* transistor. (*c*) The *IC* *npn* transistor automatically has a parasitic *pnp* transistor (dotted circle) attached to it because of the fabrication process.

undue parasitic loss and transient misbehavior in these cases. This is another point where the buried layer is valuable. By increasing the base charge of the parasitic *pnp* transistor and by lowering lifetime in this region, it is not difficult to reduce α_F in the parasitic *pnp* transistor to something less than 0.05 in typical cases. The parasitic *pnp* transistor also includes laterally injected holes that travel through the *n*-epitaxial zone to be collected at the reverse-biased, *p*-isolation-region diffusion. In practice, it is necessary to keep the outer edge of the base more than $\sim 8 \ \mu m$ removed from the isolation junction to reduce this component to an acceptable value.

Lateral injection of electrons can also occur from the emitter into the base. We have not explicitly mentioned this injection loss in the discussion of transistor gain, but the dimensions of modern *IC* transistors have been so reduced that it can be a detectable loss mechanism. The transport factor for electrons injected laterally is considerably lower than that for electrons injected downward because the effective lateral base width is large and also because some of the electrons are in the vicinity of the ohmic base contact where they will be lost to recombination. Injection laterally is inhibited naturally by the fact that ϕ_i increases with doping [Eq. (3.2.10)] and electron injection at a given bias is greater where ϕ_i is smaller. It is also inhibited by the lack of any aiding field in the lateral direction across the base. The designer can keep this component small by keeping the emitter shallow and the lateral base dimension large.

Transistors for integrated circuits can be made in smaller surface areas than are needed for typical *IC* resistors and capacitors. It is thus advantageous to design circuits that make use of transistors whenever possible, and to incorporate other devices only when absolutely necessary. If bipolar transistors are used for the active devices, they will typically make up the overwhelming fraction of total devices in an integrated circuit. Transistors are very often used in place of simpler *pn*-junction diodes, as described in Chapter 4, even though they take up slightly more surface area. This is because an *IC* process that is needed to produce transistors will yield diodes with lower series resistance and shorter switching times when those diodes are made up of the transistor structure in a diode connection. There are a total of five ways to connect a transistor as a diode because only two of three possible regions need to be accessed. Of the various diode connections, shown in Fig. 5.20,

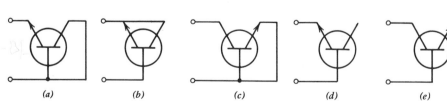

| (a) | (b) | (c) | (d) | (e) |

Figure 5.20 Five ways in which a transistor structure can be connected to obtain diode behavior.

the lowest series resistance and fastest switching times are obtained in the connection of Fig. 5.20a, which is used most often.[7]

SUMMARY

An understanding of the operation of bipolar transistors can be attained by focusing on the behavior of minority carriers at biased *pn* junctions. Because the space-charge region in a *pn* junction is a barrier to majority-carrier flow, it acts as a collector of minority carriers. Thus, when two *pn* junctions are located close together, bias on one of them can influence the minority-carrier populations in the vicinity of the second, and markedly alter its electronic behavior. This idea, which is referred to as *bipolar transistor action*, has far ranging and significant consequences. Making use of transistor action, one can build effective switches and amplifying elements. A very useful formulation of the equations for transistor action follows from the fact that majority-carrier currents in the transistor *base* between the two junctions must be nearly zero because a barrier to majority-carrier flow is always encountered along this path. The current linking the two junctions consists therefore of minority carriers. The size of the linking current depends inversely on the total majority-carrier charge in the base. Currents are nonlinear functions of terminal voltages in transistors. The nonlinearity is very strong, and practical design of transistor circuits is frequently aided by considering approximating equations that are valid over selected regions of base-emitter and base-collector bias. Under active bias, the collecting junction is continuously maintained under reverse bias while the emitting junction is continuously forward biased. Current gain in the transistor in the active-bias mode is a function of the efficiency of the injection of minority carriers into the base region, where they can be collected, as well as of their loss to recombination in the base. These mechanisms depend strongly on processing procedures and on device geometry. The Ebers-Moll model describes the operation of bipolar transistors over all ranges of bias. It is a very important first-order model and is the basis for more refined descriptions of the transistor. A reciprocity condition is part of the Ebers-Moll model. The reciprocity condition predicts that a transistor biased in the reverse-active mode will deliver the same output current as it will when biased in the forward-active mode provided that the respective terminal voltages are interchanged. This result applies to asymmetric as well as to symmetric transistors because loss mechanisms that cause a reduction in the current gain simultaneously cause an increase in the junction-saturation currents. Provided that the losses have the same voltage dependence as does the linking current, there is an exact inverse relationship and reciprocity is valid. Transistors in integrated circuits for both switching and amplification are typically built in an epitaxial layer with buried layers beneath the collector regions. The chief difference between devices for these two applications is in the dimensions perpendicular to the surface. Switching transistors are usually built in thinner epitaxial regions, and more emphasis is placed on reducing the effects of series collector resistance and charge storage in these devices than in the design of amplifying transistors. The engineering trade-offs for the two classes of devices differ. In integrated circuits, conventional *npn* transistors are shunted by parasitic *pnp* transistors in which the substrate is the collector. Although the *pnp* transistor is cut off when the *npn* transistor is under active bias, it can become active during switching. Because they are smaller than other devices in bipolar integrated circuits, bipolar transistors are used by circuit designers whenever possible.

REFERENCES

1. W. Shockley, *Bell Syst. Tech. J.*, *28*, 435 (1949).
2. J. L. Moll and I. M. Ross, *Proc. IRE*, *44*, 72 (1956).
3. H. K. Gummel, *Proc. IRE*, *49*, 834 (1961).
4. H. Kroemer, *Arch. Elek. Übertrag.*, *8*, May, August, November (1954).
5. J. J. Ebers and J. L. Moll, *Proc. IRE*, *42*, 1761 (1954).
6. H. J. DeMan, *IEEE Trans. Elect. Devices*, ED-18, 833 (1971).
7. C. S. Meyer, D. K. Lynn and D. J. Hamilton, *Analysis and Design of Integrated Circuits*, McGraw-Hill., New York, 1968, pp. 248–258.
8. H. Camenzind, *Electronic Integrated Systems Design.* Copyright 1972 by Litton Educational Publishing, Inc. Reprinted by permission of Van Nostrand Reinhold Company.
9. P. E. Gray, D. DeWitt, A. R. Boothroyd, and J. F. Gibbons, *Physical Electronics and Circuit Models of Transistors*, Wiley, New York, 1964, p. 145.

PROBLEMS

5.1   Show by using Eq. (5.1.2) that there will be a constant field in a region in which the doping density varies exponentially with distance. If a constant field of -500 V cm^{-1} exists in the base of a transistor having a base width of 0.3 μm and the dopant density is 10^{17} at the emitter-base edge, what is the dopant density at the edge of the collector-base space-charge region?

5.2   Apply Eq. (5.1.5) to an *npn* transistor under active bias. Consider that $I_C = -J_n A_E$ and prove, therefore, that

$$I_C = \frac{q A_E \tilde{D}_n N_a(x) n(x)}{\int_x^{x_B} N_a \, d\xi}.$$

5.3† Use the expression derived in Problem 5.2 to find the x dependence of the electron density in the base of the transistor of Problem 5.1. Sketch plots of $n(x)$ versus x in the base under active-bias conditions for two cases of *npn* transistors that are biased so that both pass the same collector current. Use a single set of axes for the two transistors: (i) the transistor described in Problem 5.1, and (ii) a transistor having a constant base doping of 10^{17} cm^{-3} but otherwise identical to that in Problem 5.1. Note that both plots have the same gradient at the edge of the collector space-charge zone. Why is this the case?

5.4† Derive expressions for the total stored minority charge in the base in the two transistors described in Problem 5.3. Use this result to compare the values of β_F in the two devices under the assumption that base current results only from recombination in the base.

5.5 (a) If the data plotted in Fig. 5.4 were taken on an *npn* transistor having an active area of 10 by 10 μm, what is Q_{BO}, the base-majority charge in the quasi-neutral region?
(b) What is the base width x_B if the base doping averages 10^{16} atoms cm^{-3}? (Take $\tilde{D}_n = 25$ cm^2 s^{-1}.)

5.6† Use approximating techniques to determine the net dopant atoms per unit area between the base-emitter junction plane and the base-collector junction plane in the switching transistor of Fig. 5.16 and the amplifying transistor of Fig. 5.17. Compare these values with the dopant atoms per unit area in the quasi-neutral base [Gummel number in Eq. (5.2.3)] for the transistor of Problem 5.5 and explain any differences between the results.

5.7 Using the theory of Chapter 3, justify the two forms for q_V given on Fig. 5.8. Derive expressions for the coefficients labeled K on the figure for each case.

5.8† The accompanying table gives values of the net dopant density $(N_d - N_a)$ as a function of position relative to the junction plane between the emitter and base on a transistor with the doping profile of Fig. 5.17.

Net dopant density $(N_d - N_a)$ (cm^{-3})	7.36×10^{17}	-1.25×10^{16}	-4.0×10^{16}	-1.9×10^{16}	-6.4×10^{15}	-1.5×10^{15}
Distance from base-emitter junction (μm)	-0.25	$+0.09$	$+0.23$	$+0.30$	$+0.41$	$+0.50$

Assume that the transistor has an active area of 2×10^{-6} cm^2. (a) Plot these values to a linear scale and argue that the profile can be roughly approximated by a one-sided step junction having $N_a \approx 2 \times 10^{16}$ cm^{-3}. (b) Make a plot similar to that of Fig. 5.8 to show q_{VE} versus total voltage for the junction over the bias range from breakdown to $+0.3$ V. Find values for ϕ_i and K as shown on Fig. 5.8.

5.9 Use the result of Problem 5.1 and the data of Problem 5.8 to estimate the size of the field in the quasi-neutral base of the transistor having the doping profile sketched in Fig. 5.17.

5.10 Draw an equivalent circuit similar to Fig. 5.12 and write equations similar to Eqs. (5.4.1) and (5.4.2) to represent the Ebers-Moll model for a *pnp* transistor.

5.11 Show that Fig. 5.14b represents the Ebers-Moll model for a transistor driven by base current.

5.12 Derive Eq. (5.4.13) for the voltage drop across a saturated transistor as predicted by the Ebers-Moll model. Evaluate $V_{CE\,sat}$ for $I_C/I_B = 10$, $\alpha_F = 0.985$, and $\alpha_R = 0.72$.

5.13 Use the Ebers-Moll equations (*npn* transistor) to find the ratio of the two currents I_{CEO} to I_{CBO} where I_{CEO} is the current flowing in the reverse-biased collector with the base open circuited, and I_{CBO} is the current flowing in the reverse-biased collector with the emitter open circuited. Explain the cause for the difference in the currents in terms of the physical behavior of the transistor in the two situations.

5.14[†] Consider an *npn* transistor that is biased in the active mode. At time $t = 0$, an intense visible light is focused on the collector-base space-charge region. The light produces G hole-electron pairs per unit time inside the space-charge zone. (Consider that qG is roughly of the same order as the dc base current I_B.)

(a) If the emitter-base voltage and collector-base voltage are both held constant, what will be the values of base, emitter, and collector currents for $t > 0$?

(b) Repeat part (a) if the base is driven by a current source so that the illumination does not alter I_B.

5.15 Near room temperature, the collector current I_C and the base current I_B are both normally positive for an *npn* transistor biased in the active mode. Assume that an *npn* transistor is under active bias and I_C is held constant while the temperature is increased. If one measures I_B, he will find its magnitude to decrease and ultimately to change sign. What physical effects account for this behavior? (Consider all currents at the base lead.)[9]

5.16 Consider an *npn* transistor in which both the base and emitter regions are nonuniformly doped and in which recombination of reverse-injected holes into the emitter takes place at the emitter contact. Show that the emitter efficiency γ can be written as $\gamma = (1 + Q_{BO}\tilde{D}_{pE}/Q_{EO}\tilde{D}_{nB})^{-1}$ where Q_{EO} is the emitter dopant in the quasi-neutral region of the emitter, Q_{BO} is the base dopant in the quasi-neutral region of the base, and $\tilde{D}_{pE}$ and $\tilde{D}_{nB}$ are the effective minority-carrier diffusion constants in these regions. Show the consistency of this result with that given for the prototype transistor when it is subjected to the same recombination constraints (Eq. 5.2.20).

5.17 An *npn* transistor has a cross-sectional area of 10^{-5} cm^2 and a quasi-neutral base that is uniformly doped with $N_a = 4 \times 10^{17}$ cm^{-3}, $D_{nB} = 18$ cm^2 s^{-1}, and $x_B = 0.5$ μm.

(a) If conditions in the emitter are similar to those described in Problem 5.16 and the emitter has a total dopant (Q_{EO}/q) of 8×10^9 atoms and $\tilde{D}_{pE}$ is 2 cm^2 s^{-1}, calculate γ.

(b) Estimate α_T, the transport factor if the base lifetime is 10^{-6} s.

(c) Calculate β_F for this transistor. What is the percentage error involved in taking $\beta_F \approx Q_{EO}\tilde{D}_{nB}/Q_{BO}\tilde{D}_{pE}$? This simplified form is sometimes used to estimate β_F.

5.18[†] Assume that the transistor of Problem 5.17 is placed in a radiation environment where the electron lifetime in the base decays according to the equation: $\tau_n = \tau_{n0} \exp - (t/t_d)$. In this equation t is measured in days and t_d is 10 days. Plot β_F as a function of time in days and determine the time interval until β_F drops to unity. What is the base lifetime when this occurs?

5.19[†] Consider the transistor structure shown in Fig. P5.19.

(a) Derive an expression for the electron distribution in the base as a function of x assuming that recombination takes place. [This will demand solving an equation similar to Eq. (4.3.10).]

(b) The slope of the distribution at $x = 0$ is proportional to the injected electron current, and the slope at $x = x_B$ is proportional to the collected electron current. The

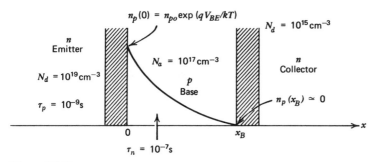

Figure P5.19

difference represents current because of recombination in the base (supplied by the base lead). Find an expression for this base recombination current as a function of L_n and x_B/L_n. $[L_n = (D_n\tau_n)^{1/2}]$.

(c) The dopant concentrations in the emitter, base, and collector are assumed to be uniform throughout each region with values of 10^{19}, 10^{17}, and 10^{15} cm^{-3}, respectively. The hole lifetime in the emitter is 10^{-9} s and the electron lifetime in the base is 10^{-7} s. The emitter and base widths are each 1 μm. Calculate the emitter efficiency and β_F. Use Fig. 1.12 to obtain values for the diffusion constants.

5.20   By considering collector current as charge in transit ($I_C = qnv_dA$), use the known distribution of injected electrons in the base of an npn prototype transistor biased in the active mode to solve for the base transit time τ_B. That is, formulate the transit time as the integrated sum of incremental path lengths divided by velocity and then carry out the integration to prove that τ_B is given by $\tau_B = x_B^2/2D_n$.

CHAPTER 6

BIPOLAR TRANSISTORS II: LIMITATIONS AND MODELS

The discussion of bipolar transistors, thus far, has been framed in terms of fundamental operation. For example, the description of a transistor biased in the active mode has emphasized that carriers are injected with high efficiency from the emitter into the base. There, the injected carriers vastly increase the minority-carrier population and flow from the base-emitter contact toward the collector. Upon reaching the base-collector space-charge zone, these added free carriers are swept across to the quasi-neutral collector region where they are majority carriers and flow as ohmic current through the collector lead.

In real transistors this simple description is remarkably accurate over a wide range of conditions. There is, however, more that needs to be considered than this basic ideal behavior. Real transistors are limited by processes that we have not specifically considered thus far. For example, a transistor as described in Chapter 5 would have a current-source output in the active mode; that is, output current would not depend in any way on the voltage applied between the base and collector terminals. In real transistors, however, both output currents and output voltages influence the device performance. Similarly, the theory that has thus far been presented would predict a current gain in the active mode that is insensitive to bias level. But this is only a rough approximation. A first objective of this chapter, therefore, is to investigate mechanisms that must be considered to obtain a more realistic picture of transistor operation.

Another major inadequacy of the theory developed thus far is that it does not deal with time-varying effects. Their consideration is thus the second major topic to be developed in this chapter. A great simplification in considering time-varying effects will accompany the modeling of transistors as charge-controlled devices. This viewpoint leads very naturally into the important topic of large-signal transient models for bipolar transistors. The amalgamation of the Ebers-Moll model with the charge-control model provides a powerful analysis technique. It will be straightforward to derive a small-signal ac equivalent circuit from the more general large-signal representation. This small-signal equivalent circuit, known as the *hybrid pi* circuit, is relatively easy to characterize because it is composed of elements that relate directly to physical mechanisms in the transistor.

The production of *pnp* transistors within the confines of standard *IC* processing is described in a final section. Two basic types of *pnp* transistors, *substrate pnp* and *lateral pnp* transistors are typically fabricated. The substrate *pnp* transistor has only limited application because its collector is not isolated. The lateral *pnp* transistor has severely limited performance compared to that of *npn* transistors. Simple models for loss mechanisms in this device explain the limited performance. An important application of lateral *pnp* transistors—a dense form of logic circuitry known as *integrated-injection logic* (I^2L)—is described at the conclusion of the chapter.

6.1 EFFECTS OF COLLECTOR BIAS VARIATION (EARLY EFFECT)

When we considered bipolar transistors under active bias in Chapter 5, the function of the voltage applied across the collector-base junction was merely to insure the efficient collection of base minority carriers and their delivery to the collector region. The magnitude of the bias only limited the range of the permissible collector voltage swing, provided that it was below the breakdown value. We have overlooked an important aspect of *pn*-junction electronics in taking this simplified view. As we noted in Chapter 3, the width of a reverse-biased *pn* junction is voltage dependent; in fact, this bias dependence made possible the operation of junction field-effect transistors. In the case of bipolar transistors, a changing collector-base bias causes variation in the space-charge layer width at the collector junction and, consequently, in the width of the quasi-neutral base region. These variations result in several effects that complicate the performance of the device as a linear amplifier. Base-width modulation under the influence of variations in collector-base bias was first analyzed by James Early,[1] and the phenomenon is generally called the *Early effect*.

The dependence of collector current on collector-base bias can be formulated directly by using the integral equations for active bias (*npn* transistor) developed in Section 5.2. In particular, from Eqs. (5.2.1) and (5.1.10), we can write

$$I_C = \frac{q\tilde{D}_n n_i^2 A_E \exp(qV_{BE}/kT)}{\int_0^{x_B} p\, dx} \tag{6.1.1}$$

where the integration is performed over x_B, the width of the quasi-neutral base region and the other terms have been defined in Section 5.1.

Variation in the base width with voltage V_{CB} causes a variation in the collector current that can be written

$$\frac{\partial I_C}{\partial V_{CB}} = \frac{-q\tilde{D}_n n_i^2 A_E \exp(qV_{BE}/kT)p(x_B)}{\left[\int_0^{x_B} p\, dx\right]^2} \frac{\partial x_B}{\partial V_{CB}} \tag{6.1.2}$$

Several of the terms in Eq. (6.1.2) can be combined to represent collector current itself, so that Eq. (6.1.2), which represents the small-signal conductance at the collector-base junction, can be written

$$\frac{\partial I_C}{\partial V_{CB}} = -I_C p(x_B)\left[\frac{1}{\int_0^{x_B} p\, dx}\right]\left[\frac{\partial x_B}{\partial V_{CB}}\right]$$

$$= -\frac{I_C}{V_A} = \frac{I_C}{|V_A|} \tag{6.1.3}$$

Since the collector is reverse biased, the derivative term in Eq. (6.1.3), $\partial x_B/\partial V_{CB}$, is negative and the Early effect results in an increase in I_C when V_{CB} is increased. This increase is evident if we examine *common-emitter* output characteristic curves such as the family shown in Fig. 6.1(*a*). These output curves are actually plots of I_C versus V_{CE} but V_{CE} is almost the same value as V_{CB} for bias in the active region ($V_{CE} \approx V_{CB} + 0.7$ V).

Equation (6.1.3) reveals that the Early effect varies linearly with collector current. The reciprocal of the factor multiplying the current has the dimension of a voltage and has been defined[2] as the *Early voltage*. The Early voltage is usually given the symbol V_A. An equation for V_A (*npn* transistor) [from Eq. (6.1.3)] is

$$V_A = \frac{\int_0^{x_B} p\, dx}{p(x_B)\,\partial x_B/\partial V_{CB}} \tag{6.1.4}$$

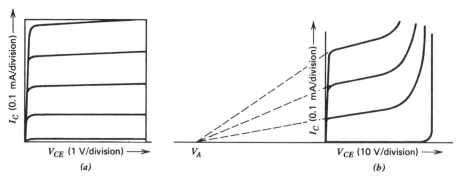

Figure 6.1 Measured output characteristics for an amplifying transistor: collector current versus V_{CE}. (a) vertical 0.1 mA/division, horizontal 1 V/division (b) vertical 0.1 mA/division, horizontal 10 V/division; tangents to the measured curves are extended to the voltage axis to determine the Early voltage V_A.

Again, the derivative in Eq. (6.1.4) is negative and, thus, the Early voltage is negative for an *npn* transistor. The analogous effect of emitter-base space-charge layer widening when the transistor is biased in the reverse-active mode can also be interpreted by using a different Early voltage, usually denoted as V_B.

Except for high-level effects, the three terms that define V_A depend only on transistor manufacturing techniques and on collector-base voltage. In practice, the collector-base voltage dependence of V_A itself is usually treated as negligible and the Early voltage is approximated by its value at a single bias, often at $V_{CB} = 0$. By using this condition to specify V_A, we can expect that a series of tangents to curves of I_C versus V_{CB} (V_{CE} in practice), drawn at the edge of the forward-active region where V_{CB} is approximately zero, should intersect the V_{CE} axis at V_A. Figure 6.1(b) shows this behavior and illustrates a practical means of obtaining the Early voltage.

A useful and informative alternative expression for V_A can be obtained by re-arranging some terms in Eq. (6.1.4). First, we use Eq. (5.1.8) to express the numerator in terms of the total base majority charge Q_B in that portion of the base where transistor action is taking place:

$$\int_0^{x_B} p \, dx = \frac{Q_B}{q A_E} \tag{6.1.5}$$

We then recognize that the denominator of Eq. (6.1.4) represents the derivative of base charge Q_B with respect to V_{CB}:

$$q A_E p(x_B) \frac{dx_B}{dV_{CB}} = \frac{dQ_B}{dV_{CB}} \tag{6.1.6}$$

The derivative on the right-hand side of Eq. (6.1.6) can be related to C_{jc}, the small-signal capacitance at the collector-base junction. The derivative is not identical to C_{jc}, however, because A_C, the area of the collector-base junction in a typical IC bipolar transistor, is substantially larger than A_E, the area of the emitter-base junction. Taking account of the area difference, we can write

$$\left| \frac{dQ_B}{dV_{CB}} \right| = C_{jc} \frac{A_E}{A_C} \tag{6.1.7}$$

Therefore, the Early voltage is

$$|V_A| = \frac{Q_B A_C}{C_{jc} A_E} \tag{6.1.8}$$

To reduce the influence of collector-base voltage on collector current, V_A should be increased in magnitude. From Eq. (6.1.8) we see that this can be accomplished in practice by increasing the ratio of total base majority charge to capacitance at the collector-base junction. Physically, this reduces the movement of the base-collector boundary into the base region.

It is very useful, for computer modeling purposes, to describe the widening of the base-collector space-charge layer through the Early voltage V_A, as we shall see in Section 6.6. It may, however, help conceptually to consider an alternative view of the Early effect that is physically more revealing for the prototype transistor introduced at the beginning of Chapter 5. For homogeneous doping and active-mode bias, the minority-carrier distribution in the base has a triangular shape (Fig. 6.2). Increasing the collector-base voltage from V_{CB} to $(V_{CB} + \Delta V_{CB})$ moves

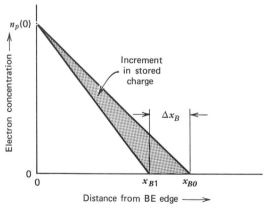

Figure 6.2 The influence of the Early effect on the minority-carrier distribution in the quasi-neutral base region of a prototype transistor. An increase in V_{CB} reduces the base width from x_{B0} to x_{B1}, increasing the gradient of n_p and decreasing the total minority charge in the base.

the base-collector junction edge a distance Δx_B from x_{B0}. A second triangular distribution specifies the new minority-carrier profile and the shaded area between the two distributions represents the decrease in stored base charge.

The collector current is thus increased in the ratio x_{B0}/x_{B1} by the change in V_{CB}. In addition, it becomes clear by looking directly at stored base charge that the Early effect reduces charge storage. This affects both the transient behavior of transistors and the dc base current since base recombination depends directly on stored base charge. We shall consider these effects in more detail in Section 6.5 when we discuss BJT models for circuit applications.

6.2 EFFECTS AT LOW AND HIGH EMITTER BIAS

Some interesting behavior is revealed when a transistor is biased into the active region and the base and collector currents are measured as functions of the base-emitter voltage. Because of the physics of transistor action, such measurements may be more easily interpreted when plotted semilogarithmically with junction voltage on the linear scale. Typical data measured on an amplifying *IC npn* transistor are given in Fig. 6.3. The excellent fit of the data for I_C and I_B to straight lines over the mid-range of currents indicates an exponential dependence on voltage, consistent with the theory that was developed in Chapter 5. Clearly, however, straight lines of constant slope do not adequately represent all of the measurements in Fig. 6.3 either for I_C or I_B. At this point, it may seem surprising that collector current is "ideal" over a wider voltage range than is base current. The reason will be clear after some discussion. First, however, we consider the variation in I_B at very low voltages where the theory given thus far obviously disagrees with experiment.

Currents at Low Emitter Bias

At low base-emitter voltage there is a reduction in the slope of the straight-line variation of log I_B with V_{BE}. The experimental data indicate that the base current asymptotically approaches a curve that can be represented by

$$I_B = I_o \exp\left(\frac{qV_{BE}}{nkT}\right) \qquad (6.2.1)$$

as V_{BE} goes to zero. In the asymptotic form [Eq. (6.2.1)], n is a parameter that is generally found to be between one and two. Furthermore, the parameter I_o in Eq. (6.2.1) is larger than is the corresponding multiplier for an exponential form that would fit the data at intermediate bias values.

The source for the added base-emitter current at low biases is clear from our discussion of current in *pn* junctions in Chapter 4; it is recombination in the base-emitter space-charge region. Space-charge-region recombination current in a diode structure was considered in Section 4.3, and Shockley-Hall-Read theory was used

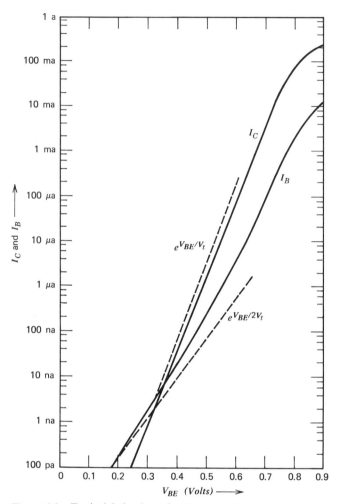

Figure 6.3 Typical behavior of collector and base currents as functions of base-emitter voltage for bias in the forward-active region.

there to derive Eq. (4.3.24). That equation for recombination current in the space-charge region has the same voltage dependence as does Eq. (6.2.1) if n is taken equal to 2. Values for n between 1 and 2 can be justified by considering possible variation of different parameters affecting recombination within the space-charge region. As was shown in Eq. (4.3.25), the importance of the recombination component relative to the injection components into the quasi-neutral regions increases as voltage is reduced. For the example taken in the diode discussion that illustrated

Eq. (4.3.25), the recombination current dominated for junction voltages less than 0.48 V.

Recombination current in the space-charge region flows only in the base-emitter leads. Its dominance does not affect the collector current, which consists almost entirely of collected electrons that are injected at the base-emitter junction. Hence, the collector current continues to be represented by Eq. (6.1.1) as V_{BE} is decreased until injection becomes so low that the collector current is dominated by generation in the space-charge region as expressed by Eq. (5.3.1). Thus, at low biases the collector current is a smaller fraction of emitter current than it is in the intermediate bias range. This behavior is brought more clearly into focus in a plot of $|I_C/I_B|$, that is β_F as defined in Eq. (5.2.22). Accordingly, the data of Fig. 6.3 has been used to make the plot shown in Fig. 6.4. The decrease in β_F as base-emitter bias decreases represents a clear limitation to the application of transistors for the amplification of low voltages.

There is great importance and enormous commercial significance to maintaining high current gain at low bias levels and therefore to reducing space-charge recombination current. Such applications for transistors and integrated circuits as hearing-aid amplifiers and *pacemaker* circuits for the medical treatment of heart patients depend on achieving adequate performance at the lowest possible current levels. For circuit applications such as these, processing efforts are usually concentrated on maintaining as high a lifetime as possible within the base-emitter space-charge region.

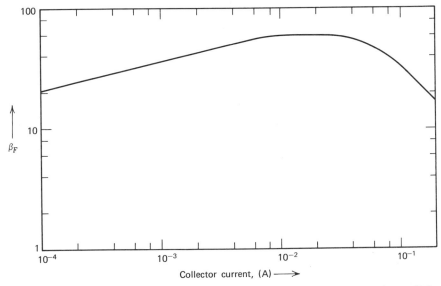

Figure 6.4 Current gain (β_F) as a function of collector current for the transistor of Fig. 6.3.

High-Level Injection

The transistor theory considered thus far has relied on the low-level injection approximation, that is, that majority-carrier populations under junction bias are essentially unperturbed from their values at thermal equilibrium. As transistors are biased more heavily in the active mode, violations of the low-level injection assumption should first become apparent either at locations where there is a high density of injected minority carriers or else where there is a low density of dopant atoms. The first situation can be expected in the base region at the emitter edge, while the second might occur in an IC transistor near the base-collector junction. Both possibilities do lead to observable effects in practical IC transistors. We shall consider them separately, concentrating first on high injection near the emitter junction.

High-Level Injection at the Base-Emitter Junction. The basic equation for collector current under active bias [Eq. (6.1.1)] was derived by considering a special case of the equation for transistor action [Eq. (5.1.7)]. The exponential dependence of current on voltage predicted by Eq. (6.1.1) is seen to be valid over nearly eight decades of current until it deviates under high bias on the base-emitter junction.

One cause for the deviation is high-level injection into the base. If the injection of electrons into the base is sufficient to cause a significant increase in the population of base majority carriers, then the full dependence of current on voltage is not stated explicitly in Eq. (6.1.1). In this case, the integrated majority charge in the denominator becomes bias dependent. (An associated, but second-order effect is a change in $\tilde{D}_n$.) Because of quasi-neutrality in the base, the expression for the integral of $p(x)$ should be written

$$\int_0^{x_B} p(x)\,dx = \int_0^{x_B} [N_a(x) + n'(x)]\,dx \tag{6.2.2}$$

where $n'(x)$ is the injected electron density and $N_a(x)$ represents the base doping profile. To proceed further, one would need to specify the x dependences in the integral and, therefore, to consider a specific dopant profile.

To continue a qualitative discussion, however, we can focus on the boundary value for the injected-electron density $n'(0)$. Since this is the maximum density, it dominates in the integral expression of Eq. (6.2.2). The boundary value can be determined by making use of Eq. (4.3.20) together with the quasi-neutrality assumption. Equation (4.3.20), which states that the pn product at the emitter-base boundary is related to the applied bias by the expression $p(0)n(0) = n_i^2 \exp (qV_{BE}/kT)$, remains valid* provided that the current levels do not become of the same order as the balanced equilibrium flow tendencies (estimated at $\sim 10^5 \, A \, cm^{-2}$

---

* The validity of Eq. (4.3.20) under high-level-injection conditions is also necessary to permit the application of the basic equation for transistor action [Eq. (5.1.7)] since the former was used in the derivation of the latter.

in Section 4.3). Currents of this size do not flow in practice without causing irreparable thermal damage to the junction.

The minority-carrier boundary value $n(0)$, which under active bias is almost the same as the injected electron density at the boundary $n'(0)$, is therefore

$$n(0) = \frac{N_a(0)}{2}\left[\left(1 + \frac{4n_i^2 \exp(qV_{BE}/kT)}{N_a^2(0)}\right)^{1/2} - 1\right] \qquad (6.2.3)$$

At low or moderate injection levels, Eq. (6.2.3) reduces to Eq. (4.3.7) as expected. Under high-injection conditions, however, the second term in the square root becomes dominant and $n(0)$ approaches an asymptote that varies as $\exp(qV_{BE}/2kT)$. Under this condition, there is a partial cancellation of the exponential increase predicted by the numerator factor of Eq. (6.1.1). Collector current tends likewise to become proportional to $\exp(qV_{BE}/2kT)$.

High-Injection Effects at the Collector (Kirk Effect).[†] The introductory comments about low injection pointed out the weakness of this assumption in locations where the material is lightly doped. Inspection of a typical doping profile for an amplifying *IC* transistor (Fig. 5.17) shows that minimal doping occurs in the collector close to the base region where the epitaxial material has not been affected by any diffusions. This region serves very useful purposes. It allows for a significant increase in the breakdown voltage of the collector-base junction, it reduces collector capacitance, and it reduces the widening of the base-collector space-charge layer into the base (Early effect). In this section, we shall see that several complicated and undesirable consequences also arise from this low collector doping.

Under amplifying conditions, the base-collector junction is reverse biased so that any minority carriers that impinge upon its edges are quickly swept across to the opposite region. We have used this physical picture to justify our assumption of essentially zero density for minority carriers at the edges of the reverse-biased collector-base space-charge region. This assumption is clearly an approximation, however; it must be violated if any carriers are to flow across the space-charge zone. Since free carriers are limited in their velocities, a current density J requires at least a density of minority carriers equal to J/qv_l, where v_l is the scattering-limited velocity (Section 1.2). This charge density will be neutralized by majority carriers in the base region, but it will add algebraically to the fixed background charge associated with the uncompensated dopant densities in the space-charge region. The presence of this charge component associated with collector current will begin to modify the theory that has been developed thus far when the density of the injected, current-carrying charge becomes comparable to the background density of dopant ions. As the current level in a transistor is increased, this mechanism can significantly alter important transistor properties.

Qualitatively, the effect may be described as follows: the free carriers entering the base-collector space-charge region modify the background charge in that region and thus affect the electric field. For a constant collector-base voltage, the integral of the field across the space-charge region is constant. Hence, any space-charge modification must be accompanied by a change in the width of the region. We shall see shortly that the width of the space-charge region tends to decrease and, therefore, the neutral base tends to widen. This mechanism is often called the *Kirk effect* after the author of the first paper to analyze it[3].

In Fig. 6.5*a* we have sketched a portion of the doping profile for an *IC* amplifying transistor from Fig. 5.17. We consider a case for which this device is biased in the active mode with 10 V applied to the collector-base junction. A plot of the electric field with little or no current flowing is sketched in Fig. 6.5*b*. Noted on the plot is

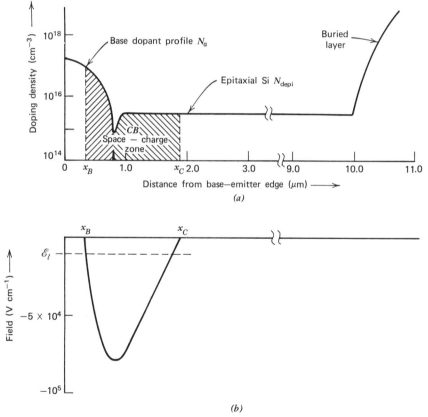

Figure 6.5 Space charge and field at the collector-base junction of the amplifying transistor shown in cross section in Fig. 5.17. (*a*) The space-charge region for 10 V collector-base reverse bias (*b*) the electric field at the collector base; the field $\mathscr{E}_l$ at which free electrons reach limiting velocity is marked on the plot.

$\mathscr{E}_l$ at which free electrons reach their limiting drift velocity v_l (Fig. 1.13). Poisson's equation for the collector-base space-charge zone of the npn transistor takes the form

$$\frac{d\mathscr{E}}{dx} = \frac{1}{\epsilon_s}\left[qN(x) - \frac{J_C}{v(x)}\right] \qquad (6.2.4)$$

where $N(x)$ is the net dopant density $(N_d - N_a)$ in the space-charge zone (negative acceptor ions in the diffused region near the base and positive donor ions in the high-resistivity epitaxial layer and in the heavily diffused buried layer). If we consider constant collector-base bias V_{CB}, we obtain a second equation involving the field $\mathscr{E}$

$$V_{CB} + \phi_i = \int_{x_B}^{x_C} - \mathscr{E}\,dx \qquad (6.2.5)$$

where ϕ_i is the built-in junction voltage [Eq. (3.2.10)]. Equation (6.2.5) implicitly specifies the width of the space-charge layer, that is, $(x_C - x_B)$. Referring again to Eq. (6.2.4), we note that the second term on the right-hand side is negligible at low currents, but will dominate at high currents. The critical current J_1 dividing the low- and high-current regions may be found by equating the two terms

$$J_1 = qN(x)v(x) \qquad (6.2.6)$$

The critical current will be reached first where $N(x)$ is a minimum; that is, in the epitaxial region, where for most bias levels $v(x)$ is at its maximum value v_l and N is N_{epi}. Thus, $J_1 = qN_{epi}v_l$. An analysis of this effect is of practical importance because in many applications the collector current density equals and exceeds the critical current J_1. For example, if $A_E \approx 10^{-5}$ cm^2 and $N_{epi} = 5 \times 10^{15}$ cm^{-3}, the critical collector current is 80 mA.

To carry out the mathematical treatment, we first multiply Eq. (6.2.4) by x and then integrate between x_B and x_C, the boundaries of the collector-base space-charge layer.

$$\int_{x_B}^{x_C} x\,\frac{d\mathscr{E}}{dx}\,dx = \frac{1}{\epsilon_s}\int_{x_B}^{x_C} x\left[qN(x) - \frac{J_C}{v(x)}\right]dx. \qquad (6.2.7)$$

The left-hand side of Eq. (6.2.7) can be integrated by parts and used with Eq. (6.2.5) to yield

$$\int_{x_B}^{x_C} x\,d\mathscr{E} = -\int_{x_B}^{x_C} \mathscr{E}\,dx = V_{CB} + \phi_i \qquad (6.2.8)$$

where $\mathscr{E}(x_C)$ and $\mathscr{E}(x_B)$ have been taken to be zero. Thus, the collector voltage can be expressed as

$$V_{CB} = \frac{1}{\epsilon_s}\int_{x_B}^{x_C} x\left[qN(x) - \frac{J_C}{v(x)}\right]dx - \phi_i \qquad (6.2.9)$$

Application of Eq. (6.2.9) to the transistor represented in Fig. 6.5 is not straightforward because of the shape of the doping profile. Mathematical complications can be avoided and a worthwhile physical picture of the Kirk effect can still be obtained by considering, for the moment, the rarely used structure shown in Fig. 6.6. For that structure, the epitaxial region is a high-resistivity p-type layer instead of the more commonly used n-type material. For a transistor made in a p-type epitaxial layer, the base-collector space-charge region would extend outward from the n^+ buried layer and the point x_C would be relatively insensitive to bias changes because of the high density of donors in the buried layer. We therefore consider x_C as constant and solve Eq. (6.2.9) to find the variation with current density of $x_{CB} = x_C - x_B$, the width of the collector-base space-charge region. For this case of acceptor doping, if we take the charge $qN(x)$ in Eq. (6.2.9) to be a constant equal to $-qN_{epi}$ up to the buried layer, and assume $v = v_l$, then

$$V_{CB} = \frac{1}{2\epsilon_s}\left[qN_{epi} + \frac{J_C}{v_l}\right]x_{CB}{}^2 - \phi_i$$

$$= \frac{qN_{epi}}{2\epsilon_s}\left[1 + \frac{J_C}{J_1}\right]x_{CB}{}^2 - \phi_i \tag{6.2.10}$$

which can be solved for x_{CB}.

At zero current the collector space-charge region has width x_{CO} where

$$x_{CO} = \left[\frac{2\epsilon_s(V_{CB} + \phi_i)}{qN_{epi}}\right]^{1/2} \tag{6.2.11}$$

The variation with current of the collector-base space-charge region x_{CB} can then be expressed as

$$x_{CB} = \frac{x_{CO}}{(1 + J_C/J_1)^{1/2}} \tag{6.2.12}$$

This variation is shown in Fig. 6.7, where we see that the Kirk effect results in a widening of the charge-neutral base region as currents approach and exceed the critical value J_1. This base-layer widening decreases the transistor current gain at high currents and also degrades the frequency response of the device. We consider

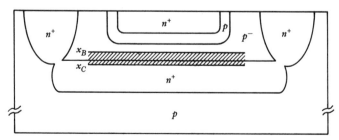

Figure 6.6   Cross section of an npn transistor made using p-type epitaxial layer. The collector-base space-charge layer extending from x_B to x_C is shown by cross-hatching.

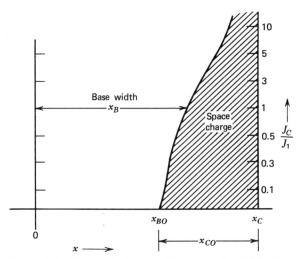

Figure 6.7   Base widening resulting from the Kirk effect in a transistor having the cross section sketched in Fig. 6.6. The variation shown has been calculated using Eq. (6.2.12).

the latter effect in the next section, but first let us return to the case of the Kirk effect in *npn* transistors with an *n*-type epitaxial layer, the most widely used structure.

The complication in the case of an *n*-type epitaxial layer arises because the dopant concentration varies considerably through the collector-base space-charge region, going from $-qN_a(x_B)$, in the tail of the *p*-type base diffusion, to $+qN_{epi}$, the fixed-donor density in the epitaxial region and finally to $+qN_d(x)$ at the outer edge of the out-diffusing buried layer. In this case, as the current level is increased above $J_1 = qN_{epi}v_l$, the sign of the space charge in the epitaxial layer changes from plus to minus. This sign reversal causes the electric field in the epitaxial layer to vary with position when current densities exceed J_1 in the opposite sense from the currentless case. Electrical flux lines that end on the negative acceptor ions in the base when currents are very low become terminated instead on the current-carrying free electrons when the Kirk effect occurs. The entire space-charge zone is thereby pushed toward the heavily doped collector region, moving from an initial position at the *pn* junction formed by the base acceptor diffusion to a final position at the edge of the buried layer at high currents. Under this very high current condition, the electric field increases linearly with distance and the space-charge region is compressed against the highly doped collector edge (Problem 6.10). Computer calculations of the Kirk effect for this case have been carried out[4] and several of the results obtained are shown in Fig. 6.8.

The doping profile of the transistor is shown in Fig. 6.8*a*, while the field and electron densities at various current levels are indicated in Figs. 6.8*b* and 6.8*c*, respectively. At low currents, the field in the collector-base region has the roughly triangular shape sketched in Fig. 6.5b. (Note that the *negative* of the field is plotted in Fig. 6.8b.) The critical current J_1 for the structure of Fig. 6.8 is of the order of 500 A cm^{-2}. In the neighborhood of this current density, the high-field region is seen in Fig. 6.8b to migrate toward the highly doped buried layer. At still higher currents, the field becomes compressed against the boundary between the epitaxial region and the buried layer. The plot of electron concentration in Fig. 6.8c has a steep gradient near the emitter to overcome the built-in opposing field in this region. The extension of the base region with increasing current is evident on the

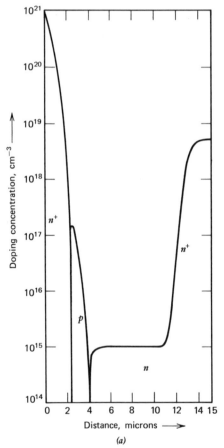

(a)

Figure 6.8   (a) Doping profile in an *npn* transistor for which a computer analysis of the Kirk effect has been carried out.[4] (b) Electric field as a function of distance for various collector currents in the transistor of (a). (c) Base minority-carrier (electron) density as a function of distance for various collector currents in the transistor of (a).

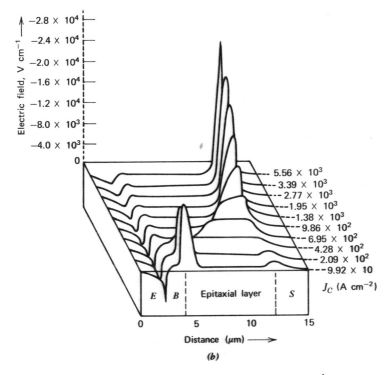

(b)

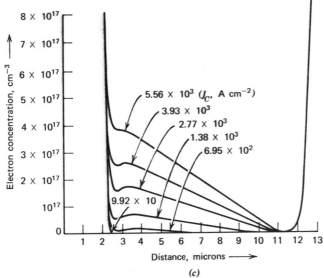

(c)

Figure 6.8   (*Continued*)

same figure. Note that at higher currents, the plots of electron density have nearly constant gradients over most of the epitaxial region.

Base Resistance

Because current gain β_F in amplifying transistors is typically so very high, one may be tempted to assume that base current is negligible and therefore that resistance in the base region is of little consequence to transistor operation. This simplified viewpoint fails to consider that small voltage differences in the base region are magnified by the exponential diode factor and thus can cause much larger differences in current densities across the emitter-base *pn* junction.

Consider the cross section of the transistor shown in Fig. 6.9. As V_{BE} is increased from zero, injection of electrons from the emitter region will be the greatest at that part of the junction where the base doping N_a is the smallest. Because the base is diffused, this will be across the bottom plane of the emitter diffusion. Majority-carrier base current will flow into this region to supply carriers for recombination and to be injected into the emitter. A typical base thickness, however, is less than one micron and therefore a sizable ohmic resistance will generally be present between the base electrode and the active area of the transistor. Since the emitter-base current is spread over the active region of the emitter (Fig. 6.9), the base current decreases continuously as one moves toward the center line of the emitter. Therefore, we cannot calculate a resistance value that will account in a straightforward way for the ohmic drop in the base region. More important, the potential dropped along the base region decreases the base-emitter bias along the same path. The density of injected-electron current therefore decreases from its highest value in the active region nearest the base electrode to a minimum at the center of the emitter. This current crowding toward the perimeter of the emitter increases with bias and causes localized heating at current levels that might be tolerable if the current

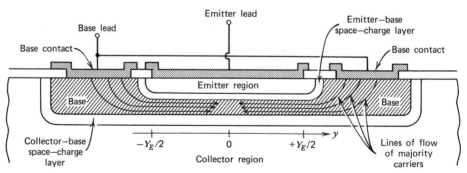

Figure 6.9   Cross section of a transistor under active bias. The base current is supplied from two side base contacts and flows toward the center of the emitter causing the base-emitter voltage drop to vary with position.

were distributed uniformly. The high-injection effects that were considered earlier in this section also occur at lower currents because of the uneven current density across the active region. In order to reduce base resistance, power transistors are made with large comblike base and emitter contacts.

A plot of collector current versus base-emitter voltage for a transistor in which base resistance is significant is shown in Fig. 6.10. Because base resistance acts to reduce the junction bias, we might try to fit the measured data from Fig. 6.10 to the expression

$$I_C = I_S \exp\left[\frac{q(V_{BE} - I_B R_B)}{kT}\right] \tag{6.2.13}$$

To do so, we recognize from our discussion of current crowding that the resistance R_B in Eq. (6.2.13) (called the *base spreading resistance*) must be variable. Using the measurements of I_C, I_B, and V_{BE}, we find that to correspond to Eq. (6.2.13), R_B must decrease with increasing current (Fig. 6.11). The initial reduction in base resistance apparent in Fig. 6.11 corresponds to the decreasing path length from the base electrode to the active region of the transistor as current increases. If only

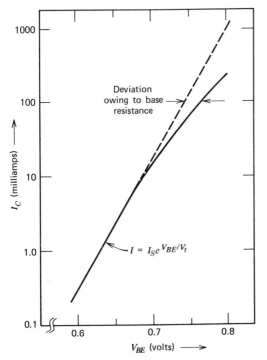

Figure 6.10   Collector current as a function of base-emitter voltage showing the deviation from ideal behavior at high currents.

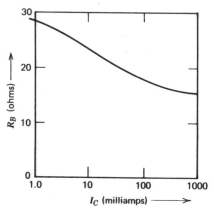

Figure 6.11   Base resistance in the transistor of Fig. 6.10. Values of R_B obtained using measured data in Eq. (6.2.13).

base spreading resistance were important, an asymptotic value for R_B would be approached when all of the current had crowded toward the perimeter of the emitter and all resistance was associated with inactive regions of the base. However, the onset of the other high-level effects, previously mentioned, may cause R_B to decrease to a lower value.

It is possible to analyze exactly the current-crowding effects that we have described by making use of a spatially distributed form of the diode equation. The mathematics become relatively tedious, however, and may obscure the relevant physical mechanisms. We shall, therefore, consider an approximate analysis for which the transistor is divided into sections. Each section is assumed to have the same current gain as the original device and to follow the ideal transistor equation, that is, to have negligible base resistance. Each section is characterized by its proportional share of the overall saturation current [I_S in Eq. (6.2.13)] and is separated from the adjacent section by a resistance corresponding to part of the physical resistance along the base region. By increasing the number of pieces into which the transistor is sliced, we can increase the accuracy of the analysis and ultimately approach the exact distributed case. It is only necessary, however, to use the four-section structure in Fig. 6.12 to illustrate both the physical mechanism of current crowding and the method of solution. The calculation of base resistance from the network sketched in Fig. 6.12 can be accomplished by assuming a current in the last leg of the ladder and then applying network laws to calculate voltage and current at the input terminal. Base resistance is then obtained by using Eq. (6.2.13) at each value of input current I_B and voltage V_{BE}. Further consideration of base resistance is found in the problems.

Summary. This section has considered several bias-dependent physical effects that were not included in the basic theory of junction transistors. The effects

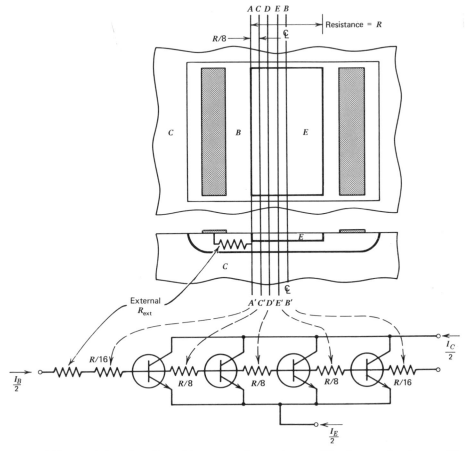

Figure 6.12 Modeling the effect of base-spreading resistance. The base region for an *IC* transistor is divided into eight sections (four on each half) and the total built-in resistance between the base contacts is divided between eight transistors.[9] Network analysis allows us to determine the effective resistance seen at the base-emitter junction as bias is varied (Problem 6.12).

described provide limits on the useful range of bias for an amplifying bipolar transistor. Analysis of these effects shows that the equation for transistor action [Eq. (5.1.7)] adequately describes high-level injection at the base-emitter junction provided that the total majority charge in the base is properly specified. The basic theory of Chapter 5 must be extended, however, in order to treat low-level injection, the Kirk effect, or base resistance. Further consequences from these effects are described in the discussion of transistor speed and models that follows.

For purposes of clarity, the discussion of high-level injection has treated effects one-by-one. In practice, of course, they occur simultaneously. Some interactions between effects, such as a reduction in base resistance because of high injection at the base-emitter junction may be surmised; others are more subtle. Computer methods have been developed that treat the high-injection problem from first principles and thus give an accurate overall analysis. The computer analysis[5] shows that the Kirk effect, also referred to sometimes as *base push-out*, plays a dominating role under high-injection conditions.

The list of limiting effects that has been treated is not all-inclusive. The mechanisms described are, however, concerned with internal transistor operation. They are thus distinguished from performance limits imposed by parasitic elements that are introduced in the *IC* fabrication process. Limiting effects resulting from these elements can usually be treated by using circuit analysis as illustrated in the discussion of lateral *pnp* transistors in Section 6.7.

6.3 BASE TRANSIT TIME

The transit time of minority carriers across the quasi-neutral base of a transistor under active bias is a significant parameter. The delay associated with this transit time was the dominant limitation on the transient behavior of the first transistors. The search for methods of reducing base transit time was responsible for many advances in semiconductor processing techniques. In the present design of bipolar transistors for integrated circuits, the base transit time has been so greatly reduced that it is just one among several important time limitations. Its discussion, however, can elucidate considerably the interrelationships between transistor design and performance.

Let us first consider the excess minority charge in the quasi-neutral base of a transistor under active bias. This charge, which we designate q_B, can be expressed

$$q_B = \int_0^{x_B} q A_E n'(x) \, dx \tag{6.3.1}$$

The injected charge q_B carries the collector current of Eq. (6.1.1) that flows because of transistor action between the emitter and collector. Accordingly, a characteristic *base transit time* τ_B for the transfer of minority carriers across the base is given by the quotient of q_B divided by I_C:

$$\tau_B = \frac{q_B}{I_C} \tag{6.3.2}$$

Calculation of the base transit time in the prototype transistor sketched in Fig. 5.1 is particularly simple. Because $n'(x)$ is linear [Eq. (5.1.11)], $q_B = \frac{1}{2}qn'(0)x_B A_E$ and I_C is given by Eq. (5.1.12). Hence, for this case

$$\tau_B = \frac{x_B{}^2}{2\tilde{D}_n} \tag{6.3.3}$$

Equation (6.3.3) gives a measure to the time required for diffusion transport across a region. Some appreciation of the low speed of diffusion transport relative to drift of carriers in an electric field can be gained by calculating τ_B for a lightly doped base ($N_a \approx 10^{16}$ cm^{-3}) having a width x_B of 1 μm. Application of Eq. (6.3.3) gives $\tau_B \simeq 144$ ps. If a field were present in the base, the carriers would experience a transit time of $x_B/\mu_n\mathscr{E}$ or $x_B^2/\mu_n V$ for a constant field. The drift transit time would be smaller than the diffusion time for any voltage drop greater than about 50 mV.

For the arbitrarily doped transistor, an expression for τ_B can be written using Eqs. (6.1.1) and (6.3.1) in Eq. (6.3.2).

$$\tau_B = \frac{\int_0^{x_B} p \, dx \int_0^{x_B} n' \, dx}{\tilde{D}_n n_i^2 \, \exp(qV_{BE}/kT)} \tag{6.3.4}$$

Equation (6.3.4) is particularly useful when considering the effects of various base dopant profiles on τ_B.

Application of Eq. (6.3.4) to the prototype transistor under high-injection conditions will also prove enlightening. For the prototype transistor under active bias, the carrier densities vary linearly with position, and the integrals in Eq. (6.3.4) are therefore particularly simple to evaluate. A plot of the minority-carrier density in the base is triangular and, because of charge neutrality, the majority carriers take on a trapezoidal distribution decreasing away from the edge of the base-emitter space-charge zone. Sketches of the densities are shown in Fig. 6.13. Under these conditions, the integrals in Eq. (6.3.4) can be evaluated and the base transit time can be written:

$$\tau_B = \frac{\left[\frac{1}{2}n'(0)x_B\right]\left[\frac{1}{2}(n'(0) + 2N_a)x_B\right]}{\tilde{D}_n[n'(0)(n'(0) + N_a)]}$$

$$= \frac{x_B^2}{4\tilde{D}_n}\left[1 + \frac{N_a}{n'(0) + N_a}\right] \tag{6.3.5}$$

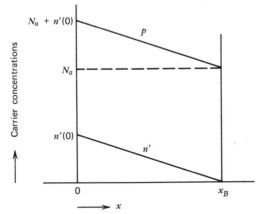

Figure 6.13 Free-carrier densities in the base of a homogeneously doped transistor. Charge neutrality causes the hole density to increase above the dopant concentration N_a.

Equation (6.3.5) shows that under low or moderate injection conditions, for which $n'(0)$ is much less than N_a, τ_B approaches $x_B^2/2\tilde{D}_n$. High bias levels actually reduce the base transit time, with τ_B approaching $x_B^2/4\tilde{D}_n$ asymptotically. This factor of two reduction in base transit time for a homogeneous doped transistor under high-level injection was first predicted by W. M. Webster[6] and is frequently called the *Webster effect*. The physical source for the reduction in τ_B is the field associated with the higher density of majority carriers that exists throughout the base under high-injection conditions. To balance a diffusion tendency of these majority carriers resulting from their nonuniform distribution, an electric field must exist along the the emitter-collector dimension. This electric field balances the diffusion tendency and prevents majority-carrier flow toward the collector. At the same time, the field aids the flow of minority carriers and acts to reduce their transit time through the base. In the case of uniform base doping, the transport of minority carriers changes from a pure diffusion process at low biases to drift-aided diffusion at higher biases.

Although *IC* transistors are not made with uniformly doped base regions, we have seen in Section 6.2 that high-injection conditions do extend the base into the uniformly doped epitaxial regions (Kirk effect). For this reason, high-injection analyses often take τ_B to be $x_B^2/4\tilde{D}_n{}^{3,4}$.

An alternative form for τ_B can be derived from Eqs. (5.1.5) and (6.3.4). If the density of electrons at the collector boundary $n(x_B)$ is taken to be zero, then Eq. (5.1.5) shows that the minority-carrier density as a function of x is

$$n(x) = -\frac{J_n}{q\tilde{D}_n p(x)} \int_x^{x_B} p(\xi)\, d\xi \tag{6.3.6}$$

Under the usual active-bias approximation that $n \approx n'$, Eq. (6.3.6) can be combined with Eq. (6.3.4) to give

$$\tau_B = \frac{1}{\tilde{D}_n} \int_0^{x_B} \frac{1}{p(x)} \left[\int_x^{x_B} p(\xi)\, d\xi \right] dx \tag{6.3.7}$$

If the distance variable x in Eq. (6.3.7) is normalized to x_B by introducing $y = x/x_B$, then Eq. (6.3.7) becomes

$$\tau_B = \frac{x_B^2}{\tilde{D}_n} \left\{ \int_0^1 \frac{1}{p(y)} \left[\int_y^1 p(\xi)\, d\xi \right] dy \right\}$$

$$= \frac{x_B^2}{v\tilde{D}_n} \tag{6.3.8}$$

where the factor v represents the reciprocal of the nested definite integral in the first form of Eq. (6.3.8). Equation (6.3.8) shows that the effect of an inhomogeneous base doping on τ_B can be incorporated into the value of v.

For a constant base doping $p(x) = N_a$, it is easy to use Eq. (6.3.8) to find that $v = 2$. When graded profiles are considered, τ_B can be reduced from $x_B^2/2\tilde{D}_n$ by about an order of magnitude. Limits on reducing τ_B are imposed by the need to maintain acceptable injection efficiency [which constrains $N_a(0)$ to be appreciably lower than the emitter doping] and the requirement of an extrinsic p-type doped material at $x = x_B$ [which forces $N_a(x_B)$ to be greater than $N_{d\ epi}$].

6.4 CHARGE-CONTROL MODEL

A useful framework for transistor equations that often simplifies analysis and design is provided by the *charge-control model* of the device[7]. For this model the controlled variable is not current or voltage; instead, equations are framed in terms of controlled charges within regions of the device.

A typical charge-control relationship for a transistor under active bias was derived in the previous section in Eq. (6.3.2). This equation, $I_C = q_B/\tau_B$ relates the minority charge stored in the quasi-neutral base q_B to the current I_C carried by transistor action between the emitter and the collector. The charge and current are linearly related with τ_B, the transit time in the quasi-neutral base, as the proportionality factor. Because Eq. (6.3.2) represents only minority-carrier transport across the base, however, it is just one portion of the charge-control model for a transistor.

Control in an amplifying *npn* transistor is exercised by the bias on the base-emitter junction. This bias affects not only q_B, but other charge components as well. The major additional components to be considered are the charges represented by holes injected into the emitter, which we designate as q_E, and the charges stored on the base-emitter and base-collector depletion capacitances, which are given the symbols q_{VE} and q_{VC}, respectively. Figure 6.14 shows these components for a prototype transistor. The two injection components, q_B and q_E, are responsible for steady-state base currents. We shall refer to their sum $(q_B + q_E)$ by the symbol q_F (because this is a case of *forward*-active bias). The other components influence the time-varying behavior of the transistor and will be treated separately. The steady-state collector current can be written in terms of q_F if a characteristic time τ_F is introduced. The equation for current, in analogy to Eq. (6.3.2), is

$$I_C = \frac{q_F}{\tau_F} \tag{6.4.1}$$

Note that all that is formally required to make Eq. (6.4.1) accurate is that I_C be linearly related to q_F (with τ_F taken to be a constant).

Because q_F represents stored minority charge in the quasi-neutral emitter and base regions, its voltage dependence is generally that of a diode and we may write

$$q_F = q_{FO}\left[\exp\left(\frac{qV_{BE}}{kT}\right) - 1\right] \tag{6.4.2}$$

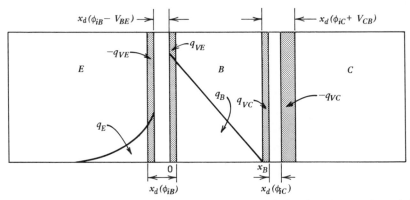

Figure 6.14 Cross section of a prototype transistor indicating the location of the charge components used in the charge-control model. The components of q_V represent base majority charge supplied to the pn junctions at the base-emitter and base-collector junctions.

where q_{FO} is a function of the doping profiles and device geometry. Since q_F is at least somewhat greater than q_B, τ_F must be greater than the base transit time τ_B. Further comments about τ_F are deferred until the charge-control model is developed more fully.

The steady-state current into the base is proportional to the rate at which q_B recombines in the quasi-neutral base plus the rate at which holes are injected into the emitter. These two rates are in turn proportional to q_F, and it is therefore possible to write a charge-control expression for the input (base) current of the transistor:

$$I_B = \frac{q_F}{\tau_{BF}} \qquad (6.4.3)$$

A considerable amount of physical analysis, involving emitter efficiency and the recombination processes for excess carriers in both the emitter and base, would be required to express either τ_{BF} or τ_F in terms of more fundamental parameters. This analysis is not of special value because both τ_{BF} and τ_F can be obtained in practice from measurements, and also because the charge-control model aims at representing the device for design, not at exploring the physical electronics of transistor models.

Equations (6.4.1) and (6.4.3) can be used to show that the steady-state current gain is simply the ratio of the two characteristic times:

$$\frac{I_C}{I_B} = \beta_F = \frac{\tau_{BF}}{\tau_F} \qquad (6.4.4)$$

For example, Eqs. (6.4.1), (6.4.3) and (6.4.4) can be applied to the prototype transistor of Fig. 5.1. If the emitter efficiency is very high in the prototype device, then $q_F \approx q_B$ with $q_B = \frac{1}{2}qn'(0)x_B A_E$. Since only base recombination is significant for this case, $\tau_{BF} = \tau_n$ and $\tau_F = x_B^2/2\tilde{D}_n$ as derived in Eq. (6.3.3). Therefore, from Eq. (6.4.4) β_F is $2L_n^2/x_B^2$ where $L_n = \sqrt{\tilde{D}_n \tau_n}$. This result for dc current gain can be compared with earlier analysis of the same problem in Sec. 5.2. There, α_F was derived as the base transport factor to be $[1 - (x_B^2/2L_n^2)]$ in Eq. (5.2.8). If we use this result for α_F in the equation $\beta_F = \alpha_F/(1 - \alpha_F)$, we find an identical value for β_F to that obtained from the charge-control analysis.

A full charge-control model for the bipolar transistor is derived by adding terms to represent the currents that flow because of time variations in stored charge. Clearly, if q_F is increasing with time, there will be a component of base current equal to dq_F/dt. Likewise, changes in the charges stored at the base-emitter and base-collector junctions (q_{VE} and q_{VC}) result in added base current. An overall expression for the base current is, therefore,

$$i_B = \frac{q_F}{\tau_{BF}} + \frac{dq_F}{dt} + \frac{dq_{VE}}{dt} + \frac{dq_{VC}}{dt} \tag{6.4.5}$$

The first three components of current in Eq. (6.4.5) flow from the base to the emitter; the last flows from the base to the collector. Combining Eq. (6.4.1) and (6.4.5) and using Kirchhoff's current law, one obtains a set of charge-control equations for the transistor under active bias:

$$i_C = \frac{q_F}{\tau_F} - \frac{dq_{VC}}{dt}$$

$$i_B = \frac{q_F}{\tau_{BF}} + \frac{dq_F}{dt} + \frac{dq_{VE}}{dt} + \frac{dq_{VC}}{dt} \tag{6.4.6}$$

$$i_E = -q_F\left(\frac{1}{\tau_F} + \frac{1}{\tau_{BF}}\right) - \frac{dq_F}{dt} - \frac{dq_{VE}}{dt}$$

We have thus derived a set of *linear* equations relating currents and charges in a bipolar transistor. These equations are in contrast to the nonlinear expressions that relate currents and voltages in the transistor.

The circuit diagram of Fig. 6.15 represents the various terms in Eq. (6.4.6). The diode from the base to the emitter passes the steady-state current and has a saturation current $I_{ES} = q_{F0}[(1/\tau_F) + (1/\tau_{BF})]$ as indicated by Eq. (6.4.2). The elements storing the charges q_F, q_{VE}, and q_{VC} are shown as capacitors with a line across them to indicate that they store charge (like capacitors), but are voltage dependent. Now that a basic set of charge-control equations has been written for a *BJT*, we are better able to gain a perspective on the limitations of this viewpoint.

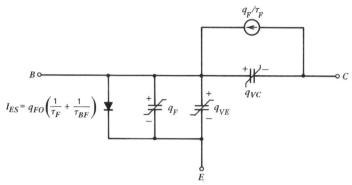

Figure 6.15 Charge-control representation of an *npn* transistor under active bias with junction charge storage and injected base charge taken into account.

The basic premise underlying charge-control analysis is the existence of a constant proportionality between quantity of charge and current. Another way of stating this is that the characteristic times in the charge-control equations must not themselves be functions of charge or of bias voltages. Under this premise, the characteristic times may be derived for dc conditions and then be used to express dynamic conditions. Strictly this premise is not correct; the characteristic times are functions of the charge. For example, to affect current at the collector, minority carriers must disperse through the base region after being injected at the edge of the base-emitter junction. Thus, during transient conditions, collector current is not in the same ratio to base charge as in the steady state.

If a dynamic problem is analyzed with the charge-control model, the solutions for charge are constrained to be a time sequence of differing "steady-state" solutions. Hence, these solutions are sometimes referred to as *quasi-static approximations*. For the greater part of the transient, however, the charge-control solution is usually a fair representation of the more exact result. The time scale over which there is significant error in the charge-control solution is of the order of the transit time in the base, that is, the order of τ_F. For most applications the time scale of interest is appreciably greater than τ_F and solution by means of the charge-control model is perfectly adequate. A specific example in the next section will make this more clear.

Applications of the Charge-Control Model

Before discussing the bipolar charge-control model further, it is worthwhile to illustrate the use of Eqs. (6.4.6) in an application. A simple circuit appropriate for this purpose is shown in Fig. 6.16*a* and the circuit plus charge-control model is sketched in Fig. 6.16*b*. What is sought is the collector-current response of an *npn*

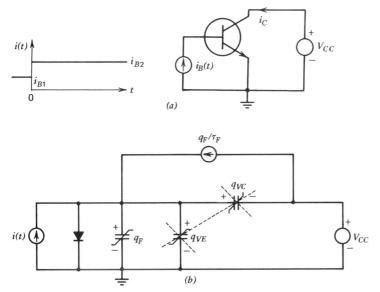

Figure 6.16 (*a*) Simple circuit for illustration of charge-control model. (*b*) Equivalent-circuit model. The cancelled elements carry negligible currents.

transistor under active bias when driven by a current source at the base. For hand calculations, several simplifications are in order. First, the current dq_{VE}/dt can be considered negligible since the base-emitter voltage varies only slightly under forward bias. This simplification is usually made for active-bias operation. Because we have chosen a simple circuit in which the collector voltage is constant, it is also possible to treat the current dq_{VC}/dt as negligible.

With these approximations the equation for the base current has only one unknown, the controlled charge q_F, and takes the form

$$i_B = \frac{q_F}{\tau_{BF}} + \frac{dq_F}{dt} \tag{6.4.7}$$

Since $i_B(t)$ is specified by

$$i_B = i_{B1}(t < 0)$$
$$= i_{B2}(t > 0)$$

the solution for q_F will contain terms associated with both the homogeneous and particular forms of Eq. (6.4.7). When the boundary values $q_F(t = 0) = i_{B1}\tau_{BF}$ and $q_F(t \to \infty) = i_{B2}\tau_{BF}$ are matched, the solution becomes

$$q_F = \tau_{BF}[i_{B2} + (i_{B1} - i_{B2}) \exp(-t/\tau_{BF})] \tag{6.4.8}$$

Under the approximations that we have made, the collector current is given by

q_F/τ_F so that its time dependence becomes [using Eq. (6.4.4)]

$$i_C = \beta_F[i_{B2} + (i_{B1} - i_{B2})\exp(-t/\tau_{BF})] \qquad (6.4.9)$$

The collector current thus changes from its initial to its final value following an exponential function with a characteristic time constant equal to τ_{BF} (Fig. 6.17).

As mentioned in the previous section, the transient solution given in Eq. (6.4.9) is in error for small values of t. At zero time, for example, Eq. (6.4.9) predicts an abrupt change in the slope of the collector current equal to $(i_{B2} - i_{B1})/\tau_F$ whereas collector current will not change until the extra electrons injected at the emitter side of the base reach the collector. A more complete analysis that takes account of the distributed nature of base charging predicts that i_C does not change at all for times close to $t = 0$; initially, the increments in the base and emitter currents are equal. The collector current first begins to change when t reaches the base transit time τ_B. The behavior of i_C then rapidly approaches the transient solution predicted by the charge-control model and essentially matches that result for times larger than $t = \tau_F$. (See inset in Fig. 6.17.)

The foregoing example may seem artificial because of the number of simplifications that have been imposed, first in the choice of circuit and second in the approximations that have been made. These physical simplifications, however, have avoided mathematical complications that might tend to obscure the use of the model.

One such complication would arise if, for example, the collector in Fig. 6.16a were not connected to an ac ground, but rather was connected to the source through

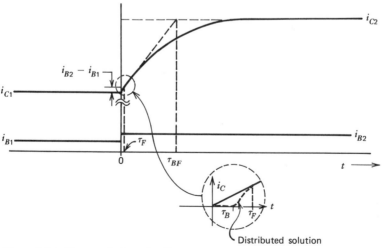

Figure 6.17 Time variation of collector current in the circuit of Fig. 6.16 as calculated from the charge-control model.

a load resistor R_L. Since V_{CB} would be variable in this case, the current required to charge q_{VC} would not be negligible. The solution for i_C can be obtained readily if we define an effective capacitance C_{jC} equal to the average of dq_{VC}/dV_{CB} over the collector voltage interval. The solution for i_C is then found to be equal to Eq. (6.4.9) provided that the time constant τ_{BF} in Eq. (6.4.9) is replaced by [Problem 6.19]:

$$\tau'_{BF} = \tau_{BF}\left(1 + \frac{R_L C_{jC}}{\tau_F}\right) \tag{6.4.10}$$

This result shows that a collector resistor lengthens the time of the current transient by the binomial factor in Eq. (6.4.10).

Large-Signal Model. The most extensive use of charge-control equations is for the solution of large-signal transient problems, and their utility is especially apparent when one is dealing with switching between regions of operation, typically between cut-off and saturation. Writing charge-control equations for large-signal switching is straightforward because all regions of operation have in common the injection and extraction of charge at the two junctions. These injection and extraction processes can be represented by adding together two sets of charge-control equations; one set that characterizes forward-active operation [Eqs. (6.4.6)] and a companion set having the same form to represent operation under reverse-active bias. The total controlled charge thus becomes a superposition of charges representing forward-active bias q_F and reverse-active bias q_R.

The full set of equations for an *npn* transistor is:

$$i_E = -\frac{dq_F}{dt} - q_F\left(\frac{1}{\tau_F} + \frac{1}{\tau_{BF}}\right) + \frac{q_R}{\tau_R} - \frac{dq_{VE}}{dt}$$

$$i_C = \frac{q_F}{\tau_F} - \frac{dq_R}{dt} - q_R\left(\frac{1}{\tau_R} + \frac{1}{\tau_{BR}}\right) - \frac{dq_{VC}}{dt} \tag{6.4.11}$$

$$i_B = \frac{dq_F}{dt} + \frac{q_F}{\tau_{BF}} + \frac{dq_R}{dt} + \frac{q_R}{\tau_{BR}} + \frac{dq_{VE}}{dt} + \frac{dq_{VC}}{dt}$$

A circuit model corresponding to these equations is shown in Fig. 6.18. Its form clearly indicates the superposed forward-active and reverse-active charge-control models. The dc components in the circuit of Fig. 6.18 have a one-to-one correspondence to the components of the Ebers-Moll large-signal representation sketched in Fig. 5.12. The Ebers-Moll model is also built up from superposed forward-active and reverse-active equivalent circuits.

In Eqs. (6.4.11), parameters associated with the reverse-active region of bias are defined analogously to those representing forward bias. For example, the controlled charge q_R is related to V_{BC} by

$$q_R = q_{R0}\left[\exp\left(\frac{qV_{BC}}{kT}\right) - 1\right] \tag{6.4.12}$$

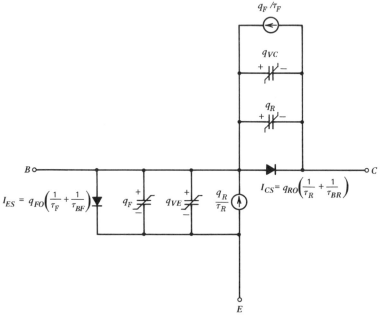

Figure 6.18   Complete bipolar charge-control model for large-signal applications.

and this charge represents storage in both the base and collector quasi-neutral regions.

The density of dopant atoms in the collector region is normally of the same order of magnitude or lower than the density in the base near the collector junction. Hence, when a transistor is under reverse-active bias, the efficiency of electron injection into the base is typically quite low (approximately 60 to 85%). Thus, a relatively large amount of charge is stored in the collector. In fact because the base is so narrow, this charge is typically the dominating component of q_R. For example, Fig. 6.19 is a sketch of the components of q_F and q_R for two cases of saturated transistors. Figure 6.19a shows the case of a homogeneously doped transistor with a lightly doped collector region and Fig. 6.19b shows a typical IC transistor. The very large amount of collector charge q_C in either transistor when it is saturated must be removed to bring it out of saturation. This can cause considerable delay. One means of speeding its removal, the incorporation of a large number of recombination centers, usually by doping the collector region with gold, was extensively employed in early designs of fast switching integrated circuits. Superior switching results can be obtained by preventing the transistor from entering the saturated mode of operation. An effective method to accomplish this was described in Section 2.6 where Schottky clamping was discussed. An IC realization of a

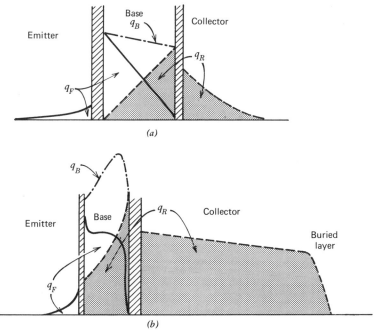

Figure 6.19 The locations of q_F and q_R (dotted lines) for saturated conditions: (a) in a homogeneously doped transistor with a lightly doped collector region and (b) in an epitaxial diffused transistor. The dot-dash lines represent the total base charge q_B.

Schottky-clamped switching transistor was shown in Fig. 5.18. Figure 6.20 shows the comparative switching times of a gold-doped switching transistor and a Schottky-clamped transistor having the same dimensions. The impressive reduction in the time delay before the output voltage rises makes clear why Schottky-clamped transistors have become so important for fast digital logic circuits.

Saturation Transient.[†] As a final application of charge-control modeling, we consider the solution for the transient behavior of a bipolar transistor within the saturated region of operation, that is, a transistor that has not been Schottky clamped. For this case all terms in Eqs. (6.4.11) need to be retained.

The circuit is shown in Fig. 6.21. We assume that $V_S \gg V_o$, the "turn-on voltage" of the base-emitter diode that was introduced in Section 2.6. We also take V_C to be much greater than the drop across the transistor in saturation [Eq. (5.4.13)] and assume that the base drive is strong enough to saturate the transistor fully; that is, that $\beta_F V_S / R_S \gg V_C / R_L$. If the switch is closed at $t = 0$, the transistor will enter the active-bias region and current will build in the collector circuit according to Eq. (6.4.9) with the modification of the time constant given in Eq. (6.4.10). The

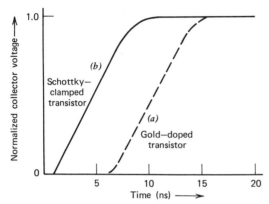

Figure 6.20 Comparisons of turn-off times of (a) gold-doped and (b) Schottky-clamped transistors. The gold-doped device uses 7 ns for recombination before it begins to change state.[11]

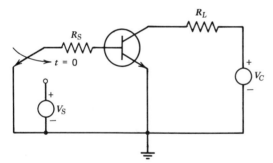

Figure 6.21 Switching circuit in which a transistor undergoes transient behavior from cut-off to saturation. The source voltage V_S is taken to be much greater than the "on voltage" V_o for the base-emitter diode. The base drive is taken to be sufficient to cause transistor saturation ($\beta_F V_S/R_S \gg V_C/R_L$).

final value of current $\beta_F i_2$ will not be reached because the transistor will enter saturation when V_{BC} becomes roughly 0.5 V. The collector current will then be limited to V_C/R_L as shown in Fig. 6.22a. Before the transistor saturates, the stored charge, q_F will increase with a wave shape similar to that of i_C (Fig. 6.22b). Although the collector current stabilizes after the transistor saturates, the charges q_F and q_R are still changing with time and the transistor is not in a steady-state "on" condition until these quantities are constant. Before saturation is reached, the collector-base junction is back biased, and q_R is essentially zero. After saturation q_R increases as does q_F.

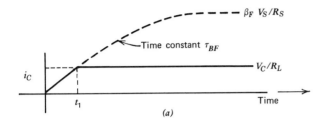

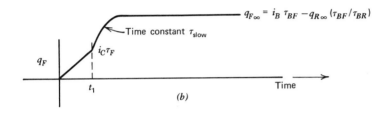

Figure 6.22 Transient behavior of (a) i_C, (b) q_F, and (c) q_R for the switching circuit in Fig. 6.21.

To analyze the saturation behavior, a set of independent simultaneous equations for q_F and q_R must be solved. The set consists of equations for i_B and i_C.

$$i_B = \frac{q_F}{\tau_{BF}} + \frac{dq_F}{dt} + \frac{q_R}{\tau_{BR}} + \frac{dq_R}{dt}$$

$$i_C = \frac{q_F}{\tau_F} - q_R\left(\frac{1}{\tau_R} + \frac{1}{\tau_{BR}}\right) - \frac{dq_R}{dt} \qquad (6.4.13)$$

In writing these equations the terms representing charging currents for q_{VC} and q_{VE} are dropped as negligible because V_{BC} and V_{BE} are roughly constant in saturation.

Equations (6.4.13) represent simultaneous differential equations for which the natural frequencies are solutions of s that satisfy the equation:

$$\left(s + \frac{1}{\tau_{BF}}\right)\left(s + \frac{1}{\tau_R} + \frac{1}{\tau_{BR}}\right) + \left(s + \frac{1}{\tau_{BR}}\right)\frac{1}{\tau_F} = 0 \qquad (6.4.14)$$

The roots of this quadratic are approximately given by

$$\left| s_1 \right| = \frac{1}{\tau_{\text{FAST}}} = \left(\frac{1}{\tau_F} + \frac{1}{\tau_R} + \frac{1}{\tau_{BR}} + \frac{1}{\tau_{BF}} \right)$$

and

$$\left| s_2 \right| = \frac{1}{\tau_{\text{SLOW}}}$$

$$= \tau_{\text{FAST}} \left(\frac{1}{\tau_F \tau_{BR}} + \frac{1}{\tau_R \tau_{BF}} + \frac{1}{\tau_{BF} \tau_{BR}} \right) \tag{6.4.15}$$

These roots represent normal modes of the transient solution for q_F and q_R in saturation. As the subscripted names imply, the root s_1 is responsible for a transient solution that is completed long before the steady state is reached, while the root s_2 leads to the transient behavior for q_F and q_R that dominates while the transistor remains saturated. Hence, a good approximation to the transient behavior of q_F and q_R in saturation (denoted by q_{FS} and q_{RS}) is

$$q_{FS} = (q_{F\infty} - q_{F1}) \left[1 - \exp \frac{-(t - t_1)}{\tau_{\text{SLOW}}} \right] + q_{F1}$$

and

$$q_{RS} = (q_{R\infty} - q_{R1}) \left[1 - \exp \frac{-(t - t_1)}{\tau_{\text{SLOW}}} \right] + q_{R1} \tag{6.4.16}$$

where t_1 is the time at which saturation begins, $q_{F\infty}$ and $q_{R\infty}$ are the final values and q_{R1} and q_{F1} are the values of the two controlled charges at $t = t_1$.

The total saturation charge $q_{ST} = (q_{FS} + q_{RS})$ is sometimes combined in terms of the forward charge $q_F = I_C \tau_F$ at the edge of saturation and an added portion that enters the channel after the transistor saturates.

$$q_{ST} = I_C \tau_F + q_S \left[1 - \exp \frac{-(t - t_1)}{\tau_{\text{SLOW}}} \right] \tag{6.4.17}$$

where

$$q_S = q_{F\infty} + q_{R\infty} - I_C \tau_F$$

For the prototype transistor in which essentially all charge is stored in the base, these definitions aid in understanding the physical mechanisms. They can be simply represented as shown in Fig. 6.23. The important feature, emphasized by Fig. 6.23, is that transistor switching is not completed when the collector current assumes a steady value (which occurs at the time of saturation), but that the steady state is only reached when q_R and q_F have reached their final values.

We have just considered the "turn-on" transient. To calculate the "turn-off" transient a similar two-part analysis would need to be carried out. We would first find solutions in the saturated region while q_R and q_F are reduced to the values

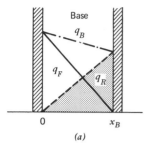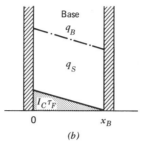

Figure 6.23 Two alternative representations for total charge at steady state in a saturated, homogeneously doped transistor with high emitter and collector doping. (a) The two charge-control variables q_F and q_R are preserved. (b) The total charge is broken up into $I_C \tau_F$, which is present at the onset of saturation, and q_S, which is added during the saturation transient.

applying at the edge of active-mode bias and then solve for the transient through the active region to transistor cut-off.

It will not be of much benefit to carry these calculations any further; they have already served to indicate the analytical means afforded by charge-control analysis. We see that substantial mathematical labor is involved in obtaining hand-calculated solutions to the charge-control equations. We also have seen that their format is a useful representation not only because they provide a set of linear simultaneous equations, but also because it is straightforward to identify the physical basis for each term and to simplify the equations in many special cases. Fortunately, the charge-control equations are a particularly simple set to implement for solution by computer routines when detailed solutions are desired. Further applications of the model to switching circuits are extensively described in reference 8.

6.5 SMALL-SIGNAL TRANSISTOR MODEL

When transistors are biased in the active region and used for amplification, it is often worthwhile to approximate their behavior under conditions of small voltage variations at the base-emitter junction. If these variations are smaller than the thermal voltage $V_t = (kT/q)$, it is possible to represent the transistor by a linear equivalent circuit. This representation can be of great aid in the design of amplifying circuits. It is called the small-signal transistor model. It can be derived readily by making use of several of the charge-control relationships that were discussed in Section 6.4.

When a transistor is biased in the active mode, collector current is related to base-emitter voltage by Eq. (5.2.1), which is repeated here for convenient reference.

$$I_C = I_S \exp\left(\frac{qV_{BE}}{kT}\right) = I_S \exp\left(\frac{V_{BE}}{V_t}\right) \tag{6.5.1}$$

Hence, if V_{BE} varies incrementally, I_C will also vary according to

$$\frac{\partial I_C}{\partial V_{BE}} = \frac{I_S}{V_t} \exp\left(\frac{V_{BE}}{V_t}\right) = \frac{I_C}{V_t} \equiv g_m \tag{6.5.2}$$

This derivative is recognized as the transconductance and given the usual symbol g_m. Notice that g_m is directly proportional to the bias current in the transistor. The variation of base current with base-emitter voltage can be found most directly by making use of the charge-control expressions for I_C and I_B [Eqs. (6.4.1) and (6.4.3)].

$$\frac{\partial I_B}{\partial V_{BE}} = \frac{\partial(q_F/\tau_{BF})}{\partial V_{BE}} = \frac{\partial(I_C\tau_F/\tau_{BF})}{\partial V_{BE}} = \frac{\tau_F g_m}{\tau_{BF}} = \delta g_m \tag{6.5.3}$$

where the ratio of τ_F to τ_{BF} is expressed as a *defect factor* δ. Referring to Eq. (6.4.4), we see that δ is equal to β_F^{-1}.

The base minority charge q_F varies with base-emitter voltage according to

$$\frac{\partial q_F}{\partial V_{BE}} = \frac{\partial(I_C\tau_F)}{\partial V_{BE}} = g_m\tau_F \equiv C_D \tag{6.5.4}$$

where the symbol C_D (often called the *diffusion capacitance*) represents the capacitance associated with incremental changes in the injected minority-carrier charge.

If the incremental voltages and currents in Eqs. (6.5.2), (6.5.3), and (6.5.4) are identified with ac signals, then the system of equations can be represented by the equivalent circuit shown in Fig. 6.24. The small-signal incremental currents and voltages are denoted by lowercase symbols. The base-emitter input circuit is a parallel RC network having a time constant τ_F/δ that is just the base time constant τ_{BF}.

The collector-emitter output circuit consists of a current source activated by the input voltage. The output current for a given v_{BE} is proportional to g_m, and therefore dependent on the dc bias I_C as shown in Eq. (6.5.1). The equivalent circuit of Fig. 6.24 emphasizes the fact that to first order the input is decoupled from the output and the output is insensitive to collector-base voltage variations. From the analysis

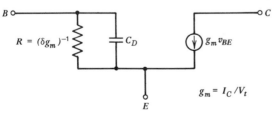

Figure 6.24   Equivalent circuit representing small-signal active bias for a bipolar junction transistor. Only first order effects have been considered.

of Section 6.1, we know that the voltage across the collector-base junction does influence collector current, chiefly as a result of the Early effect. The variation of I_C with V_{CB} was shown in Section 6.1 to be the ratio of the collector current I_C to the Early voltage V_A. In terms of the small-signal parameters,

$$\left|\frac{\partial I_C}{\partial V_{CB}}\right| \equiv \frac{I_C}{|V_A|} = \frac{g_m V_t}{|V_A|} = \eta g_m \tag{6.5.5}$$

where a new parameter $\eta \equiv V_t/|V_A|$ has been introduced to represent the ratio of the change in I_C when V_{CB} is varied to the change in I_C when V_{BE} is varied.

Variation in I_C with V_{CB} must also result in a change in the controlled charge q_F with V_{CB}. This can be calculated from the charge-control relationship of Eq. (6.4.1).

$$\left|\frac{\partial q_F}{\partial V_{CB}}\right| = \left|\frac{\partial (I_C \tau_F)}{\partial V_{CB}}\right| = \tau_F \eta g_m = \eta C_D \tag{6.5.6}$$

Any change in base minority charge results in a change in base current as well as in collector current. Thus, variation in V_{CB} results in a change in I_B that is given by

$$\left|\frac{\partial I_B}{\partial V_{CB}}\right| = \left|\frac{\partial (q_F/\tau_{BF})}{\partial V_{CB}}\right| = \frac{\eta g_m \tau_F}{\tau_{BF}} = \eta \delta g_m \tag{6.5.7}$$

The variations calculated in Eqs. (6.5.5), (6.5.6), and (6.5.7) can be incorporated into the linear equivalent circuit of Fig. 6.24 by adding three elements as shown in Fig. 6.25. The variation in I_C calculated in Eq. (6.5.5) represents a change in the current flowing from collector to emitter in response to a change in collector-base voltage. It is thus modeled as a current generator activated by collector-base voltage. The current generator is directed from collector to emitter because an increase in V_{CB} increases I_C as described in Section 6.1. The change in injected charge storage in response to a change in collector-base voltage calculated in Eq. (6.5.6) is modeled as a capacitor from collector to base. The variation in current calculated in Eq. (6.5.7) flows between the base and the emitter and is caused by a changing collector-base voltage. It is thus modeled as a current generator directed

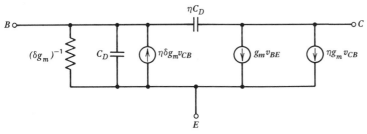

Figure 6.25 Small-signal equivalent circuit for a bipolar junction transistor with Early-effect elements.

from the emitter to the base. This direction is consistent with a reduced base current as a consequence of a reduction in q_F.

The equivalent circuit sketched in Fig. 6.25 can be simplified by employing two procedures common in circuit analysis. First, the generators involving v_{CB} can be bridged across the collector-base node pairs by making use of the equivalence that is illustrated in Fig. 6.26. In this way some generators will be activated by the voltage that appears across their terminals and can be replaced by passive elements. The second procedure involves reexpressing the activating voltages for several generators by making use of the identity: $v_{CE} \equiv v_{CB} + v_{BE}$. When these operations are performed and the admittance of elements in parallel is summed, the circuit can be simplified to the form shown in Fig. 6.27. This form of the small-signal equivalent circuit is widely referred to as the *hybrid pi circuit* for the transistor. The term *hybrid* is used because the generator is a voltage-activated current source and it therefore refers quantities of differing dimensions. The reference to *pi* denotes the general geometric shape of the circuit in the form of a Greek letter Π.

For accurate representation of the transistor small-signal behavior in some situations, two additional effects remain to be considered. The first of these is base resistance discussed in Section 6.2. There we saw that current-crowding effects cause an overall dc base resistance R_B [as defined implicitly in Eq. (6.2.13)] that is a function of collector current. The dependence of R_B on I_C for a typical *npn* transistor is shown in Fig. 6.11. To incorporate this result into a small-signal equivalent

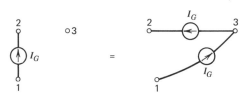

Figure 6.26   Two equivalent circuits in which generators "bridge" nodes.

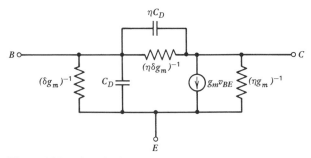

Figure 6.27   Simplified small-signal equivalent circuit including Early-effect elements.

circuit, the interdependences of the small-signal variations in voltage and current must be considered. To do this properly, it is necessary to consider the functional relationships between V_{BE}, I_B, I_C, and R_B. Total differentiation of V_{BE} with respect to I_C, for example, will result in three terms:

$$\frac{dV_{BE}}{dI_C} = \frac{\partial V_{BE}}{\partial I_C}\bigg|_{I_B,R_B} + \frac{\partial V_{BE}}{\partial I_B}\bigg|_{I_C,R_B}\left(\frac{dI_B}{dI_C}\right) + \frac{\partial V_{BE}}{\partial R_B}\bigg|_{I_C,I_B}\left(\frac{dR_B}{dI_C}\right) \qquad (6.5.8)$$

From Eq. (6.2.13) we have

$$V_{BE} = I_B R_B + V_t \ \ell n\left(\frac{I_C}{I_S}\right) \qquad (6.5.9)$$

Therefore, using Eq. (6.5.8), we obtain

$$\frac{dV_{BE}}{dI_C} = \frac{V_t}{I_C} + R_B \frac{dI_B}{dI_C} + I_B \frac{dR_B}{dI_C} = \frac{1}{g_m} + \delta R_B + I_B \frac{dR_B}{dI_C} \qquad (6.5.10)$$

where we have used the previously defined symbols g_m and δ while retaining the term involving the derivative of R_B. In practice this derivative can be obtained from a plot similar to that in Fig. 6.11.

To obtain an equivalent-circuit representation from Eq. (6.5.10), we may solve for the base input resistance R_I where

$$R_I = \frac{dV_{BE}}{dI_B} = \frac{dV_{BE}}{dI_C} \cdot \frac{dI_C}{dI_B} = \frac{dV_{BE}}{dI_C} \cdot \frac{1}{\delta} \qquad (6.5.11)$$

Using Eq. (6.5.10) in (6.5.11), we thus derive

$$R_I = \frac{1}{\delta g_m} + \left(R_B + \frac{I_B}{\delta}\frac{dR_B}{dI_C}\right) \qquad (6.5.12)$$

Thus, in order to take account of base resistance, a resistor of value

$$r_b = R_B + I_C \frac{dR_B}{dI_C} \qquad (6.5.13)$$

must be added in series with the base-emitter resistance $(\delta g_m)^{-1}$ that was previously determined. This has been done in the circuit sketched in Fig. 6.28. As is apparent from this circuit, when base resistance is taken into account the current generator in the output circuit is no longer actuated by the applied base-emitter voltage, but is rather a function of an internal node-pair voltage. It is left as a problem to show that the current gain in the presence of base resistance is also properly modeled when the circuit of Fig. 6.28 is used.

Our consideration of base resistance has been limited to dc and low-frequency effects. Base resistance can lead to more complex behavior at higher frequencies because the base resistance and junction capacitances behave like distributed

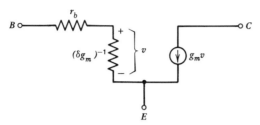

Figure 6.28   Low frequency, small-signal equivalent circuit including base resistance.

transmission lines. These effects can be modeled with fair accuracy by using an equivalent shunt RC network in place of r_b. A full discussion of this topic is provided in reference 9.

The last complication that we shall add to the hybrid pi circuit is the capacitance associated with the junction space-charge regions. This capacitance is in parallel with the base-emitter and base-collector capacitors shown in Fig. 6.27. The added capacitance is usually denoted by C_{je} and C_{jc} and is calculated from the junction capacitance equations already derived in Chapter 3. For increased accuracy, it is sometimes necessary to divide the total collector junction capacitance into a portion that bridges the base impedance element (referred to as C_{sc}) and a portion C_{jc} that is returned to the intrinsic transistor base node, that is, the node connected to δg_m. This division implies that some parts of the collector capacity will not be charged through the base resistance. The overall equivalent circuit including these effects is shown in Fig. 6.29.

The circuit taken in its entirety may appear formidable. Fortunately, it is seldom necessary to deal directly with the overall hybrid pi circuit in hand calculations. Either it is true that several elements in the circuit have negligible effect under given

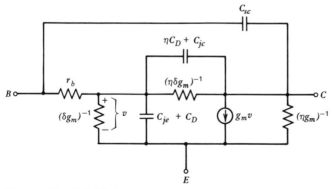

Figure 6.29   Hybrid pi circuit with space-charge capacitances and base resistance.

conditions and the circuit can therefore be simplified, or else the calculations are carried out by a computer.

Equivalences between Models

Three models to represent bipolar transistors have now been discussed: the Ebers-Moll Model in Section 5.4, the charge-control model in Section 6.4 and the hybrid pi model in Section 6.5. The first two of these models are valid for all ranges of bias, while the third is a small-signal equivalent circuit for use only in the region of active bias. Since the models have regions of overlapping validity, a series of relationships between their parameters can be expected. For example, in Problem 6.20 it is found that the Ebers-Moll parameters α_F and α_R are related to the charge-control parameters τ_F and τ_{BF} by

$$\alpha_F = \frac{\tau_{BF}}{\tau_F + \tau_{BF}}$$

and

$$\alpha_R = \frac{\tau_{BR}}{\tau_R + \tau_{BR}} \tag{6.5.14}$$

respectively. In the Ebers-Moll representation, the six parameters of the basic charge-control model (q_{F0}, q_{R0}, τ_F, τ_{BF}, τ_R, and τ_{BR}) are reduced to only four parameters α_F, α_R, I_{ES}, and I_{CS}).

Relationships such as Eq. (6.5.14) are useful to obtain model parameters by measurements and to establish properties of the models. For example, the reciprocity condition of the Ebers-Moll model $\alpha_F I_{ES} = \alpha_R I_{CS}$ (Eq. 5.4.7) can be used on the charge-control model to derive

$$\frac{q_{F0}}{\tau_F} = \frac{q_{R0}}{\tau_R} \tag{6.5.15}$$

A useful relationship that allows the experimental determination of the charge-control parameter τ_F can be derived by amalgamating the small-signal equivalent circuit with the charge-control model under active bias (Fig. 6.15). Consider that only dc bias is applied to the collector. Hence, the ac small-signal equivalent circuit has the collector shorted to ground. Under this condition, it is usually a good approximation to neglect the Early-effect elements in the circuit of Fig. 6.29 and also to consider base resistance as insignificant. The circuit of Fig. 6.29 can then be simplified to the form shown in Fig. 6.30. The current gain i_C/i_B in this circuit is

$$\frac{i_C}{i_B} = \frac{(1/\delta)(1 - j\omega C_{jc}/g_m)}{1 + j\omega[(C_{je} + C_{jc})/g_m\delta + \tau_F/\delta]}$$

$$\approx \left(\frac{\tau_{BF}}{\tau_F}\right)\left[1 + j\omega\left(\tau_{BF} + \frac{(C_{je} + C_{jc})\tau_{BF}}{g_m\tau_F}\right)\right]^{-1} \tag{6.5.16}$$

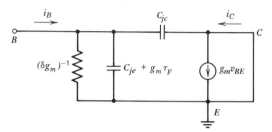

Figure 6.30   Equivalent circuit for obtaining the interrelationship between f_T and τ_F.

where we have used Eq. (6.4.4) and omitted the frequency-dependent term in the numerator. This numerator term is only of consequence at frequencies appreciably above those at which the imaginary term in the denominator is dominant.

As the frequency is increased, the current gain decreases and, from Eq. (6.5.16), the magnitude of the current gain will be unity at a frequency f_T which is approximately given by

$$f_T = \frac{1}{2\pi \left(1 + \dfrac{C_{je} + C_{jc}}{g_m \tau_F} \right) \tau_F} \tag{6.5.17}$$

Solving this equation for τ_F, we have

$$\tau_F = \frac{1}{2\pi f_T} - \frac{(C_{je} + C_{jc})}{g_m} \tag{6.5.18}$$

Thus, measurements of short-circuit current gain (i.e., current gain with the collector in an ac short-circuit connection) as a function of frequency provide a means to obtain the charge-control parameter τ_F. The parameter f_T can be obtained by extrapolating a plot of gain versus frequency to unity gain. A value for τ_F is then obtained by using f_T in Eq. (6.5.18). The low frequency gain β_F can then be used to calculate $\tau_{BF} = \beta_F \tau_F$ (Eq. 6.4.4).

If these measurements are repeated for the transistor under reverse-active bias, the parameters β_R and τ_{BR} can be obtained in a similar manner. Measurements of leakage current under active bias permit the calculation of I_{CS} and I_{ES} by using Eq. (5.4.11) and the corresponding relationship for reverse-active bias. Thus, all parameters for the basic transistor models can be extracted from a series of measurements when the equivalences between parameters of the various transistor models are considered.

6.6 BIPOLAR TRANSISTOR MODEL FOR COMPUTER SIMULATION[†]

For computer simulation of transistors, precision takes precedence over conceptual or computational simplicity. To maximize the usefulness of computer programs, models for transistors should be accurate for both large- and small-signal applications and they should also be readily characterized by parameters that are relatively easy to obtain and to verify. These requirements have been met most successfully thus far by simulations that are based on the Ebers-Moll equations, which were introduced in Section 5.4.

The starting point for our discussion is the so-called "transport version" of the Ebers-Moll equations. This consists of Eqs. (5.4.5) and (5.4.6), which specify transistor currents in terms of the linking current between the emitter and the collector and additional base-emitter and base-collector diode components.

In Eq. (5.1.9) we derived an expression for the linking current I_n which we repeat here

$$I_n = I_S \left[\exp\left(\frac{V_{BC}}{V_t}\right) - \exp\left(\frac{V_{BE}}{V_t}\right) \right] \tag{6.6.1}$$

This equation plus the forms derived in Eqs. (5.4.5) and (5.4.6) allow the Ebers-Moll equations to be written in the following form:

$$
\begin{aligned}
I_C &= -I_n - \frac{I_S}{\beta_R}\left[\exp\left(\frac{V_{BC}}{V_t}\right) - 1\right] \\
I_E &= I_n - \frac{I_S}{\beta_F}\left[\exp\left(\frac{V_{BE}}{V_t}\right) - 1\right] \\
I_B &= \frac{I_S}{\beta_F}\left[\exp\left(\frac{V_{BE}}{V_t}\right) - 1\right] + \frac{I_S}{\beta_R}\left[\exp\left(\frac{V_{BC}}{V_t}\right) - 1\right]
\end{aligned}
\tag{6.6.2}
$$

In this formulation, the three parameters I_S, β_F, and β_R suffice to characterize the basic Ebers-Moll relationships. Additional terms must be added to the equations in this set to represent effects not included in the Ebers-Moll model, for example, the phenomena described earlier in this chapter. Gummel and Poon[10] have shown relatively straightforward methods by which Eqs. (6.6.2) can be modified to incorporate three important second order effects: (1) recombination in the emitter-base space-charge zone at low emitter-base bias, (2) the current-gain decrease experienced under high current conditions, and (3) effects of space-charge-layer widening (Early effect) on the linking current between the emitter and the collector. The consequences of these second order effects lead to the deviations from ideal performance that are revealed in the sketches in Figs. 6.31a and 6.31b.

Recombination in the Space-Charge Regions. As we saw in Chapter 4, recombination in the space-charge zone leads to modified diode relationships for the junction currents. These can be modeled by adding four parameters to the Ebers-Moll model in order to define base current in terms of a superposition of ideal diode and nonideal diode components.

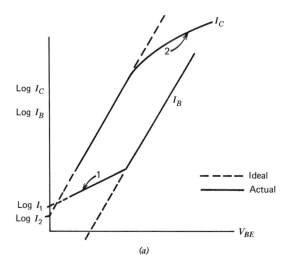

(a)

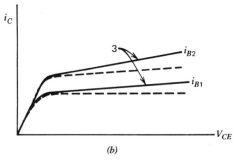

(b)

Figure 6.31   The results of second order effects on bipolar transistor characteristics in the active mode. The numbers on the figures refer to the effects enumerated in the text. The base current extrapolated to zero base-emitter voltage is I_1 in Eq. (6.6.3).

$$I_B = \frac{I_S}{\beta_F}\left[\exp\left(\frac{V_{BE}}{V_t}\right) - 1\right] + I_1\left[\exp\left(\frac{V_{BE}}{n_e V_t}\right) - 1\right]$$

$$+ \frac{I_S}{\beta_R}\left[\exp\left(\frac{V_{BC}}{V_t}\right) - 1\right] + I_2\left[\exp\left(\frac{V_{BC}}{n_c V_t}\right) - 1\right] \qquad (6.6.3)$$

The new parameters I_1, I_2, n_e, and n_c are found in practice by measurements made at low base-emitter biases. For example I_1 is obtained from the intercept of a plot of log I_B versus V_{BE} extrapolated to $V_{BE} = 0$ (Fig. 6.31).

Early Effect and High-Level Operation.   Both the high-current effect (2) and the Early effect (3) can be incorporated by modifying the value of I_S, the multiplier for the linking current between the emitter and the collector. In Section 5.1 it was shown that I_S depends inversely on the total base majority charge Q_B. For convenience we repeat here the two pertinent equations, Eqs. (5.1.10) and (5.1.8).

$$I_S = \frac{q^2 A_E^2 n_i^2 \tilde{D}_n}{Q_B} \qquad (5.1.10)$$

where

$$Q_B = q A_E \int_0^{x_B} p(x)\, dx \qquad (5.1.8)$$

In the Gummel-Poon model Q_B is represented by components having a bias dependence that can be calculated readily. First, there is the "built-in" base charge Q_{B0} where

$$Q_{B0} = q A_E \int_0^{x_B} N_a(x)\, dx \qquad (6.6.4)$$

In addition to this term there are emitter and collector charge-storage contributions (q_{VE} and q_{VC}) plus the charge associated with forward and reverse injection of base-minority carriers. These are all summed to represent Q_B by the equation:

$$Q_B = Q_{B0} + C_{je} V_{BE} + C_{jc} V_{BC} \frac{A_E}{A_C} + \frac{Q_{B0}}{Q_B} \tau_F I_S \left[\exp\left(\frac{V_{BE}}{V_t} \right) - 1 \right]$$

$$+ \frac{Q_{B0}}{Q_B} \tau_R I_S \left[\exp\left(\frac{V_{BC}}{V_t} \right) - 1 \right] \qquad (6.6.5)$$

By defining several parameters, Eq. (6.6.5) can be put into a more manageable format.

$$q_b \equiv \frac{Q_B}{Q_{B0}}; \qquad I_{KF} \equiv \frac{Q_{B0}}{\tau_F};$$

$$I_{KR} \equiv \frac{Q_{B0}}{\tau_R}; \qquad |V_A| \equiv \frac{Q_{B0}}{C_{jc}} \frac{A_C}{A_E}$$

$$|V_B| \equiv \frac{Q_{B0}}{C_{je}} \qquad (6.6.6)$$

The key variable, total base charge Q_B, is normalized in Eq. (6.6.6) to Q_{B0}, and its dimensionless counterpart is designated as q_b. The two charge-control time constants τ_F and τ_R together with Q_{B0} define "knee currents" I_{KF} and I_{KR} having a significance that will shortly become apparent. The definition of the Early voltage V_A and the equivalent Early voltage for reverse operation V_B is the same as was derived in Eq. (6.1.8).

In terms of the normalized parameters, Eq. (6.6.5) can be written in the following form.

$$q_b = q_1 + \frac{q_2}{q_b} \qquad (6.6.7)$$

where q_1 and q_2 are auxiliary variables as defined by

$$q_1 = 1 + \frac{V_{BC}}{|V_A|} + \frac{V_{BE}}{|V_B|}$$

$$q_2 = \frac{I_S}{I_{KF}}\left[\exp\left(\frac{V_{BE}}{V_t}\right) - 1\right] + \frac{I_S}{I_{KR}}\left[\exp\left(\frac{V_{BC}}{V_t}\right) - 1\right] \qquad (6.6.8)$$

These new variables provide convenient indication of the significance of the second order effects. If the Early effect is negligible, q_1 will approach unity. If high-injection effects are not important, q_2 will be small.

Thus, base-width modulation effects have been modeled through the introduction of the two Early voltages while high-level bias effects are specified through the knee currents I_{KF} and I_{KR}. The Gummel-Poon model thus requires the specification of three variables I_S, β_F, and β_R for the basic Ebers-Moll model and then adds four more, I_1, I_2, n_e, and n_c to model space-charge recombination effects. The Ebers-Moll parameters that are specified should be valid in the mid-bias range, and no high-level effects should be present.

Finally, base-width and majority-charge modulation are modeled by specifying a variable q_b that depends on the values of four additional variables I_{KF}, I_{KR}, V_A, and V_B. The overall model is thus specified by 11 parameters plus the temperature (to enable a calculation of V_t). The collected equations making up the model for an *npn* transistor are:

$$I_B = \frac{I_S}{\beta_F}\left[\exp\left(\frac{V_{BE}}{V_t}\right) - 1\right] + I_1\left[\exp\left(\frac{V_{BE}}{n_e V_t}\right) - 1\right]$$

$$+ \frac{I_S}{\beta_R}\left[\exp\left(\frac{V_{BC}}{V_t}\right) - 1\right] + I_2\left[\exp\left(\frac{V_{BC}}{n_c V_t}\right) - 1\right]$$

$$I_C = \frac{I_S[\exp(V_{BE}/V_t) - \exp(V_{BC}/V_t)]}{q_b} - \frac{I_S}{\beta_R}\left[\exp\left(\frac{V_{BC}}{V_t}\right) - 1\right]$$

$$- I_2\left[\exp\left(\frac{V_{BC}}{n_c V_t}\right) - 1\right] \qquad (6.6.9)$$

$$q_b = \frac{q_1}{2} + \frac{\sqrt{q_1^2 + 4q_2}}{2}$$

$$q_1 = 1 + \frac{V_{BE}}{|V_B|} + \frac{V_{BC}}{|V_A|}$$

$$q_2 = \frac{I_S}{I_{KF}}\left[\exp\left(\frac{V_{BE}}{V_t}\right) - 1\right] + \frac{I_S}{I_{KR}}\left[\exp\left(\frac{V_{BC}}{V_t}\right) - 1\right]$$

To illustrate the validity of this equation set we might consider low-level, active-mode operation for which $q_2 \simeq 0$ (because the collector current in Fig. 6.32 is much less than the knee current I_{KF}). For this case

$$I_C \simeq \frac{I_S \exp{(V_{BE}/V_t)}}{1 + V_{BC}/|V_A|} \simeq I_S \exp{\left(\frac{V_{BE}}{V_t}\right)}\left(1 - \frac{V_{BC}}{|V_A|}\right) \tag{6.6.10}$$

and

$$\frac{\partial I_C}{\partial V_{CB}} = \frac{I_C}{|V_A|} \tag{6.6.11}$$

as was derived from fundamental considerations in Eq. (6.1.3).

For operation at high current levels for which the Early effect is of far lesser consequence than are high-injection effects, we have $q_2 > q_1$. Under this condition, the normalized base charge q_b has a high-bias asymptotic behavior of the form

$$q_b = \sqrt{\frac{I_S}{I_{KF}}} \exp{\left(\frac{V_{BE}}{2V_t}\right)} \tag{6.6.12}$$

Thus, collector current will vary as (cf. Fig. 6.32)

$$I_C = \sqrt{I_S I_{KF}} \exp{\left(\frac{V_{BE}}{2V_t}\right)} \tag{6.6.13}$$

It is left as a problem to show that the intercept of the asymptotic behavior of the Gummel-Poon model at low bias with the high-bias asymptote (Eq. 6.6.13) occurs at the "knee" current $I_C = I_{KF}$. The physical basis for the form represented by Eq. (6.6.13) is that sufficiently high injection into the base region leads to a bias

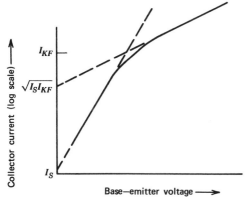

Figure 6.32 Logarithm of collector current versus V_{BE} in the active mode to illustrate the Gummel-Poon model for high-bias effects. The asymptotes for low and high bias intersect at the "knee" current $I_C = I_{KF}$.

dependence for the base majority-carrier concentration as was described in the discussion of Eq. (6.2.3).

6.7 DEVICES: *pnp* TRANSISTORS

There is a major advantage in circuit yield and economy if an integrated-circuit process is directed toward the production of only one type of transistor. Basically because of the higher mobility of electrons than of holes, IC processes are centered on the production of high quality *npn* transistors. If a *pnp* transistor is needed in a circuit, it is usually made without complicating the IC process. Two types of *pnp* transistors have been designed that meet this constraint: *substrate* and *lateral pnp transistors*.

Substrate *pnp* Transistors

One reason for the excellent performance of *npn* transistors is that the region in which transistor action takes place in these devices has been designed to be away from the surface and uniform over the broad area of the junction that is parallel to the surface plane. A *pnp* transistor with these features can be obtained using a standard planar process by making the *pnp* emitter from the p-type diffusion that is normally used for the *npn* base region. The structure of the *pnp* transistor is shown in cross section in Fig. 6.33.

The epitaxial region serves as the *pnp* base and the grown *np* junction at the interface between the epitaxial layer and the substrate is the collector junction. The collector region is therefore the substrate of the integrated circuit; hence, it is not isolated from other *pnp* transistors formed in the same way. For this reason,

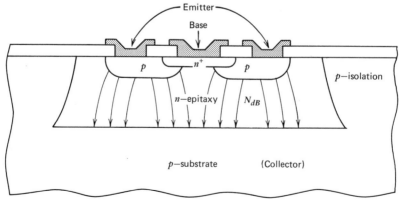

Figure 6.33   Cross section of a substrate *pnp* transistor. Flow lines for the linking current are sketched on the figure.

these so-called *substrate pnp transistors* can only be used in an integrated circuit when the collector junction is an ac ground as in an emitter-follower circuit. This is a valuable component for many integrated circuits, but it obviously cannot be used in every case for which a *pnp* transistor is desirable. Although substrate *pnp* transistors do not have the built-in base field resulting from a graded base that is found in double-diffused *npn* transistors, they can be made with values of β up to about 100 at 1 mA. Dependent on the process used, vertical *pnps* have been designed to operate in the range 1 μA to roughly 10 mA and to have values of f_T as high as 10 MHz.

Lateral *pnp* Transistors

There is a convenient way to make a *pnp* transistor with an isolated collector using standard processing. This is to employ the standard *npn* base *p*-type diffusion for both the emitter and the collector by spacing two *p*-regions close together as shown in the cross-sectional view in Fig. 6.34. The device formed by this construction is known as a *lateral pnp transistor* because transistor action takes place laterally— that is, parallel to the surface between the emitter and collector regions. This design sacrifices the advantages normally gained by removing transistor action from the surface region. As a result, the performance of lateral *pnp* transistors is markedly inferior to that of standard *npn* transistors. Nevertheless, lateral *pnp* transistors are frequently used in both analog and digital integrated circuits.

The two collector regions for the lateral *pnp* transistor shown in Fig. 6.34 are joined; typically, the collector completely surrounds the emitter region to improve current gain in the transistor. A buried layer is also usually included, as shown in Fig. 6.34. The buried layer improves transistor gain and frequency response in two

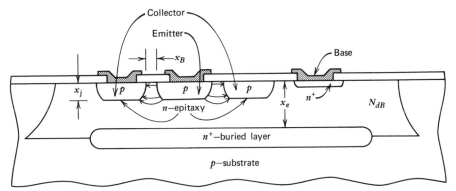

Figure 6.34 Cross section of a typical lateral *pnp* transistor for *IC* applications. The diffused collector region completely surrounds the emitter. Flow lines for the linking current are sketched on the figure.

ways: (1) by reducing base resistance and (2) by suppressing the collection of holes at the junction between the epitaxial layer and the substrate. One way of understanding the second improvement is to note that the built-in field that results from the doping gradient in the buried layer repels any holes that are incident from the epitaxial region. Alternatively, one can consider that the inclusion of a buried layer causes an increase in the base doping Q_B of the parasitic substrate *pnp* transistor. By Eq. (5.1.10), this reduces the loss of holes to the parasitic device and, thereby, improves the gain of the lateral *pnp*.

Collector Current. The current linking the emitter and the collector in a lateral *pnp* transistor follows a two-dimensional path as can be seen from the flow lines sketched on Fig. 6.34. Because the epitaxial region is uniformly doped, the boundary value for the injected hole density in the base at a given emitter-base bias will be uniform along the edge of the emitter-base junction regions. The density gradient that causes the injected holes to diffuse toward the collector will be a maximum near the surface where the spacing between the junctions is a minimum. If one moves away from the surface along the emitter-base junction, the gradient in the hole density decreases slowly as the distance between the diffused *p*-regions increases. Thus, the base width for the lateral *pnp* transistor is only approximated by the spacing between the junctions at the surface (x_B in Fig. 6.34), and the hole current is nonuniform along the emitter-base junction. The configuration of the flow lines for the linking current is also influenced by the thickness of the epitaxial layer and by the geometry of the buried layer.

An analysis of lateral *pnp* transistors[12] has considered these effects in detail and has given an empirical means of relating the linking current I_p to the flow in a one-dimensional transistor of base width x_B and emitting area $P_E x_j$ where P_E is the perimeter of the emitter and x_j is the depth of the diffused junction. The actual current is written

$$I_p = F \frac{q P_E x_j D_p n_i^2}{N_{dB} x_B} \exp\left(\frac{q V_{EB}}{kT}\right) \tag{6.7.1}$$

where F can be shown to be a function of the two dimensionless ratios: x_e/x_j and x_B/x_j.[12] As seen in Fig. 6.34, x_j is the depth of the emitter and collector diffusion, x_e is the thickness of the epitaxial region up to the edge of the buried layer, and x_B is the separation of the emitter and collector at the oxide-silicon interface. Curves showing $F(x_e/x_j, x_B/x_j)$ from reference 12 are given in Fig. 6.35. A typical value for both of these arguments might be 2, in which case F from Fig. 6.35 would be roughly 1.8. Thus, the total current is nearly doubled from that predicted by one-dimensional analysis.

The spacing x_B is determined by the separation of the emitter and collector windows in the photolithographic mask minus the lateral diffusions under the oxide toward one another by the acceptors forming the emitter and the collector.

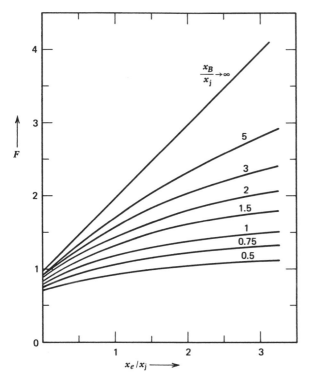

Figure 6.35 Geometry-dependent factor F relating actual collector current to that obtained from a one-dimensional model.[12]

The photolithographic spacing is typically limited to about 6 μm, and the sideways diffusion roughly equals the diffusion in the vertical direction. Thus, x_B is generally of the order of 4 μm although by diffusing deeper, one can attain spacings between 2 and 3 μm. By reducing x_B, one improves gain and frequency response, but suffers in reproducibility and reliability.

The epitaxial layer is typically lightly doped (between 10^{15} and 10^{16} donors cm^{-3}). Hence, only moderate forward bias at the emitter-base junction is required to bring the base into a high-level injection condition (Problem 6.9). Chou[12] has shown that high-level injection results in a fall-off of the gain of lateral *pnp* transistors because of three effects: (1) a lessened dependence of $p'(0)$—the excess hole density at the emitter-base junction—on emitter-base voltage, which can be accounted for by writing the analogous equation to Eq. (6.2.3) for holes, (2) an effective variation in the diffusion constant D_p because of the *Webster effect*, (Section 6.3), and (3) voltage drops in series with the applied emitter-base bias because of resistances in the base and the emitter.

Base Current. When the *npn* transistor was considered, it was only necessary to take account of three components of base current for an accurate device model. The three components were caused by injection of minority-carrier holes into the emitter (the dominant source for base current under most conditions of bias for the *npn* transistor), recombination of injected electrons in the base, and recombination of injected electrons in the emitter-base depletion region (important at low emitter-current levels). Currents analogous to these three components, in which the roles of electrons and holes are interchanged, are also important in lateral *pnp* transistors. Because of the presence of the oxide-silicon surface and the lateral geometry of the device, however, two other significant causes for base current are present in the lateral *pnp* transistor. These other components are caused by extra recombination at the oxide-silicon interface and in the neighborhood of the buried layer. An adequate representation of base current in the lateral *pnp* transistor can be obtained when these five current components are accounted for. The five components of base current enumerated above are indicated schematically in Fig. 6.36.

The three contributions to I_B that are analogous to the significant components of base current in *npn* transistors can be expressed by equations similar to those

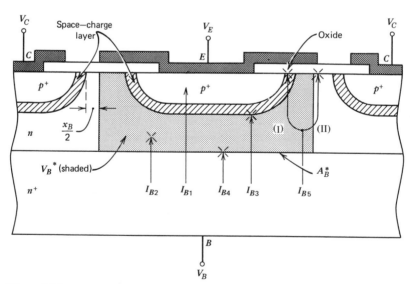

Figure 6.36 Schematic illustration of the physical origins of the base current in a lateral *pnp* transistor.[12] I_{B1} represents electron injection into the emitter, I_{B2} represents base recombination, I_{B3} represents space-charge zone recombination away from the surface, I_{B4} accounts for recombination at the buried layer and collection by the substrate, and I_{B5} represents surface recombination of holes.

already presented. An exception is the formulation of an analogous equation to Eq. (5.2.4) for base recombination. The two-dimensional flow pattern in the lateral *pnp* transistor complicates a proper definition for the volume of the base. Chou[12] has shown that base recombination can be written

$$I_{B2} = \frac{qn_i^2[\exp{(qV_{EB}/kT)} - 1]}{N_{dB}\tau_p} V_B^*$$ (6.7.2)

where τ_p is the lifetime of holes in the epitaxial region and V_B^* represents the volume defined in Fig. 6.36 by a surface that bisects the base width x_B.

It has proven convenient to account for the flow of holes to the buried layer (I_{B4}) by assigning a recombination velocity s_{nn^+} to the interface between the undoped epitaxial region and the buried layer. As introduced in Eq. (4.2.20) the recombination velocity is multiplied by the incident excess carrier density to express the total recombination rate at a surface. Assuming a relatively long lifetime in the epitaxial layer, one can use the boundary value for excess holes at the base-emitter junction to write

$$I_{B4} = qs_{nn^+} A_B^* \frac{n_i^2}{N_{dB}} \left[\exp\left(\frac{qV_{EB}}{kT}\right) - 1 \right]$$ (6.7.3)

where A_B^* is the area at the lower surface of V_B^* (Fig. 6.36). In the lateral *pnp* transistor, there is really no plane having the recombination velocity s_{nn^+}. Rather, the recombination velocity is an effective parameter that can be used to account for all the hole current into the buried layer. The incident holes may recombine at the interface, recombine within the buried layer, or else be collected across the junction between the buried layer and the substrate. Recombination at the oxide-silicon interface is also treated by defining a recombination velocity s_{os} and expressing current by equations similar to Eq. (6.7.3).

Once all the expressions for base current are written, the current itself can be calculated if appropriate values for s_{nn^+}, s_{os}, and the hole lifetimes in the bulk and space-charge zones can be obtained. In general, special test structures are needed to determine these parameters.[12] Experiments on lateral *pnp* transistors have shown that for typically low values of s_{os} ($s_{os} \sim 1$–5 cm s^{-1}), the recombination current at the oxide-silicon surface I_{B5} is generally negligible. All of the other components have significance in one or another range of useful emitter-base bias. There appears to be a fairly wide range for s_{nn^+} the recombination velocity at the buried layer. Values between 10 and 2000 cm s^{-1} have been reported. This variation apparently arises from differing dopant densities, buried-layer geometries and processing schedules. When s_{nn^+} becomes roughly 100 or greater, the vertical hole current generally becomes an important component.

The current gain β in lateral *pnp* transistors is substantially lower than that exhibited by *npn* transistors. Values of 20 or less are common although, with care, values of β up to about 100 have been achieved. In the low microampere

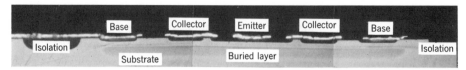

Figure 6.37   Stained angle-lap section of a lateral *pnp* transistor. (Courtesy Signetics Corporation.)

range, β is typically lower than one. It increases with current until I_C is roughly 1 mA (for an emitter area $\sim 10^{-5}$ cm^2). At higher currents, β decreases strongly as collector current increases. The increasing β at low biases corresponds to the lessening importance of recombination in the emitter-base space-charge region. The decrease in β at higher currents accompanies the onset of high injection effects.

Figure 6.37 is a composite photograph (enlarged ~ 1500 times) of a cross section through a lateral *pnp* transistor. It was made by using angle-lapping and staining techniques that delineate diffused regions in integrated circuits. Such photographs are of great value in assessing *IC* process control and in locating errors.

Integrated-Injection Logic.   The lateral *pnp* transistor is an important component in a family of very densely packed digital integrated circuits. This circuit family, usually called *integrated-injection logic* or I^2L has been described as "superintegrated" because it merges together *pnp* and *npn* transistors. The collector of one device functions simultaneously as the base of the other. Because this is the case, much of the surface area that is normally required for contacts, interconnections, and isolation diffusions in digital circuits can be saved.

An understanding of the ideas underlying I^2L can be gained by considering the basic I^2L gate shown in cross section in Fig. 6.38*a*. A circuit diagram that corresponds to the gate is shown in Fig. 6.38*b*. Note that the emitter of the lateral *pnp* transistor in the circuit is adjacent to the *p*-base of what appears in the cross-sectional view to be a conventional *npn* transistor. In I^2L, the *npn* transistor operates in the inverted mode, that is, with the uppermost n^+ region (normal emitter) functioning as a collector. When the *npn* transistor is connected in this way, many digital circuits can be crowded into a small surface area on the chip.

When bias is applied to the gate shown in Fig. 6.38, the holes delivered by the lateral *pnp* transistor to the base of the *npn* device cause the latter to saturate unless base current is drawn through electrode *B*. The gate is thus bistable, with the *npn* transistor either biased in cut-off or else in saturation depending upon the presence or absence of current through electrode *B*. The gates shown in Fig. 6.38 can be interconnected to obtain flip-flops or to carry out binary logic functions. A full analysis of the I^2L cell can be carried out by enumerating all of the physical origins for currents in the structure. The procedure is similar to that employed in our analysis of the lateral *pnp* transistor.

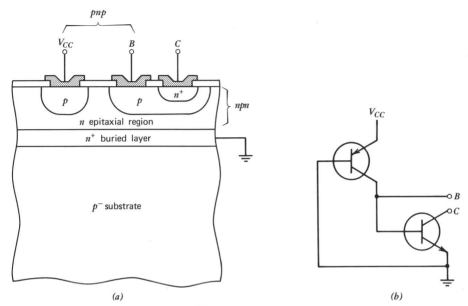

Figure 6.38   (*a*) Cross section of an $I^2\tilde{L}$ gate. (*b*) Equivalent circuit for the I^2L gate.

SUMMARY

The basic theory of transistor action introduced in Chapter 5 must be augmented by considering several important physical effects to obtain acceptable accuracy for many design situations. For transistors biased in the active mode, one such effect is the variation in collector current as a result of variation in collector-base bias. This phenomenon generally called the *Early effect* can be conveniently described by the introduction of an *Early voltage*, which indicates the output current variation at a given bias point. Other effects result in limits to the application of transistors at different ranges of bias. The operation of bipolar transistors at low current levels is generally limited by recombination within the base-emitter space-charge region. This recombination reduces the injection of minority carriers into the base and results in a decrease in current gain (β_F) as the quiescent bias in the devices is reduced to very low levels.

When transistors are biased at high current levels, several important effects can take place. One of these is a reduction in emitter efficiency when the base minority-carrier density begins to approach the dopant density at the edge of the base-emitter space-charge region. Another effect at high current levels is the modification of the space-charge configuration within the collector-base depletion region. When this condition is reached, the boundaries of the quasi-neutral base region are affected; in general, widening of the base region takes place. To analyze this so-called *Kirk effect*, it is necessary to consider in detail the doping profile of the transistor and to obtain a consistent set of solutions of Poisson's equation and the equations expressing mobile space charge in terms of bias currents. Base spreading resistance is another source for the complication of transistor performance at higher current levels. The effect of base spreading resistance is to reduce the bias of those parts of the base-emitter junction that are remote from the base contact because of the ohmic voltage drop associated with the delivery of base majority carriers to the active base region. In a given device all of the high-current effects may occur simultaneously at a given bias level. Understanding them in detail is more important in the modification of device design than it is in making use of the transistors in circuits. For circuit design, an empirical model of high level performance is usually all that is necessary.

An equation for the transport of injected carriers across the quasi-neutral base of a transistor (the *base transit time*) can be formulated in the general case of arbitrary base doping. An understanding of this equation and of the steps in its derivation is useful in developing the charge-control representation of the transistor. Consideration of the effect of high-level operation on the base transit time leads to an understanding of the *Webster effect* in which minority-carrier transit time across the base is reduced by the field associated with excess majority carriers introduced into the base.

The charge-control model of the transistor comprises a set of linear differential equations that provide an extremely useful description of the transistor for circuit-design purposes. The model is useful when considering the device from its terminals outward; that is, the charge-control model does not give accurate information about any of the distributed effects within the device. When considered on an incremental voltage basis, the charge-control model can be used to derive a small-signal equivalent circuit (*hybrid pi model*) that is especially helpful for the design of amplifying circuits.

Interrelationships between the parameters of the charge-control model, the Ebers-Moll model, and the hybrid pi model are useful both in gaining an understanding about the various models and in selecting experimental means to obtain model parameters. A model that successfully represents most of the significant physical effects in bipolar transistors and that is suitable for computer analysis of circuit behavior has been presented (*Gummel-Poon* model). In this model, several parameters are added to the basic Ebers-Moll transistor representation. Because of the processing requirements for integrated circuits, lateral *pnp* transistors—that is, transistors in which the separation between the emitter and collector is parallel to the surface—are often employed. The performance of lateral *pnp* transistors is frequently dominated by effects different from those that are significant for *npn* transistors. Lateral *pnp* transistors play an important role in the operation of *integrated-injection logic* circuits (I^2L).

REFERENCES

1. J. M. Early, *Proc. IRE*, *40*, 1401, (1952).
2. F. A. Lindholm, and D. J. Hamilton, *Proc. IEEE*, *59*, 1377 (1971).
3. C. T. Kirk, *IRE Trans. Electron Devices*, *ED-9*, 164 (1962).
4. H. C. Poon, H. K. Gummel, and D. L. Scharfetter, *IEEE Trans. Electron Devices*, *ED-16*, 455, (1969). Reprinted by permission.
5. O. Manck, H. H. Heimeier, and W. L. Engl, *IEEE Trans. Electron Devices*, *ED-21*, 403 (1974).
6. W. M. Webster, *Proc. IRE*, *42*, 914 (1954).
7. R. Beaufoy, and J. J. Sparkes, *Automat. Teleph. Elect. J.*, *13*, Reprint 112 (1957).
8. P. E. Gray, and C. L. Searle, *Electronic Principles: Physics, Models and Circuits*, Wiley., New York, 1969.
9. P. E. Gray, D. DeWitt, A. R. Boothroyd, and J. F. Gibbons, *Physical Electronics and Models*, SEEC Volume II, Wiley, New York, 1964.
10. H. K. Gummel, and H. C. Poon, *Bell Syst. Tec. J.*, *49*, 827 (1970).
11. R. N. Noyce, et al, *Electronics*, July 21, 1969, page 74.
12. S. Chou, *Solid-State Electron* 14, 811 (1971).
13. A. S. Grove, *Physics and Technology of Semiconductor Devices*, Wiley, New York, 1967, pp. 229, 240.

PROBLEMS

6.1   Show that the relationship for the Early voltage V_A that is derived in Eq. (6.1.3) properly specifies $\partial I_C / \partial V_{CB}$ for the prototype transistor by considering collector current to be carried purely by diffusion between the emitter and the collector.

6.2   Calculate the value of V_A at $V_{CB} = 0$ for a prototype transistor in which the base is doped to 10^{17} atoms cm^{-3} of boron and the collector is doped to 10^{16} atoms cm^{-3} of phosphorus if the neutral base width is 2.5 μm. Consider the junction to be a step between the two concentrations. What is the indicated slope of I_C owing to the Early effect?

6.3   Compare the Early voltages for the two transistors that are considered in Problem 5.3. Treat the base-collector junction as a step junction and take the collector doping to be 10^{15} cm^{-3} in both cases.

6.4[†]   Consider an *npn* transistor in which the base doping varies *linearly* across the quasi-neutral region, going from 10^{17} cm^{-3} at the emitter side to 10^{16} at the collector side. The base width is 1 μm and both emitter and collector have $N_d = 10^{19}$ cm^{-3}.
   (a) Sketch the minority-carrier densities (i) at thermal equilibrium, and (ii) under low-level, active-bias conditions.
   (b) Sketch the "built-in" electric field in the base.
   (c) Derive an expression for the field and give its maximum value.
   (d) Determine the approximate ratio between the two Early voltages (i) V_A that is encountered in forward-active bias and (ii) V_B that characterizes reverse-active bias.

6.5   Using Eq. (6.1.4), discuss qualitatively the dependence of V_A on collector-base bias for (a) the prototype transistor and (b) an *IC* amplifying transistor.

6.6   One criterion of the onset of current crowding is a drop in the transverse base voltage exceeding kT/q. Estimate the corresponding collector-current level for a transistor that has a β_F of 50, the impurity distribution shown in Fig. P6.6, and a stripe geometry with $Z_E = 0.1$ cm and $Y_E = 2 \times 10^{-3}$ cm (see Fig. 5.3). It can be shown that, for the stripe geometry, the base spreading resistance is

$$R_B = \frac{\bar{\rho}_B Y_E}{6 x_B Z_E}$$

The average resistivity of the base region $\bar{\rho}_B$ is given by

$$\bar{\rho}_B \simeq \frac{A_E x_B}{\mu_p Q_{B0}}$$

where μ_p is the mobility that corresponds to $N_a = Q_{B0}/q A_E x_B$.[13]

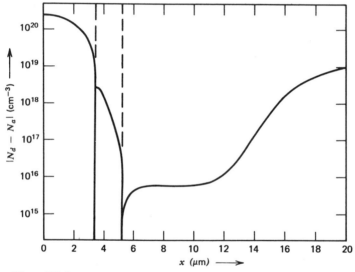

Figure P6.6

6.7 Consider a *pn* junction having the properties described in Problem 4.12. If the junction is to be used as the emitter-base of a *pnp* transistor, what is the maximum value of β_F that can be attained?

6.8[†] Reconsider Problems 5.17 and 5.18 in which β_F was to be found in a transistor and degradation due to radiation was considered. In the present case, assume that the transistor is to be used at a voltage such that $J_t/J_r = 10$ in Eq. (4.3.25).
(a) Compute β_F for this case.
(b) If lifetime in the space-charge zone varies in the same way as base minority lifetime resulting from the radiation damage, calculate the information required in Problem 5.18.

6.9 Derive an expression and make a plot of β_F/β_{FO} versus collector current for an *npn* transistor in which high-injection effects at the emitter edge of the base (as described in

Section 6.2) degrade current gain. The current gain β_{FO} is the value obtained in the middle bias range. Consider a prototype transistor with $A_E = 5 \times 10^{-5}$ cm^2, $N_a = 5 \times 10^{16}$ cm^{-3}, $D_n = 20$ cm^2 s^{-1}, and $x_B = 5$ μm. Consider that the base current is dominated by reverse injection into the emitter. Under this condition, base current will remain proportional to $\exp(V_{BE}/V_t)$. Thus, beta will fall off as $n(0)$ begins to depart from its middle-range dependence on V_{BE}.

6.10† Consider the extreme case of the Kirk effect (described in Section 6.2) in which the collector space-charge region is moved to the edge of the n^+ buried layer. Take the case that the negative space charge is completely due to electrons in transit and the positive space charge is provided by a very high donor concentration that starts abruptly at the edge of the buried layer. Calculate the field and space-charge-layer width as a function of current under the assumption that the electron velocity v is given by the following two cases: (a) $v = \mu\mathscr{E}$ and (b) $v = v_l$. These are two cases of space-charge-limited currents in solids.

6.11 (a) Show that R_B in Eq. (6.2.13) can be obtained from a plot of I_C versus measured V_{BE} such as in Fig. 6.10. In particular, if I_{CA} is the *actual* collector current and I_{CI} is the collector current that would flow in the absence of base resistance (the *ideal* current), show that

$$R_B = \frac{V_t \beta_F}{I_{CA}} \ln\left(\frac{I_{CI}}{I_{CA}}\right)$$

(b) Use the following data for the transistor of Fig. 6.10 to obtain a plot of R_B versus I_{CA}. $I_S = 3 \times 10^{-14}$ A, $V_t = 0.0252$ V, $\beta_F = 100$ (assumed constant).

V_{BE} (V)	I_{CA} (mA)
0.70	11.25
0.72	22.4
0.75	56.2
0.80	200

6.12† If the total resistance R in the transistor shown in Fig. 6.12 is 150 Ω (0.75 square at 200 Ω/square), and the external resistor is 20 Ω, use the network shown in the inset of the figure to investigate the crowding of current due to base resistance. Assume that each of the transistor segments has $\frac{1}{8}$ of the I_S given in Problem 6.11, but that each has $\beta_F = 100$.

(a) Assume currents of 1 and 10 mA in the innermost transistor segment. Work through the network to calculate the total current flowing in the overall transistor.

(b) Use the results of Problem 6.11 to calculate R_B and the total applied bias between the external base and emitter leads.

(c) Assuming that base resistance is the only important high current effect, what value does R_B approach at the highest current levels?

(d) Show that R_B approaches the sum of the external resistance plus $11R/128$ at very low currents.

(You will need to maintain about four significant figures for V_{BE} to derive accurate results; a calculator is highly desirable.)

6.13   Prove the statements made in the paragraph containing Eq. (6.3.3).

6.14[†] Use Eq. (6.3.8) to calculate τ_B for a transistor having a constant (built-in) base field resulting from an exponential variation in base doping. Specifically, take the case that the built-in voltage drop between $x = 0$ at the emitter side to $x = x_B$ at the collector side of the base is κV_t where V_t is the thermal voltage kT/q.
   (a) Show that $v = \kappa^2/(\kappa - 1 + e^{-\kappa})$ and that v behaves properly as $\kappa \to 0$.
   (b) Calculate τ_B for $\kappa = 20$, $x_B = 0.5\ \mu$m, $D_n = 20\ \text{cm}^2\ \text{s}^{-1}$, and explain why this value of κ is about as large as can be realized practically.

6.15[†] Consider the influence of the Kirk-effect results that are sketched in Fig. 6.8 on τ_B. Use Eq. (6.3.8) to make a semiquantitative plot of τ_B versus collector current. The indicated fall-off in high-frequency performance is an important consequence of the Kirk effect.[3]

6.16   Find an expression for τ_{BF} (as introduced in Eq. 6.4.3) in terms of transistor geometry and minority-carrier lifetime in the case of a prototype transistor for which the emitter efficiency is given by Eq. (5.2.20). Formulate the conditions required in order that τ_{BF} will become nearly equal to τ_n, the electron lifetime in the transistor base.

6.17   Show that the set of charge-control equations for a *pnp* transistor is of the form:

$$i_C = -\frac{q_F}{\tau_F} + \frac{dq_R}{dt} + q_R\left(\frac{1}{\tau_R} + \frac{1}{\tau_{BR}}\right) + \frac{dq_{VC}}{dt}$$

$$i_E = \frac{dq_F}{dt} + q_F\left(\frac{1}{\tau_F} + \frac{1}{\tau_{BF}}\right) - \frac{q_R}{\tau_R} + \frac{dq_{VE}}{dt}$$

$$i_B = -\frac{dq_F}{dt} - \frac{q_F}{\tau_{BF}} - \frac{dq_R}{dt} - \frac{q_R}{\tau_{BR}} - \frac{dq_{VE}}{dt} - \frac{dq_{VC}}{dt}$$

6.18   Consider the physical nature of the space charges represented by q_{VE} and q_{VC} to argue that the signs for the derivatives of these quantities are correct in Eq. (6.4.11), and in the equation set of the preceding problem (for a *pnp* transistor). *Hint: Consider the sign of the charges supplying the base current and the sign of the q_V terms. Reference to Fig. 6.14 may be helpful.*

6.19   Carry through the steps necessary to show the validity of Eq. (6.4.10).

6.20   Analyze the circuit shown in Fig. 6.18 [which corresponds to Eq. (6.4.11)] under dc conditions. Show that if one defines

$$\alpha_F = \frac{\tau_{BF}}{\tau_F + \tau_{BF}}$$

$$\alpha_R = \frac{\tau_{BR}}{\tau_R + \tau_{BR}}$$

and

$$I_{ES} = q_{FO}\left(\frac{1}{\tau_F} + \frac{1}{\tau_{BF}}\right)$$

$$I_{CS} = q_{RO}\left(\frac{1}{\tau_R} + \frac{1}{\tau_{BR}}\right)$$

Eqs. (6.4.11) will reduce to the Ebers–Moll equations (Eqs. 5.4.1).

6.21[†] Show that τ_{SLOW} in Eq. (6.4.15) can be expressed by

$$\tau_{SLOW} = \frac{(\beta_F + 1)\tau_{BR} + (\beta_R + 1)\tau_{BF}}{1 + \beta_F + \beta_R}$$

This form for τ_{SLOW} is frequently used in practice.

6.22 (a) Show that q_{FO} as defined in Eq. (6.4.2) is linearly related to $n_{po}(0)$ and write its value for a prototype transistor. *Hint: Consider Eqs.* (6.1.1) *and* (6.4.1)

(b) Show that V_{CE} for a saturated transistor is given by

$$V_{CE} = \frac{kT}{q} \ell n \left[\frac{q_{RO}(q_F + q_{FO})}{q_{FO}(q_R + q_{RO})} \right]$$

6.23 A transistor has the following charge-control parameters:

$$\tau_F = 12 \text{ ns}, \qquad \beta_F = 100, \qquad \tau_R = 36 \text{ ns}, \qquad \beta_R = 10$$

(a) Evaluate the forward stored charge q_F if the collector current $I_C = 2$ mA and the transistor operates just on the boundary of the saturation region with $V_{CB} = 0$.

(b) Determine the base-charge components q_F and q_R if the base current is now made $I_B = 0.5$ mA with I_C remaining at 2 mA.

(c) Compare the charge stored for cases (a) and (b).

6.24 For a calculated variation in charge with voltage to be modeled by a capacitance, not only must the variation be activated by the terminal voltage associated with the charge but also the sign of the charge variation must be consistent with that of a capacitor. Show that this condition is fulfilled when the variation in q_F [as calculated in Eq. (6.5.6)] is modeled in the capacitor ηC_D of Fig. 6.27.

6.25 Show by considering the signs of v_{BE} and the generator that the equivalent circuit shown in Fig. 6.27 is valid either for *npn* or for *pnp* transistors.

6.26 Discuss the limitation of the small-signal equivalent circuit to variations in base-emitter voltage that are less than V_t. What steps in the derivation of the circuit require this limitation?

6.27 Carry out the steps necessary to reduce the equivalent circuit of Fig. 6.25 to the form sketched in Fig. 6.27.

6.28 Verify Eq. (6.5.15).

6.29 An increment of emitter current dI_E is applied to a transistor under active bias with quiescent current I_E and base-emitter voltage V_{BE}. Assume that the simplified hybrid pi circuit of Fig. 6.30 (with C_{jc} negligibly small) is applicable. Show that a time

$$\tau_E \simeq \tau_F + (C_{je}V_t/I_C)$$

is required in order to bring the base-emitter voltage to a new steady-state value. This is one of the delays that affects the measured f_T in a transistor.

6.30 (a) Obtain the hybrid pi representations for the two transistors described in Problems 5.1 and 5.3. Take $I_C = 2$ mA and $\phi_i + V_{CB} = 10V$ for both transistors. In the uniform base transistor, take $\tau_n = 100$ ns.

(b) Comment on the relative performance of the two transistors when they are used as small-signal amplifiers.

6.31[†] Consider an *npn* transistor that is biased in the active mode and illuminated in the collector-base space-charge region. The radiation produces hole-electron pairs at a rate *r* pairs per unit time. Consider *r* to be a sinusoidal function of time.
(a) Briefly indicate the flow of the generated carriers.
(b) Indicate how the effects of radiation might be incorporated in the low frequency hybrid pi circuit (take the Early effect to be negligible).
(c) Use the circuit of part (b) to compute i_C as produced by the radiation if v_{BE} is zero (base-emitter ac shorted).
(d) Repeat part (c) if the base-emitter is ac open-circuited (i.e., $i_B = 0$).

6.32 Show that a symmetrical equivalent circuit diagram like that in Fig. P6.32 represents the "transport version" of the Ebers-Moll equations [Eqs. (5.4.5) and (5.4.6)], provided that we define

$$I_{AA} = I'_{ES}\left[\exp\left(\frac{qV_{BE}}{kT}\right) - 1 \right]$$

$$I_{BB} = I'_{CS}\left[\exp\left(\frac{qV_{BC}}{kT}\right) - 1 \right]$$

where I'_{ES} and I'_{CS} differ from I_{ES} and I_{CS} in the original *E-M* equations.

Advantages for this representation over that of the more conventional Ebers-Moll equivalent circuit (Fig. 5.12) are discussed by J. Logan (*Bell System Tec. J.*, April 1971).

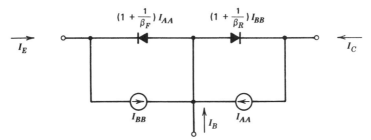

Figure P6.32

6.33[†] Derive Eqs. (6.6.2).

6.34[†] Derive the solution for q_b in Eq. (6.6.7).

6.35[†] Using the Gummel-Poon equations derived and discussed in Section 6.6 consider active-bias and low-level conditions to show that a decrease in $\beta_F(I_C/I_B)$ is predicted as current decreases. Find the variation in β_F as a function of I_C at low levels. Show that the parameter n_e can be obtained from the behavior noted and plot a reasonable sketch for β_F in a transistor having $\beta_F = 100$ at currents above 0.9 mA with a dropoff asymptote that intersects the midrange β_F at $I_C = 0.5$ mA. How might the curve of β_F be used to obtain a value for I_1?

6.36[†] (a) Show that the intercept of the two asymptotic forms expressing collector current as a function of base-emitter bias in the Gummel-Poon model occurs at the "knee" current $I_C = I_{KF}$ (Fig. 6.32).

(b) Show that the Gummel-Poon model predicts that in the high-current region, β_F becomes proportional to I_C^{-1}.

CHAPTER 7

PROPERTIES OF
THE OXIDE-SILICON SYSTEM

In our discussion of the electronics and technology of materials in Chapter 1, we noted that silicon had become overwhelmingly important as a semiconductor material because of some very special properties. The chief reason for its prominence among possibly competitive materials to be used for semiconductor devices is the ability to produce by compatible technologies both a semiconductor (single-crystal silicon) and an insulator (amorphous silicon oxide) that have superb electrical and mechanical properties. This ability has made planar technology possible and, in turn, the reliable production of large-scale integrated circuits. Thus, properties of the oxide-silicon system are fundamental to the performance of integrated-circuit devices. Knowledge of these properties and of their control has been responsible for many advances in device design and performance. Despite years of work in the area, research on the oxide-silicon system is still ongoing, and new applications are continually being found.

A useful starting point for the consideration of the oxide-silicon system is the construction of an energy-band diagram. The utility of the band diagram is

304

enhanced considerably, however, by adding to the figure a third component, a metal overlay above the oxide. The metal provides an electrode at which the voltage can be fixed, and the resultant three-component, metal-oxide-silicon (MOS) system is useful in understanding several very important integrated-circuit structures: most notably the metal-oxide-silicon field-effect transistor (MOSFET), which is often called* the insulated-gate field-effect transistor (IGFET).

The insulated-gate field-effect transistor is a device of such major importance that the entire succeeding chapter is devoted to it. The advent of dense, large-scale integrated circuits has made the IGFET roughly of equal importance to the bipolar transistor among the devices used in integrated circuits. It may well become the most significant integrated-circuit device because of certain attributes that we will consider later. The discussion of the MOS system in this chapter will help greatly to focus on the physical electronics that underlies the operation of the IGFET.

After we have incorporated the effects of oxide charge into MOS theory, we can deduce the electronic behavior of oxide-silicon systems without a metal overlay. Conditions at the interface between silicon and its oxide influence the properties of *pn* junctions. Since most *pn* junctions made by the planar process intersect the silicon surface at an oxide-silicon interface, the topics of this chapter are necessary additions to our discussion of the properties of *pn*-junction devices.

A direct application of MOS electronics has been in the fabrication of precise capacitors for integrated circuits. Arrays of MOS capacitors can be used to make charge-coupled devices (CCDs). The CCD was invented because of a detailed understanding of the oxide-silicon system. Although charge-coupled devices are still the subject of intensive research, they have already been applied to integrated circuits for signal processing, optical imaging, and computer memory.

Acceptor-doped silicon is considered throughout the chapter to establish a uniformity in the presentation of material. The oxide-silicon system with donor doping is the subject of several problems, and results for this system are included in an equation summary. In most cases, the modifications to be made upon changing doping type are clear.

Throughout the discussion that follows, we use the term silicon oxide instead of silicon dioxide, although complete oxidation of silicon results in SiO_2. In practice, the oxide encountered is almost entirely SiO_2 and the physical properties mentioned in the text refer to this material.

7.1 THE MOS STRUCTURE

To derive an energy-band diagram for the metal-oxide-silicon system, we apply the basic principles that we have used previously in studying systems of metals

---

* Strictly speaking, IGFET denotes a wider class of devices than those made from metal-oxide-silicon structures. This distinction is unimportant to us, and we will use the terms interchangeably.

and semiconductors, and of p-type and n-type silicon. The starting point is recognition that systems at thermal equilibrium are characterized by a constant Fermi energy. In this case of a three-component system, the Fermi level is constant throughout all three materials: the metal, the oxide, and the silicon.

Thermal-Equilibrium Energy-Band Diagram

For the present, we shall idealize the MOS system by considering that the interfaces between the materials consist of planes and are free of charges. The Fermi levels in the various materials become equalized by the transfer of negative charge from the materials with the higher Fermi levels (smaller work functions) across the interfaces to the materials with lower Fermi levels (greater work functions). As discussed in Chapter 2, the vacuum level is a continuous function of position and, hence, knowledge of the electron affinities of the insulator and semiconductor, and of the work functions for the semiconductor and metal permit the construction of a unique energy-band diagram. In Fig. 7.1, aluminum (work function = 4.1 eV), silicon oxide (electron affinity ~ 0.95 eV), and uniformly doped p-type silicon (electron affinity 4.15 eV) having a work function of 5.0 eV are considered.* The vacuum level is designated E_0 and the various energies when the materials are separated are indicated on the figure.

When the materials form a system at equilibrium, negative charge has been transferred from the aluminum into the silicon because the work function of the metal is 0.9 eV less than the work function of the silicon. The insulator, which is incapable of transferring charge (since it possesses essentially zero mobile charge), sustains a voltage drop because of the charge stored on either side of it. That charge, in turn, consists of a very thin sheet of positive charge (a plane in the ideal case of a perfect conductor) at the surface of the metal and of uncompensated acceptors extending into the semiconductor from its surface. The voltage corresponding to this energy difference therefore divides across the oxide and the space-charge zone at the surface of the silicon. It may be puzzling to consider charge transfer in the MOS system with reference to an oxide that is an almost perfect insulating material. In fact, if the system being considered were fabricated without any path for charge flow between the metal and the silicon other than the oxide, the materials could exist in a condition of nonequilibrium (i.e., with Fermi levels unequal) for very long periods. Nearly every MOS system of interest does, however, have some alternative path for the transfer of charge that is much more transmissive to charge flow than is the oxide. Hence, we can assume that there is thermal equilibrium between the metal and the semiconductor. Under these assumptions

---

* There is considerable variation in tabulated values for work functions and electron affinities. The values given here have provided accurate correspondence to theory in a number of MOS experiments.

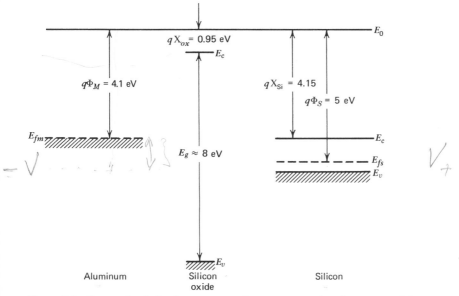

$qX_{ox}= 0.95$ eV

E_c

$q\Phi_M = 4.1$ eV

$qX_{Si} = 4.15$

$q\Phi_S = 5$ eV

E_o

E_{fm}

$E_g \approx 8$ eV

E_c

E_{fs}

E_v

E_v

Aluminum   Silicon   Silicon
                  oxide

Figure 7.1   Energy levels in three separated components that form an MOS system: aluminum, thermally grown silicon oxide, and p-type silicon containing $N_a \sim 2 \times 10^{15}$ cm^{-3}. (Although recent measurements indicate E_g in SiO$_2$ to be ~ 9 eV, many experimental data appear to be consistent with the 8 eV value that we will use.)

the band diagram for the MOS system formed with the materials of Fig. 7.1 is sketched in Fig. 7.2. The vacuum level, which plays no significant role in most device analysis, has been omitted from Fig. 7.2. In terms of charge flow, the diagram of Fig. 7.2 would result from the transfer of holes in the p-type silicon to an ohmic contact (not shown) where the holes are freely converted into electrons. These electrons are then supplied from the aluminum of the MOS system. The charge imbalance in the aluminum because of the transfer of these electrons leaves a sheet of positive charge at the metal surface near the oxide (as close as it can get to the equal quantity of negative charge stored on the silicon surface). The drop of 0.6 eV across the oxide shown in Fig. 7.2 is typical. The exact value of the energy difference between the metal and the surface of the semiconductor depends upon specific oxide properties that have not been given in this example.

The band diagram of Fig. 7.2, in which the p-type silicon is depleted of holes at its surface, can be compared to the diagram for gold and n-type silicon sketched in Fig. 2.5. The space charge at the surface of the silicon in both cases consists of ionized dopant atoms, and the energy bands in the silicon are curved in the region of the space charge with a spatial curvature that can be calculated by solving Poisson's equation. The presence of the oxide in the case of Fig. 7.2 acts to reduce

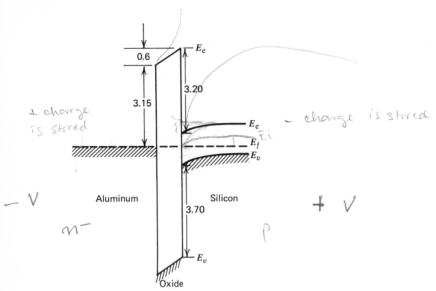

Figure 7.2 Energy-band diagram at thermal equilibrium for an MOS system composed of the materials indicated in Fig. 7.1. The oxide is assumed to be free of any charge.

the surface field by separating the surface charges, but there is otherwise no important difference in the band diagram within the silicon itself.

In terms of electrical characteristics, there are, of course, major differences in the two situations. In the MOS system, electrons cannot pass freely in either direction across the oxide. This distinction shows itself on the band diagram by the presence of abrupt steps in the energy of the allowed states for electrons in the MOS system. Referring to Fig. 7.2, we note that electrons in the metal that exist at the Fermi level are 3.15 eV lower in energy than the "conduction-band edge" in the silicon oxide.* Because of this separation of the allowed energies for free electrons, it is proper to characterize the metal-oxide interface by a 3.15 eV barrier for electron emission into the oxide. By the same reasoning, there exists a barrier of 3.20 eV at the silicon-oxide interface for electrons in the conduction band of the silicon and a barrier of 4.30 eV for electrons in the valence band of the silicon. The validity of these barrier heights has been confirmed directly by measurements of the photon energies required to emit electrons from the metal and semiconductor into silicon oxide.

---

* As mentioned previously, there are theoretical objections to describing an amorphous material like silicon oxide in terms of energy bands. Nonetheless, within the context of our present discussion, the concept is meaningful.

The Effect of Bias Voltage

We have seen that for an energy-band diagram of the idealized MOS system at thermal equilibrium, the metal and the semiconductor form two plates of a charged capacitor. The capacitor is charged to a voltage that represents the energy equivalent of the difference in the metal and semiconductor work functions. The application of a bias voltage between the metal and the silicon causes the system to depart from thermal equilibrium and to change the amount of charge that is stored on the capacitor. For the particular case considered in Fig. 7.2, if a negative voltage is applied to the metal with respect to the silicon, its effect is in opposition to the built-in voltage on the capacitor. It therefore tends to reduce the charge stored on its plates at equilibrium.

In a particular case of interest the applied voltage is set to a value that exactly compensates the difference in the work functions of the metal and the semiconductor. The stored charge on the MOS capacitor is then reduced to zero and the fields in the oxide and the semiconductor vanish. In this situation, the energy bands in the silicon are level or flat in the surface regions as well as in the bulk (Fig. 7.3). Because of the effect of the applied voltage on the band diagram, the voltage applied to achieve flat bands in the silicon is called the *flat-band voltage* and usually designated V_{FB}. The flat-band voltage varies with the dopant density in the silicon as well as with the specific metal used for the MOS system. Notice that the MOS system is not at thermal equilibrium under the flat-band condition; therefore, the

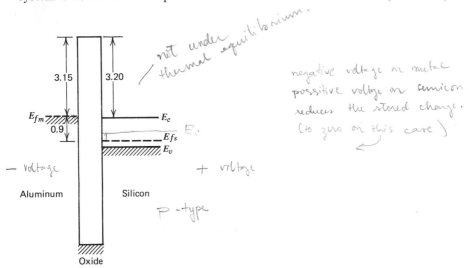

Figure 7.3 Energy-band diagram of the MOS system of Fig. 7.2 under flat-band conditions. A voltage equal to V_{FB} has been applied between the metal and the silicon to achieve this condition, which does not correspond to thermal equilibrium.

Fermi energy is different in the metal than in the silicon (Fig. 7.3). The voltage applied to the ideal MOS system to bring it to the flat-band condition equals the difference in the work functions of the metal and the silicon.

$$V'_{FB} = \Phi_M - \Phi_S \equiv \Phi_{MS} \tag{7.1.1}$$

(The result is written V'_{FB} in Eq. (7.1.1) because the true flat-band voltage V_{FB} is a function of additional parameters.)

Continuing with our consideration of the system of Fig. 7.2, if the silicon is held at ground and the voltage applied to the metal is kept negative but increased in magnitude above V'_{FB}, the MOS capacitor will begin to store positive charge at the silicon surface. This positive charge is made up of an increase in the hole population at the surface. The surface therefore has a greater density of holes than N_a, the acceptor density. This condition is called *surface accumulation*, and the zone at the surface containing the increased hole population is known as the *accumulation layer*.

The surface accumulation layer is a space-charge layer composed of free carriers. Hence, the solution of Poisson's equation within the accumulation layer is the same as the solution for a Schottky ohmic contact, which was described in Section 2.4. From consideration of Eq. (2.4.2) it was shown that half of the space charge of free carriers exists within $\sqrt{2}$ times the Debye length L_D of the surface, and the total extent of the accumulation layer is therefore a few times the value of L_D at the surface. To grasp the order-of-magnitude of these figures, consider that a p-type silicon wafer with $N_a = 10^{15} \text{ cm}^{-3}$ is accumulated so that the surface density is 10 times the bulk density. The Debye length at the surface for this case is about 400 Å [from Eq. (2.4.3)], roughly one third of the typical oxide thickness. Sketches of the energy-band diagram and of the charge configurations for the case of MOS surface accumulation are given in Figs. 7.4a and 7.4b, respectively.

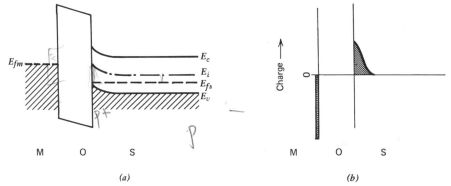

(a)   (b)

Figure 7.4 (a) Energy-band diagram of an MOS system with p-type silicon biased into accumulation, (b) charge in the same MOS system.

We have already seen that the MOS system of Fig. 7.2 with zero voltage applied between the metal and the silicon stores negative charge at the silicon surface and positive charge on the metal. This is consistent with a built-in positive voltage between the metal and the silicon. If this built-in voltage is aided by applying a positive voltage between the metal and the silicon, the silicon will become further depleted as more acceptors become exposed at its surface. Correspondingly, the positive charge on the metal increases. Because of the behavior of the silicon surface charge, this condition is called *surface depletion*. The energy-band diagram and the charge configuration for depletion-mode bias in the MOS system of Fig. 7.2 are sketched in Fig. 7.5. Under depletion-mode bias, the energy-band diagram behaves very similarly to the back-biased Schottky-barrier diode. Figure 7.5a can be compared to Fig. 2.6b to demonstrate this similarity (with due allowance for the fact that the silicon is *n*-type in Fig. 2.6 and *p*-type in Fig. 7.5).

If the voltage applied to the metal in the MOS system is continuously increased, however, behavior unlike that of a metal-semiconductor diode occurs. In the MOS case, as the metal voltage is increased, the field at the surface of the silicon increases and the energy bands bend considerably away from their levels in the bulk of the silicon. In the surface region, the majority carriers have been depleted and generation of carriers will therefore exceed recombination [Eq. (4.2.9)]. The generated hole-electron pairs are separated by the field, the holes being swept into the bulk and the electrons moving to the oxide-silicon interface where they are held because of the energy barrier between the conduction band in the silicon and that in the oxide. If the voltage at the metal is changed slowly enough, this generation process can bring the populations of free carriers at the silicon surface into localized equilibrium with the bulk of the silicon. In that case a constant Fermi level can be drawn from the bulk to the oxide interface and Fermi-Dirac statistics can be used to calculate the concentrations of carriers in the silicon. In fact, all of the energy-band diagrams drawn in this section have assumed this to be true. We shall have occasion later to consider alternative behavior. If the Fermi level remains constant

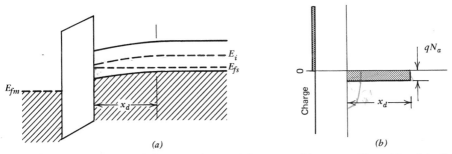

Figure 7.5 (a) Energy-band diagram of an MOS system with *p*-type silicon biased in the depletion mode, (b) charge in the same MOS system.

in the silicon while the bands bend with applied voltage, at sufficiently high voltages, the intrinsic Fermi level E_i at the silicon surface will cross the Fermi level.
The conduction-band edge at the oxide-silicon interface will then be closer to the
Fermi level than the valence-band edge is to the Fermi level. The applied voltage
has therefore created an *inversion layer*, so-called because the surface contains
more electrons than holes, even though the material has been doped with acceptor
impurities. The applied voltage between the metal and the silicon has induced a
pn junction near the surface. A sketch of the energy-band diagram for the MOS
system of Fig. 7.2 when biased to inversion is shown in Fig. 7.6a.

When E_i is just slightly below E_f at the surface, there is a low electron density
(order n_i) in the inversion layer, and the MOS system is said to be biased in the
weak inversion region. On the other hand, when $(E_c - E_f)$ at the surface is less than
$(E_f - E_v)$ in the bulk, the electron density in the inversion layer is greater than the
hole density in the bulk, and the system is in the *strong inversion* region. The
dividing point between the two cases is conveniently taken when the electron
density at the surface equals the acceptor concentration. While this division be-

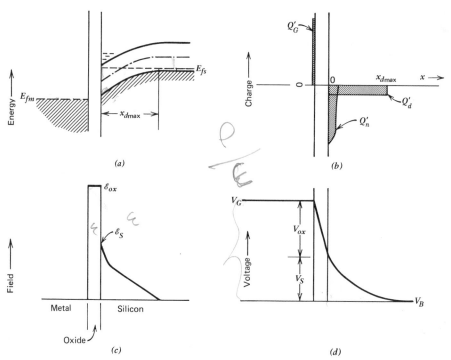

Figure 7.6   (*a*) Energy-band diagram, (*b*) space-charge configuration, (*c*) field, and (*d*) potential
distribution for an MOS system with *p*-type silicon biased into inversion.

tween weak and strong inversion is somewhat arbitrary, it is a convenient distinction for many purposes.

In the depletion and inversion regions of bias, the applied voltage induces negative charge in the semiconductor by repelling holes from the surface to create the depletion region and inducing electrons to form the inversion layer. Figure 7.6b shows the charge storage in the MOS system of Fig. 7.2 when it is biased into inversion. The free electrons in the inversion layer are very close to the surface because of the attractive force of the surface field. The acceptor charge is distributed throughout the depletion region, which extends away from the surface toward the bulk of the silicon. Once the silicon is biased into strong inversion, the free-electron population at the surface is nearly an exponential function of the potential at the surface so that there is little change in surface potential with increasing metal voltage. Thus, the total potential drop across the depletion region and the depletion-layer width in the silicon are both relatively constant when the surface is inverted. The maximum width of the depletion layer is usually denoted $x_{d \, max}$ as in Fig. 7.6b. The field and potential variations in the inverted MOS structure are shown in Figs. 7.6c and 7.6d, respectively. The field at the interface between the oxide and the silicon is discontinuous, falling from $\mathscr{E}_{ox}$ to $\mathscr{E}_s$ because of the change in the permittivity of the two materials. The total voltage across the structure $(V_G - V_B)$ is made up of a drop across the oxide and a drop across the space-charge region in the silicon. Equations for these two voltages are derived in Section 7.3.

This qualitative discussion of surface charge, voltage, and fields has served to introduce the important effect of the voltage applied to the overlying metal in determining the properties of the silicon surface. Although the system is basically only a capacitor, the various forms that the surface charge in the silicon can assume cause very significant differences in the electrical properties of the silicon surface. For example, the surface can be highly conducting and electrically connected to the bulk when it is accumulated; it can be highly insulating when it is depleted of free carriers, or else it can be highly conducting, but disconnected from the bulk, when it is inverted. These three conditions can be controlled by the bias applied between the metal and the bulk of the silicon. Because of this control, the metal layer is usually referred to as the *gate*, and the voltage at the metal is most often denoted by V_G.

Silicon Gate. For some device applications, it is very desirable to construct the gate from heavily doped silicon rather than from a metal (Problem 7.2). The silicon may be conveniently deposited over the gate oxide in an epitaxial reactor as discussed in Chapter 1. Because the silicon layer is stable at high temperatures, dopant atoms can be diffused into the substrate after the gate has been deposited and the whole system can be passivated with an overcoating oxide. The advantages of this technology become more apparent in the discussion of practical MOS transistor circuits in Chapter 8. Although the silicon may be deposited over the gate oxide in an epitaxial system, it is not a single crystal. Rather, it consists of a film of connected crystallites that are typically submicron in diameter. Because it is composed of many crystals, the layer is called *polycrystalline silicon*. Even though the gate is silicon, its electrical function in the MOS system is identical to that of the metal, and silicon-gated oxide systems are frequently described as MOS systems.

7.2 CAPACITANCE OF THE MOS SYSTEM

As with the Schottky barrier and the *pn* junction, analysis of the behavior of the small-signal capacitance measured between the two output electrodes provides very valuable insight into the electrical behavior of the MOS system. In the case of the MOS system, this analysis has been central to research that has resulted in the present well-developed state of understanding of the oxide-silicon system and its technology. The qualitative discussion of the previous section provides a valuable framework on which to build an understanding of the behavior of the small-signal capacitance as the applied gate voltage is changed.

Consider first that the MOS system is biased with a steady voltage that causes the silicon surface to be accumulated. For a *p*-type silicon sample such as that of Fig. 7.2, this corresponds to a negative applied voltage and would result in a charge configuration like that sketched in Fig. 7.4. The excess holes at the surface are pulled very close to the oxide. If a small ac voltage is superposed on the dc bias, it causes small variations in the charges stored on the metal gate and at the silicon surface. If the system is now connected to an instrument that measures the capacitance associated with these variations, the capacitance measured will be very close to that of the oxide itself because the spatial extent of the modulated charge in the silicon is small compared to the oxide thickness. The more the surface is accumulated, the thinner will be the accumulated layer (effectively, the Debye length at the surface will be reduced by the added carriers); hence, the capacitance will approach asymptotically the capacitance associated with the pure oxide. Thus, the capacitance per unit area* C' in accumulation approaches

$$C'_{ox} = \frac{\epsilon_{ox}}{x_{ox}}$$

accumulation and inversion at low frequency (7.2.1)

where x_{ox} is the oxide thickness. When the gate voltage is changed in the direction of its flat-band value, the surface accumulation decreases to zero and the capacitance decreases as the Debye length at the surface increases. To obtain an exact formulation for capacitance in this bias range, it is necessary to solve Poisson's equation under the condition that free electrons, free holes, and dopant atoms all contribute to the total space charge at the surface. This analysis was first carried

---

* Primes are used on symbols for capacitance and charge to indicate that they are on a per-unit-area basis.

out by Kingston and Neustadter[1] and their results have been widely applied to obtain various electrical properties of MOS systems. A particular result that is readily calculated from the Kingston and Neustadter theory is the capacitance per unit area when $V_G = V_{FB}$. If this quantity is called C'_{FB}, it can be expressed[2] by

$$C'_{FB} = \frac{1}{x_{ox}/\epsilon_{ox} + [kT/q^2\epsilon_s N_a]^{1/2}}$$

$$= \frac{1}{1/C'_{ox} + L_D/\epsilon_s} \qquad \text{Flat band -} \qquad (7.2.2)$$

where L_D, the extrinsic Debye length has been defined in Eq. (3.2.14).

   When the gate voltage becomes more positive than the flat-band voltage, holes are repelled from the surface of the silicon and the system is in the depletion-bias condition. Under this condition, relatively straightforward electrostatic analysis shows that the overall capacitance corresponds to the capacitance that is obtained by a series connection of the oxide capacitance and the capacitance across the surface depletion region (Problem 7.3):

$$C' = \frac{1}{1/C'_{ox} + x_d/\epsilon_s} \qquad \text{Depletion -} \qquad (7.2.3)$$
$$\text{bias condition \&}$$

where x_d is the width of the surface depletion layer, which depends upon gate bias as well as the doping and oxide properties. From Eq. (7.2.3) we see that the capacitance of the system decreases as the depletion region widens. *high frequency bias*

   When the gate bias is increased sufficiently to invert the surface, a new feature must be considered to describe the MOS capacitance behavior. We recall that the inversion layer at the MOS surface results from the generation of minority carriers. Hence, the population of the inversion layer can change only as fast as carriers can be generated within the depletion zone near the surface. This limitation causes the measured capacitance to be a function of the frequency of the ac signal that is used to measure the small-signal capacitance of the system.

   The simplest case arises when both the dc gate-bias voltage and the small-signal measuring voltage are changed very slowly so that the silicon can always approach equilibrium. In this case, the signal frequency is low enough so that the inversion-layer population can "follow" it. The capacitance of the MOS system is just that associated with charge storage on either side of the oxide; its value is therefore approximately C'_{ox}. Under this circumstance, a plot of measured capacitance versus gate bias follows the dashed curve marked "low frequency" in Fig. 7.7: going from C'_{ox} in the accumulation region of bias through a decreasing region as the surface traverses the depletion region and moving back up to C'_{ox} when the surface becomes inverted.

   The results of Problem 7.6 show that a characteristic time to form an inversion layer at the surface of an MOS system biased to inversion is of the order of

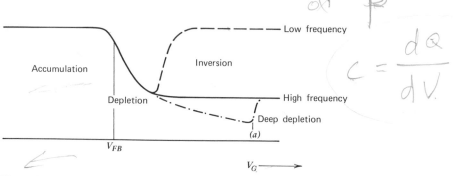

Figure 7.7 Small-signal capacitance of an MOS system with p-type silicon. Low-frequency behavior: both the bias voltage and the ac measuring signal vary slowly (less than ~ 10 Hz). High-frequency behavior: bias voltage varies slowly, but the ac measuring signal varies rapidly (typical circuits use 1 MHz); deep depletion variation: both the gate bias voltage and the ac measuring signal vary rapidly. The behavior at point (a) is described in the text.

$(2N_a\tau_0/n_i)$ where τ_0 is the minority-carrier lifetime at the surface.* For typical values of lifetime (1 μs) and dopant concentrations (10^{15} cm^{-3}), this time is roughly 0.2 s. Therefore, the small-signal measuring voltage must be changed very slowly to observe the low-frequency C-V curve (dashed curve of Fig. 7.7).

The presence of some means, such as surface illumination, that can stimulate the surface generation and recombination rates will increase the range of the low-frequency behavior. If the inversion layer extends to contact a source for electrons, the low-frequency curve can be observed at frequencies extending into the MHz range. This is the case because the electrons in the inversion layer can then be supplied and withdrawn rapidly through the ohmic connection.

When the ac measuring signal is changed rapidly while the dc bias voltage is varied slowly, the inversion layer cannot respond to the measuring signal. The number of charges in the silicon space-charge layer is modulated instead by the movement of holes at the far edge of the depletion region. The capacitance then corresponds to the series combination of the oxide capacitance and the depletion-region capacitance, as was true in depletion-mode bias. Since the depletion region reaches a maximum width $x_{d\,\text{max}}$ when the system goes into strong inversion, the measured capacitance approaches a value corresponding to the series connection of the oxide capacitance and to that associated with the maximum depletion-region width. It remains constant at this value as the bias voltage is increased further. This high-frequency C-V curve is shown by the solid line of Fig. 7.7.

---

* Generation from surface states is considered negligible. (See Section 7.5.) Note that τ_0 can vary considerably in practice.

A final capacitance behavior is sketched in Fig. 7.7. This is the curve shown dot-dash and marked *deep depletion*. It corresponds to the experimental situation in which both the gate bias voltage V_G and the small-signal measuring voltage vary at a faster rate than can be accommodated by generation-recombination in the surface depletion region. Since the inversion layer cannot form, the depletion region becomes wider than $x_{d\,max}$ and the name *deep depletion* is properly descriptive. The capacitance in this mode is given by Eq. (7.2.3) as was true in the normal depletion mode; here, however, x_d exceeds the inversion value $x_{d\,max}$ and C' does not reach a minimum. One means of generating a "deep depletion" capacitance curve is to sweep the gate bias V_G with a low-frequency triangular wave upon which is superposed the sinusoidal ac measuring frequency. The generation of carriers increases as the depletion layer is widened, however, and the deep-depletion curve is frequently observed to relax to the high-frequency curve at higher biases. The relaxation is indicated schematically at point (*a*) on the deep-depletion curve of Fig. 7.7.

7.3 MOS ELECTRONICS

The previous two sections have given a qualitative introduction to the charge induced at the silicon surface by the voltage applied between the overlying gate and the substrate. The particular application of this theory to the IGFET in the next chapter demands a more quantitative background, especially in the representation of conditions under inversion-mode bias. To provide this background, an approximate analysis based on the depletion approximation that helped to simplify the theories of Schottky barriers and *pn* junctions will be carried out. The analysis avoids an exact representation of the free-carrier densities in the inversion layer because solutions of the exact case cannot be written in explicit form. The complexity of the exact forms is not needed to develop a model of the MOS system that adequately fits our purposes. The complete solutions have been published[1,3] and are available for reference.

Thermal-Equilibrium Analysis

For the first level of analysis, it is advantageous to consider the silicon surface region to be at thermal equilibrium with the bulk. Although this assumption will be removed shortly, the results obtained for thermal equilibrium will be enlightening in the more general case. They will also be readily generalized for the nonequilibrium case. We define potential in the silicon as in Eq. (3.1.2):

$$\phi(x) = \frac{E_f - E_i(x)}{q} \tag{7.3.1}$$

Since we consider thermal equilibrium, E_f is constant, but E_i can vary with position. The potential ϕ_p in the neutral bulk of the silicon of Fig. 7.8 is thus negative because

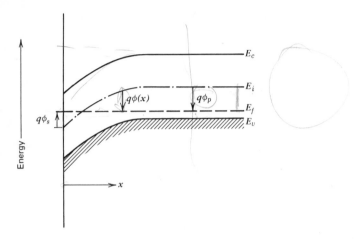

Figure 7.8 Energy-band diagram showing the potential as defined in Eq. (7.3.1) in the vicinity of the silicon surface in an MOS system. The case sketched corresponds to a positive value of surface potential (ϕ_s).

the material is p-type and E_f is less than E_i. At the surface, the potential ϕ_s is expressed as

$$\phi_s = \frac{E_f - E_i(0)}{q} \tag{7.3.2}$$

The carrier densities are related to $\phi(x)$ by Eqs. (1.1.26) and (1.1.27)

$$p = n_i \exp\left(-\frac{q\phi}{kT}\right)$$

$$n = n_i \exp\left(\frac{q\phi}{kT}\right) \tag{7.3.3}$$

From these equations and the definitions of ϕ_p and ϕ_s, we can express the surface free-carrier densities n_s and p_s in terms of the potential drop ($\phi_s - \phi_p$) across the depletion region at the silicon surface.

$$p_s = N_a \exp\left[\frac{q(\phi_p - \phi_s)}{kT}\right]$$

$$n_s = \frac{n_i^2}{N_a} \exp\left[\frac{q(\phi_s - \phi_p)}{kT}\right] \tag{7.3.4}$$

These equations can be used to give convenient "benchmarks" for the surface potential at various free-carrier densities. Table 7.1 lists these ϕ_s values for p-type

Table 7.1   **MOS Surface-Charge Conditions for *p*-type Silicon**

$(V_G - V_{FB})$	ϕ_s	Surface Charge Condition	Surface Carrier Density
Negative	Negative $\lvert\phi_s\rvert > \lvert\phi_p\rvert$	Accumulation	$p_s > N_a$
0	Negative $\phi_s = \phi_p$	Neutral	$p_s = N_a$
Positive (small)	Negative $\lvert\phi_s\rvert < \lvert\phi_p\rvert$	Depletion	$n_i < p_s < N_a$
Positive (larger)	0	Intrinsic	$p_s = n_s = n_i$
Positive (larger)	Positive $\lvert\phi_s\rvert < \lvert\phi_p\rvert$	Weak inversion	$n_i < n_s < N_a$
Positive (larger)	Positive $\phi_s = -\phi_p$	Onset of strong inversion	$n_s = N_a$
Positive (larger)	Positive $\lvert\phi_s\rvert > \lvert\phi_p\rvert$	Strong inversion	$n_s > N_a$

silicon together with corresponding gate-bias ranges and descriptions of the surface-charge conditions at each value of ϕ_s.

The strong inversion condition, the last entry in Table 7.1, holds the greatest importance for device applications of the MOS system. Once strong inversion is reached, the surface potential ϕ_s remains relatively constant at $-\phi_p$ because n_s is such a sensitive function of ϕ_s [cf. Eq. (7.3.4)]. For example, if n_s is increased from N_a to $10 \times N_a$, ϕ_s is only increased by 58 mV at room temperature (Problem 7.10). For uniform doping, we can relate the surface potential at the onset of strong inversion to the maximum width of the depletion layer through the use of the depletion approximation as was done for the Schottky barrier in Chapter 2 [Eq. (2.2.3)].

$$\phi_s = -\phi_p = \frac{q}{2\epsilon_s} N_a x_{d\,max}^2 + \phi_p \tag{7.3.5}$$

Equation (7.3.5) can be solved for the maximum width of the depletion layer.

$$x_{d\,max} = \sqrt{\frac{4\epsilon_s \lvert\phi_p\rvert}{qN_a}} \tag{7.3.6}$$

The total charge stored in the depletion layer (per unit area) is designated as Q_d' where

$$Q_d' = -qN_a x_{d\,max} = -\sqrt{4\epsilon_s qN_a \lvert\phi_p\rvert} \tag{7.3.7}$$

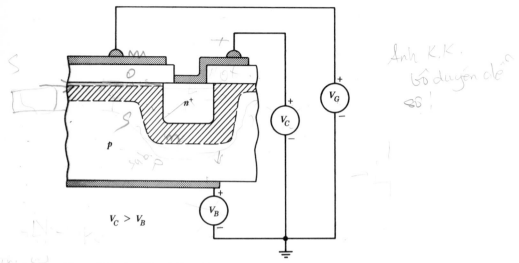

Figure 7.9   A diffused junction in the vicinity of an MOS capacitor can be used to bias the induced junction between the bulk of the silicon and an inversion layer formed at the oxide-silicon interface. The cross-hatching indicates the extent of the space-charge region in the depleted silicon.

Nonequilibrium Analysis

Once the MOS system has been biased into the inversion mode, a *pn* junction exists between the surface and the bulk of the silicon. If there is a nearby *n*-type diffused region that contacts the inverted surface as shown in Fig. 7.9, it is possible to apply a bias to the *pn* junction. Application of bias in this manner corresponds to a nonequilibrium condition within the silicon, and some current will flow between the inversion layer at the surface and the bulk. However, in practical applications of MOS systems, the junction will be reverse biased, and the currents will be small.

An energy-band diagram for the case of bias applied to an inverted surface is characterized by two *quasi-Fermi levels*,* one for the *p*-region and one for the *n*-region. Just as for the *pn* junction under reverse bias (Chapter 3), the two quasi-Fermi levels are separated by the applied reverse bias. This situation is sketched in Fig. 7.10.

---

* The concept of *quasi-Fermi levels* to be used under nonequilibrium conditions was introduced by Shockley[4] and has proven very useful in representing the densities of carriers that are nearly at thermal equilibrium (and hence whose densities can properly be calculated once the Fermi energy is specified). A useful discussion of quasi-Fermi levels and their proper application has been given by Blakemore[5].

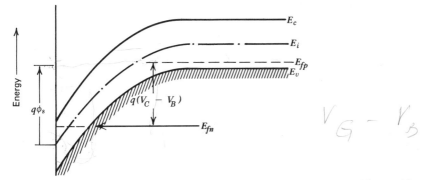

Figure 7.10   Energy-band diagram for an inverted surface on *p*-type silicon with a voltage $(V_C - V_B)$ applied between the inversion layer and the substrate.

An applied reverse bias between the induced surface *n*-region and the bulk increases the charge Q'_d in the depletion layer. Since the negative charge induced by $V_G - V_B$ is shared between the depletion and inversion layers, an increase of the charge in the depletion layer means that there is less charge available to form the inversion layer for a given gate voltage. Looked at another way, more gate voltage must be applied to induce the same number of electrons in the inversion layer when there is a reverse bias. With reverse bias present, the surface potential at the onset of strong inversion becomes $\phi_s = -\phi_p + (V_C - V_B)$ rather than $\phi_s = -\phi_p$, where $V_C - V_B$ is the reverse voltage applied between the inversion layer (or *channel*) and the bulk. Consequently, the change of the surface potential between flat-band and strong inversion is $2|\phi_p| + V_C - V_B$ rather than $2|\phi_p|$. The corresponding maximum depletion-region width $x_{d\,max}$ and the depletion-layer charge Q'_d (per unit area) become

$$x_{d\,max} = \sqrt{\frac{2\epsilon_s(2|\phi_p| + V_C - V_B)}{qN_a}} \tag{7.3.8}$$

and

$$Q'_d = -\sqrt{2\epsilon_s qN_a(2|\phi_p| + V_C - V_B)} \tag{7.3.9}$$

These two quantities are plotted in Fig. 7.11 as functions of the dopant concentration for several values of bias between the inversion layer and the bulk.

The previous discussion about charge in the MOS system showed that flat band $(V_G - V_B = V_{FB})$ corresponds to the condition of charge neutrality. Therefore, $(V_G - V_B) - V_{FB}$ is the effective voltage tending to charge the MOS capacitor. In this context, V_{FB} for the MOS system is analogous to ϕ_i, the built-in voltage for the *pn* junction; that is, both act like offsets of the zero level in equations relating stored charge to an applied bias. To express the charge in terms of applied voltage, we

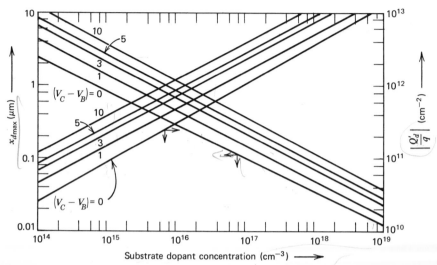

Figure 7.11 Maximum depletion-region width $x_{d\,max}$ and corresponding area density of charge in the depletion region Q'_d/q as functions of the substrate dopant concentration with the applied channel-to-substrate bias $(V_C - V_B)$ as a parameter. The curves are solutions to Eqs. (7.3.8) and (7.3.9).

carry out the following analysis. The charging voltage $[(V_G - V_B) - V_{FB}]$ is the sum of a drop V_{ox} across the oxide and a drop in the silicon $(\phi_s - \phi_p)$ (cf. Fig. 7.6d).

$$V_G - V_B - V_{FB} = V_{ox} + \phi_s - \phi_p \qquad (7.3.10)$$

The field in the insulating oxide is constant in the absence of any oxide charge. In terms of the applied voltage and oxide thickness, this field $\mathscr{E}_{ox}$ is

$$\mathscr{E}_{ox} = V_{ox}/x_{ox} = [(V_G - V_B - V_{FB}) - (\phi_s - \phi_p)]/x_{ox} \qquad (7.3.11)$$

Just inside the silicon and adjacent to the oxide (before any silicon charge is encountered), the normal displacement D will be constant and the field $\mathscr{E}_s$ will be

$$\mathscr{E}_s = \frac{\epsilon_{ox}\mathscr{E}_{ox}}{\epsilon_s} \qquad (7.3.12)$$

If Eq. (7.3.12) is combined with Eq. (7.3.11) and the definition for oxide capacitance (per unit area) $C'_{ox} = \epsilon_{ox}/x_{ox}$ [Eq. (7.2.1)] is used, we find

$$\epsilon_s\mathscr{E}_s = C'_{ox}[(V_G - V_B - V_{FB}) - (\phi_s - \phi_p)] \qquad (7.3.13)$$

Gauss' law states that the charge contained in a volume equals the permittivity times the electric field emanating from the volume. Applying Gauss' law to a volume extending from just inside the silicon at the oxide-silicon interface to the field-free bulk region, we can write

$$-\epsilon_s \mathcal{E}_s = Q'_s = Q'_n + Q'_d \qquad (7.3.14)$$

where Q'_s, the total charge induced in the semiconductor, is composed of the mobile electron charge Q'_n and the depletion-region charge Q'_d (all per unit area). These quantities are shown in Fig. 7.6b. Using Eq. (7.3.14) in Eq. (7.3.13), we can express the mobile charge Q'_n as

$$Q'_n = -C'_{ox}[(V_G - V_{FB} - V_B) - (\phi_s - \phi_p)] - Q'_d \qquad (7.3.15)$$

We insert the values of ϕ_s and Q'_d into Eq. (7.3.15) in order to relate the mobile charge Q'_n to the applied voltages. In strong inversion with an applied reverse bias between the channel and the bulk, $\phi_s = -\phi_p + (V_C - V_B)$, and Eq. (7.3.15) becomes

$$Q'_n = -C'_{ox}(V_G - V_{FB} - V_C - 2|\phi_p|) + \sqrt{2\epsilon_s q N_a(2|\phi_p| + V_C - V_B)} \qquad (7.3.16)$$

Equation (7.3.16) reduces to the simpler expression

$$Q'_n = -C'_{ox}(V_G - V_{FB} - V_B - 2|\phi_p|) + \sqrt{4\epsilon_s q N_a|\phi_p|} \qquad (7.3.17)$$

with no applied bias between the channel and the bulk. Note that the first terms on the right-hand side of either Eq. (7.3.16) or (7.3.17) are negative while the second terms in both equations are positive. The positive terms are, however, smaller in magnitude, (since $|Q'_s| > |Q'_d|$) so that the difference Q'_n is negative, as must be the case for the inversion layer on a p-type substrate.

From these equations, we can express directly the gate voltage necessary to induce a conducting channel at the surface of the semiconductor. This voltage, known as the threshold voltage V_T, is defined as the gate voltage that results in $Q'_n = 0$. From Eq. (7.3.16), an expression for V_T can be written as

$$V_T = V_{FB} + V_C + 2|\phi_p| + \frac{1}{C'_{ox}} \sqrt{2\epsilon_s q N_a(2|\phi_p| + V_C - V_B)} \qquad (7.3.18)$$

Taken one by one, the terms in Eq. (7.3.18) are readily seen to have the proper qualitative behavior. First, V_T contains V_{FB} because a gate voltage equal to V_{FB} is necessary to bring the silicon to a charge-neutral condition. (In the system of Fig. 7.2, V_{FB} is negative, tending to reduce the size of V_T.) Second, increasing the channel voltage V_C increases the gate voltage necessary to induce a given charge near the silicon surface. Third, $2|\phi_p|$ volts must be applied to cause the silicon energy bands to be bent to an inverted condition. Finally, the square-root term accounts for the uniform distribution of space charge in the depletion region. This term is inversely proportional to the oxide capacitance. It increases with $V_C - V_B$ to reflect the redistribution of semiconductor charge Q'_s from the inversion layer (where it contributes to Q'_n) into the depletion layer (where it forms part of Q'_d).

Table 7.2 **Formulas for the Oxide-Silicon System**

p-type substrate (n-channel)	n-type substrate (p-channel)
Flat-band voltage	

$$V_{FB} = \Phi_{MS} - \frac{Q'_{ss}}{C'_{ox}} - \frac{1}{C'_{ox}} \int_0^{x_{ox}} \frac{x}{x_{ox}} \rho(x)\, dx$$

p-type substrate (n-channel)	n-type substrate (p-channel)
Bulk potential	
$\phi_p = -\dfrac{kT}{q} \ln\left(\dfrac{N_a}{n_i}\right)$	$\phi_n = \dfrac{kT}{q} \ln\left(\dfrac{N_d}{n_i}\right)$
Surface potential for strong inversion Thermal equilibrium $\phi_s = |\phi_p|$ $\phi_s - \phi_p = 2|\phi_p|$	$\phi_s = -|\phi_n|$ $\phi_s - \phi_n = -2|\phi_n|$
With bias $(V_C - V_B) = V_{CB}$ $\phi_s = |\phi_p| + V_{CB}$	$\phi_s = -|\phi_n| - |V_{CB}|$
Maximum depletion width, $x_{d\,max}$ Thermal equilibrium $\sqrt{\dfrac{4\epsilon_s|\phi_p|}{qN_a}}$	$\sqrt{\dfrac{4\epsilon_s|\phi_n|}{qN_d}}$
With bias V_{CB} $\sqrt{\dfrac{2\epsilon_s(2|\phi_p| + V_{CB})}{qN_a}}$	$\sqrt{\dfrac{2\epsilon_s(2|\phi_n| + |V_{CB}|)}{qN_d}}$
Work-function difference, Φ_{MS} $\Phi_M - (X + E_g/2q + |\phi_p|)$	$\Phi_M - (X + E_g/2q - |\phi_n|)$
Threshold voltage V_T (arbitrary reference) $V_{FB} + V_C + 2|\phi_p|$ $+ \dfrac{1}{C'_{ox}} \sqrt{2\epsilon_s q N_a(2|\phi_p| + V_C - V_B)}$	$V_{FB} + V_C - 2|\phi_n|$ $- \dfrac{1}{C'_{ox}} \sqrt{2\epsilon_s q N_d(2|\phi_n| + V_B - V_C)}$

The charge in the inversion layer can be expressed very simply in terms of the difference between the applied gate voltage and the threshold voltage. From Eqs. (7.3.16) and (7.3.18)

$$Q'_n = -C'_{ox}(V_G - V_T) \tag{7.3.19}$$

Equation (7.3.19) should be used with caution for V_G approximately equal to V_T, because it is derived on the assumption that there are no electrons at the surface until the onset of strong inversion when the surface potential $\phi_s = -\phi_p + V_C - V_B$. This is, of course, an approximation that can be removed by considering the exact solutions of Poisson's equation for the MOS system.[1,3]

The more frequently used equations that we have derived for the MOS system have been collected in Table 7.2 for ready reference. Results for n-type substrates as well as p-type substrates are included.

7.4 OXIDE AND INTERFACE CHARGE

The theory presented thus far has neglected consideration of an important characteristic of the oxide-silicon system. This is the influence of charge within the oxide and at its surfaces. The presence of charges in the oxide and at the oxide-silicon interface is unavoidable in practical systems. To appreciate the importance of oxide charge, it is worthwhile to estimate the order of magnitude of the charge densities that have been considered in discussing the MOS system. Just above the threshold voltage where the system enters the inversion region, for example, the surface density of electrons Q'_n/q will be of the same order of magnitude as the density of dopant atoms (per unit area). For homogeneously distributed dopant atoms, the area density is $N_a^{2/3}$, or 10^{10} cm^{-2} if N_a is 10^{15} cm^{-3}. When this density is compared to the area density of silicon atoms (5×10^{22})$^{2/3}$ or 1.35×10^{15} cm^{-2}, it becomes clear that a surface-charge density that is only about 10^{-5} times as large as the atom density can cause the MOS system to depart from the ideal analysis. Fortunately, when they are formed carefully, interfaces between thermally grown, amorphous silicon oxide and single-crystal silicon can contain charge densities in the order of 10^{10} cm^{-2} and lower. This property of the oxide-silicon system is unique and is one of its outstanding attributes. We shall consider specific sources for oxide charge after we have calculated its influence on MOS properties.

Theoretical Analysis. Consider that a density of charge Q'_{ox} is located at the plane $x = x_1$ within the oxide as shown in Fig. 7.12a. The charges at x_1 will induce equal and opposite charges that are divided, in general, between the silicon and the metal gate. The closer is x_1 to x_{ox} the oxide-silicon interface, the greater will be the fraction of induced charge within the silicon. Because this induced charge changes the charge stored in the MOS system at thermal equilibrium, it correspondingly

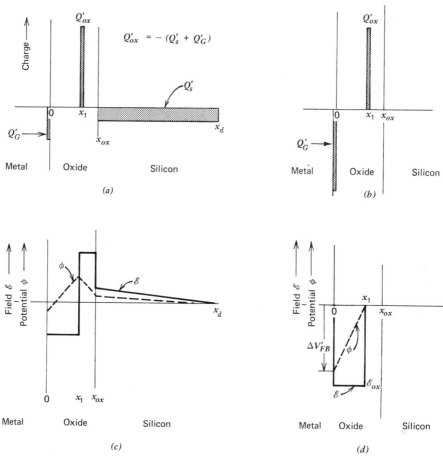

Figure 7.12 The effects of a fixed oxide-charge density Q'_{ox} on the MOS system (a) Charge configuration at zero bias: $Q'_{ox} = Q'_s + Q'_G$; (b) charge at flat band: $Q'_{ox} = Q'_G$; (c) field (solid line) and potential (dotted line) at zero bias, (d) field (solid line) and potential (dotted line) at flat band. The silicon bulk is taken as the reference for potential in the diagrams for (c) and (d).

alters the flat-band voltage from the value predicted in the ideal MOS analysis [Eq. (7.1.1)].

The size of the shift in the flat-band voltage is readily found by using Gauss' law to obtain the value of gate voltage that causes all of the oxide charge Q'_{ox} to be mirrored in the gate electrode—so that none is induced in the silicon. This condition is illustrated in Fig. 7.12b. Referring to the figure and employing Gauss' law, we note that the field is constant between the metal (at $x = 0$) and Q'_{ox} (at x_1) and zero between x_1 and the silicon surface (at $x = x_{ox}$). Its value between the gate and x_1, $\mathscr{E}_{ox}$ is

$$\mathscr{E}_{ox} = -\frac{Q'_{ox}}{\epsilon_{ox}} \qquad 0 < x < x_1 \qquad (7.4.1)$$

The gate voltage resulting from the presence of Q'_{ox} is the negative integral of $\mathscr{E}_{ox}$. Because it contributes to the flat-band voltage, we call it $\Delta V'_{FB}$ (i.e., the change in V'_{FB} from the ideal MOS case because of a planar sheet of oxide charge). The field and potential variations for the zero-bias case (Fig. 7.12a) are shown in Fig. 7.12c; those for flat band (Fig. 7.12b) are shown in Fig. 7.12d. An expression for $\Delta V'_{FB}$ is

$$\Delta V'_{FB} = x_1 \mathscr{E}_{ox} = -\frac{x_1 Q'_{ox}}{\epsilon_{ox}} \qquad (7.4.2)$$

Equation (7.4.2) can be rewritten by using Eq. (7.2.1) to express $\Delta V'_{FB}$ in terms of C'_{ox} the oxide capacitance per unit area

$$\Delta V'_{FB} = -\frac{Q'_{ox} x_1}{C'_{ox} x_{ox}} \qquad (7.4.3)$$

The maximum value for $\Delta V'_{FB}$ occurs expectedly when Q'_{ox} is situated at the oxide-silicon interface ($x_1 = x_{ox}$), while a charge situated adjacent to the gate ($x_1 = 0$) contributes nothing to the flat-band voltage.

The results given for the sheet of charge at $x = x_1$ can be generalized to account for the shift in flat-band voltage by an arbitrary distribution of charge $\rho(x)$ by superposing and integrating the increments that result from charges distributed throughout the oxide. The overall result is

$$\Delta V_{FB} = -\frac{1}{C'_{ox}} \int_0^{x_{ox}} \frac{x}{x_{ox}} \rho(x) \, dx \qquad (7.4.4)$$

Fixed charge at the oxide-silicon interface is frequently treated separately from charge that is incorporated in the oxide itself, even though surface charge can be accounted for in the formulation of Eq. (7.4.4). The interface charge density is often called Q_{ss}. Because it represents charge per unit area, however, we shall use the designation Q'_{ss}. The contribution of Q'_{ss} to the flat-band voltage is

$$\Delta V'_{FB} = -\frac{Q'_{ss}}{C'_{ox}} \qquad (7.4.5)$$

An expression for V_{FB} that includes the effects of differing work functions in the gate and in the silicon as well as the influence of oxide charge can be written by combining Eqs. (7.1.1), (7.4.4) and (7.4.5).

$$V_{FB} = \Phi_{MS} - \frac{Q'_{ss}}{C'_{ox}} - \frac{1}{C'_{ox}} \int_0^{x_{ox}} \frac{x}{x_{ox}} \rho(x) \, dx \qquad (7.4.6)$$

From Eq. (7.4.6), we see that the effect of oxide charge is to shift the flat-band voltage from its value in the ideal case. If the oxide charge is stable, the shift in flat band produces a corresponding shift in the threshold voltage V_T [Eq. (7.3.18)]. Experimentally, this threshold shift would cause the capacitance versus gate-voltage curves to be translated along the V_G axis. A typical result for a high-frequency capacitance curve is sketched in Fig. 7.13 (dashed curve).

In some cases, oxides and oxide-silicon interfaces can contain charge densities that are unstable and can be influenced by the applied voltage. In this case, the threshold voltage is itself dependent on the gate voltage. The capacitance versus voltage curve will then be distorted as in the dotted curve in Fig. 7.13. To understand this behavior as well as the influence of fixed charge, it is worthwhile to consider the physical sources of oxide charge.

Origins for Oxide Charge

A discussion of various sources for extra allowed energy levels at a silicon surface was given in Section 2.5 in conjunction with our consideration of metal-semicon-ductor contacts. That section, which would be profitably reviewed at this time, pointed out that even clean surfaces can be expected to have extra allowed energy levels at energies different from those in the bulk. Also, the inevitable presence of impurities incorporated during device processing adds others. Electrons in these extra levels and ions associated with them both contribute to interface charge.

To see the origin for voltage-dependent charge at the interface, such as that which leads to the distorted C-V curve shown dotted in Fig. 7.13, consider an oxide-silicon interface characterized by allowed levels at the energy E_s shown in Fig. 7.14. If the gate voltage causes the Fermi level at the surface to cross E_s as the MOS

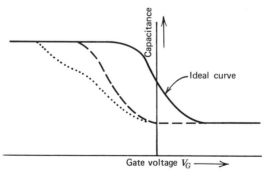

Figure 7.13 Fixed charge in the oxide causes the capacitance-voltage curve to translate along the V_G axis without distortion (dashed curve); charge that is influenced by the gate voltage causes distortion (dotted curve).

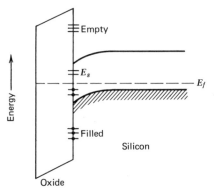

Figure 7.14   Energy levels at the oxide-silicon interface.

system is biased, then the charge state of these levels will change. This causes the second term in Eq. (7.4.6) to be voltage dependent and leads to the distorted curve. As was noted in Section 2.5, energy levels that can be charged readily from the silicon are associated with so-called *fast surface states*. The presence of fast surface states in densities that approach normal Q'_n/q values, (i.e., on the order of 10^{10} cm^{-2} or higher) is generally unacceptable for reliable device design. Modern MOS technology has successfully reduced these states to tolerable levels, although higher densities seemed unavoidable for many years and were a major cause for the lagging development of MOS devices. The density of fast surface states is typically reduced by baking the oxidized silicon wafer in hydrogen or forming gas (a hydrogen-nitrogen mixture).

Fixed charge in the oxide can be tolerated unless its density becomes so high that threshold voltages cannot be reached without applying fields that lead to breakdown within the oxide or at the silicon surface. Part of the fixed charge at the interface [Q'_{ss} in Eq. (7.4.5)] arises from uncompleted silicon-to-silicon bonds and is thus sensitive to the orientation of the silicon wafer. Surface charge Q'_{ss} from this source is higher for $\langle 111 \rangle$-oriented silicon than for $\langle 100 \rangle$-oriented silicon because more bonds are unfilled in the transition from silicon to silicon oxide in the $\langle 111 \rangle$ direction. The fixed oxide charge is, however, also dependent on the nature of the high temperature processing that the silicon undergoes, especially near the end of the fabrication process. Some of the unfilled bonds can generally be completed by annealing.

A troublesome charge in the MOS system arises from mobile ions that may be present in the oxide. The alkali metals, particularly sodium have been especially difficult to control. Sodium is a widely distributed impurity in many metals and laboratory chemicals and is easily transmitted by human contact. Equation (7.4.4) shows that the distribution of the charges, as well as their total concentration, influences ΔV_{FB}. The alkali ions have sufficient mobility to drift in the oxide when relatively low voltages are applied. Their mobility increases with temperature and the problem of flat-band instability is likewise magnified as temperature is raised. Since the metal ions are positively charged, negative V_G causes the ions to migrate to the metal-oxide interface where they do not affect flat-band voltage. Positive voltage, however, can move the ions to the silicon interface where their effect is maximal. Consequently, the characteristics of an MOS structure with mobile ions in the oxide are unstable. For threshold voltage stability of about 0.1 V, less than 2×10^{10} cm^{-2} mobile ions can be tolerated in the oxide. This source of oxide charge was another vexing problem in the development of practical MOS systems. It is avoided by careful processing and by the introduction of impurities that immobilize the alkali ions. Chlorine and its compounds, particularly HCl have been introduced into the thermal oxidation process for silicon and these have led to the production of very stable MOS systems. Another technique is to add phosphorus to the oxide.

Charge can also be introduced into the MOS system by irradiating it. The radiation may consist of energetic electrons incident on the device during its fabrication or of high energy particles or photons encountered during operation (as in a space environment). Both fixed charge in the oxide and fast surface states can be increased by radiation.

As an example, high energy photons (those with energies greater than the approximately 8 or 9 eV band gap of silicon oxide) can generate electron-hole pairs in the oxide, just as light with energy greater than the band gap of silicon can create electron-hole pairs in the silicon. After electron-hole pairs are generated in the oxide, however, their behavior differs from that of carriers that are generated in a semiconductor. Since the oxide is thin and contains so few free carriers, the probability of recombination is small. Instead, most of the electrons are swept out of the oxide by any field that may be across it. There are typically a large number of hole traps within the oxide, however, and many holes are therefore immobilized and increase the positive charge in the oxide. This shifts the flat-band voltage as predicted by Eq. (7.4.4).

As can be seen in Fig. 7.2, electrons can be excited into the oxide by photons of appreciably lower energy than the silicon-oxide band gap. These electrons may be photoemitted from the metal if the photon energy exceeds the energy barrier (3.15 eV in the case of aluminum) or 3.2 eV if the emission is from the conduction band of the silicon. In practice, appreciable emission from the silicon only occurs when photon energies are sufficient to excite electrons from the silicon valence band (which requires photons having 4.30 eV of energy). If an oxide has been charged by trapped holes, that charge can be reduced by causing the photoemission of electrons from either the metal or the silicon. Some of the photoemitted electrons will recombine with the trapped holes, thereby reducing the positive oxide charge.

As was described in Section 3.4, energetic electrons are created in silicon when an avalanching field is present. Thus, avalanche within the silicon provides another means for electrons to gain sufficient energy to surmount the barrier at the oxide interface. This mechanism is especially significant for device applications because it is under electrical control. In Chapter 8, more specific device effects are described. Within the framework of the present discussion, avalanche near the silicon surface can simply be regarded as an alternative to photoemission for providing a supply of energetic electrons that can enter the oxide.

7.5 SURFACE EFFECTS ON *pn* JUNCTIONS[†]

Several important device effects can occur when a *pn* junction exists in the vicinity of an oxide that is covered by an overlying gate (as in the structure sketched in Fig. 7.9). We have already seen in the discussion of Fig. 7.9 (in Section 7.3) that the bias applied to the junction can modify the charge Q'_n stored in the channel and the charge Q'_d stored in the depleted regions of an inverted MOS system. This is an effect of the junction on the MOS system. There is also a strong need for the *IC* designer to understand the effect of the MOS system on the properties of the junction. Most of the *pn* junctions in devices that are made by the planar process intersect the oxide-silicon interface. The only exceptions are those formed to buried layers. Thus, the properties of the oxide-silicon system can exert a significant influence on the circuit performance of *pn*-junction devices such as bipolar junction transistors.

Figure 7.15 shows the cross section of a planar $n^+ - p$ integrated-circuit diode. If this diode were, for example, situated above a second *pn* junction, it might represent the emitter and base regions of a bipolar transistor. In Section 4.3, it was shown that the generation of electron-hole pairs in the depletion region of a *pn* junction is generally the dominant source for the reverse leakage current of the junction. Likewise, at low forward bias, recombination in the space-charge region is the major current component. Generation and recombination in the space-charge region not only causes departures from the ideal diode law but, more seriously, as discussed in Chapter 5, these processes degrade transistor performance by adding components of base current that are not delivered to the collector. Some special features of these processes in surface space-charge regions deserve our attention.

To be more specific, consider section *A-A'* through the diode in Fig. 7.15*a*. The ideal *pn*-junction analysis considered the oxide-silicon surface as being free of charge, which would correspond to a neutral oxide-silicon surface, that is, one characterized by flat bands as sketched in Fig. 7.15*b*. The discussion of the MOS system, however, has made clear that the flat-band condition does not correspond to thermal equilibrium. It must usually be imposed by applying a gate-to-substrate bias equal to V_{FB}. If there is no electrode over section *A-A'*, then the most significant

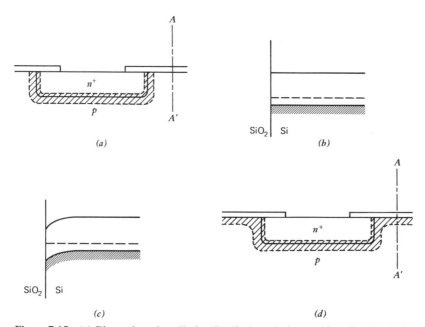

Figure 7.15 (*a*) Planar junction diode. (*b*) Ideal analysis considers flat band along section *AA'*. (*c*) Real oxide-to-*p*-silicon interface is usually depleted because of positive oxide charge. (*d*) Surface depletion zone is appended to junction depletion region.

influence on the surface is that of oxide charges. These are almost always positive causing either depletion or even inversion of *p*-type silicon (Fig. 7.15*c*). Conversely, positive charge in the oxide tends to cause accumulation in *n*-type silicon. (The consequence of these effects on the reliable production of bipolar transistors is considered in Problem 7.13.) Returning to consideration of the *p*-region in Fig. 7.15, we see that surface depletion in the vicinity of section *A-A'* will enlarge the overall junction depletion zone because it connects to the zone at the diffused junction. If the surface is inverted, there is an extension of the *n*-region along the oxide surface as well as an enlargement of the depletion region.

One effect of the enlarged depletion zone is to provide more volume for the generation of current under reverse bias. Of greater significance, however, is the fact that the oxide-silicon surface will generally contain generation-recombination sites in greater density than in the bulk. Furthermore, the activity of these sites is dependent on the surface potential at the oxide-silicon surface as was shown in the Shockley, Hall, Read (*SHR*) theory of recombination (Section 4.2).

Gated-Diode Structure. To consider the effect of the surface potential on the *pn* junction, we introduce the concept of a *gated diode*.[6] This is the structure shown in

Fig. 7.16*a* in which a metal gate overlays both the *p* and *n* regions of a diode. To focus attention on the physical mechanisms in the gated diode, we consider that the bulk of the silicon is at ground potential ($V_B = 0$), and we take the *n*-region to be biased positively to a voltage V_R (reverse bias on the junction).

At gate voltages V_G that are negative with respect to flat band, the surface of the *p*-type region is accumulated, and the depletion region at the surface is small. The depletion region near the surface in the *n*-type region is hardly changed by the gate voltage because it is heavily doped. Thus, for negative V_G, the leakage current of the diode is approximately determined by generation of holes and electrons in the fabricated or "metallurgical" junction region. From Eq. (4.3.26), this current, which we denote by I_M, is given by

$$I_M = \frac{qn_i}{2\tau_0} x_i A_M \tag{7.5.1}$$

where τ_0 is the lifetime, A_M is the area of the metallurgical junction, and x_i is the active region for generation as described in Section 4.3. In practice, x_i can be taken to be equal to the bulk depletion-layer width x_{db} and, hence, to vary with V_R in the same way as does x_{db}. The current is therefore nearly insensitive to V_G for this bias condition.

If V_G is made more positive than the flat-band voltage, the surface of the *p*-region under the gate electrode will be depleted as shown in Fig. 7.16*b*. Two components of current will now flow in addition to I_M as given in Eq. (7.5.1). First, there will be a current because of the generation of carriers in the depletion region induced by the gate. Again from Eq. (4.3.26), this component, which we call I_F, (arising from the field-induced junction) will be given by

$$I_F = \frac{qn_i}{2\tau_0} x_{ds} A_F \tag{7.5.2}$$

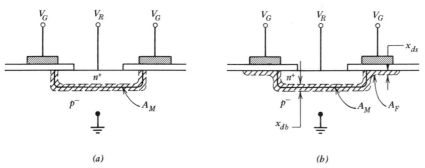

(a)   (b)

Figure 7.16 Gated-diode structure showing depletion regions when the gate voltage is V_{FB} for the *p*-region and (*b*) when the gate causes a depletion region of width x_{ds} at the silicon surface.

where x_{ds}, the width of the depletion region at the surface, is a function of the applied gate voltage V_G; and A_F, the area of the surface depletion region, is determined by the coverage of the gate electrode over the p-region. The second component of current that arises when the p-region becomes depleted results from the activity of surface generation sites. This component, which we call I_S, is best described in terms of a parameter called the surface recombination velocity s that we introduced in Section 4.2. There, it was shown that the surface recombination velocity is directly proportional to N_{st} the density of generation-recombination sites at the surface [Eq. (4.2.20)]. If we consider generation-recombination sites that have energies near E_i the mid-gap energy, then from Eq. (4.2.17) we can write that I_S is equal to the electronic charge times the generation rate or

$$I_S = \frac{q n_i s_o A_F}{2} \tag{7.5.3}$$

where s_o has been defined in Eq. (4.2.21), $s_o = N_{st} v_{th} \sigma$, v_{th} is the thermal velocity, and σ is the capture cross section associated with the generation-recombination site. The value of s_o is directly proportional to the number of surface generation-recombination centers—and it is therefore strongly influenced by device processing and annealing procedures.

If the gate voltage is increased until the surface of the p-type silicon becomes inverted, I_F in Eq. (7.5.2) rises to a maximum when x_{ds} increases to $x_{d\,max}$. Once inversion occurs, however, the surface electron density n_s becomes much greater than the intrinsic density n_i, and the surface recombination velocity decreases markedly from s_o following the prediction of Eq. (4.2.20). For typical values (Problem 7.16) I_F is smaller than I_S when the surface is depleted; after inversion I_S, the surface generation component, is negligible and the reverse-bias current becomes the sum of I_M and I_F. Typical measured behavior for reverse leakage current in a gated diode is sketched in Fig. 7.17. When the pn junction is under

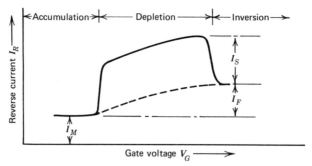

Figure 7.17 Reverse current in the gated diode as a function of V_G, showing the marked increase in leakage when the surface is depleted. The currents I_S and I_F are formulated in Eqs. (7.5.2) and (7.5.3).

forward bias at low voltages, the surface space-charge region also affects the recombination current, but we shall not discuss that condition in detail here.

In calculating the behavior of the surface-charge layer, the dependence of the threshold voltage on the reverse bias across the junction must be considered. Quantitatively, the presence of a reverse bias on the junction can be incorporated into the threshold voltage calculation by using Eq. (7.3.18) and letting the channel voltage V_C be equal to the diode reverse-bias voltage V_R while holding the bulk voltage V_B at zero.

In the absence of a gated-diode structure, the surface condition is determined by the oxide charge. In many cases of *IC* processing for bipolar transistors, the surface-charge condition does impose a wide depletion zone over the *p*-region surfaces. In these cases, *pn* junctions turn out to be leaky. Often a slow annealing step can be carried out to reduce s_o and the oxide charge content. This step usually improves the junction characteristics significantly.

7.6 MOS CAPACITORS AND CHARGE-COUPLED DEVICES   (CCDs)

Probably the most straightforward device use of the oxide-silicon system is to make high-quality, precisely controlled capacitors. One particular example of this application is shown in Fig. 7.18a, which is an enlarged view of an integrated circuit whose function is the conversion of analog signals to digital representation.[7] The A/D conversion is accomplished by a sequential comparison of the signal with fractions of a reference voltage. The reference is divided by applying the comparison through an array of capacitors having capacitances that are successively reduced by powers of two.

The capacitors in the A/D conversion circuit are MOS devices that are visible as square structures in Fig. 7.18a. For accurate conversion of analog signals, precise control over the ratios of the capacitance values in the circuit is necessary. This control is successfully achieved with MOS capacitors having the structure shown in Fig. 7.18b. The floating metal strips in Fig. 7.18b serve only to assure reliable precision during etch steps for capacitor fabrication and do not have any circuit function.[7]

Charge-Coupled Devices. Charge-coupled devices (CCDs)[8] consist almost entirely of closely spaced arrays of MOS capacitors. The capacitors are built close enough to one another so that the free charge stored in the inversion layer associated with one MOS capacitor (the channel) can be transferred to the channel region of the adjacent device. The exchange of charge is under the control of the voltages applied to the gates of the MOS capacitors.

In CCDs, the electrical signal is represented by the channel charge. The basic CCD operation depends upon the fact that the channel does not form instantaneously under a gate to which the threshold voltage has been applied; a relatively

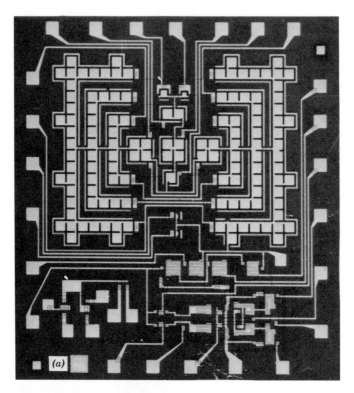

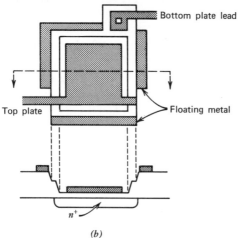

Figure 7.18   (a) Integrated circuit for the conversion of analog signals to a digital representation.[7] The circuit makes use of many precision MOS capacitors having the structure shown in (b).

long time period must elapse before sufficient generation has occurred to charge the channel (Problem 7.6). This time delay needed to establish thermal equilibrium between the silicon surface and the bulk is responsible for the different capacitance versus V_G variations that were sketched in Fig. 7.7. If the gate of the MOS structure is biased into the strong inversion region, the capacitance can range from C'_{ox} when the semiconductor surface is in equilibrium with the bulk to a very much lower value when all of the semiconductor space charge is contained in the depletion layer at its surface (deep depletion). At thermal equilibrium, the Fermi level in the silicon is constant and the band diagram is sketched in Fig. 7.8. Under non-equilibrium conditions, the Fermi level in the silicon is not constant and a band diagram similar to that sketched in Fig. 7.10 applies.

The significant idea that is exploited in CCDs is that if the gate bias is greater than the threshold voltage, the channel can be charged to any inversion density between Q'_n equals zero, which corresponds to the deep-depletion mode, and a value of Q'_n equal to the inversion density at thermal equilibrium. The channel charging can be accomplished either by transferring Q'_n from an adjacent gate or, in imaging applications, by localized stimulation of the generation rate. There is thus a lower limit for the speed of CCD operation. The signal must be transferred into and out of a stage at a rate that is fast compared to the background generation rate at a depleted silicon surface. No stage in a CCD line can be held for a long time in a deep-depletion condition, and noise will be added continuously because of thermal generation over the period during which the signal traverses a CCD circuit.

These constraints impose very severe requirements on the quality of the oxide-silicon system. As we saw from the discussion of surface generation rates, the requirements imposed can be met if the density of generation-recombination sites at the surface is held to a very low value. Manufacturers of CCD circuits routinely hold these state densities to values in the low 10^9 cm^{-2} range, which results in surface generation currents that are in the range of 1 to 10 nA cm^{-2}. A very small amount of charge is involved in many CCD circuits. Because CCD gates are typically a few $(\mu m)^2$ in area, values of detected charge range from roughly 10 electrons to about 10^7 electrons. By way of comparison, the thermal generation current of 1 nA cm^{-2} corresponds to roughly 60 electrons s^{-1} under a 1 $(\mu m)^2$ gate.

To discuss the basic operation of a CCD, we assume that charge is transferred to and held on MOS stages in times that are short compared to the time that would be necessary for thermal generation of an inversion layer. Consider the MOS system shown in Fig. 7.19 in which three capacitors are situated side by side on a silicon surface. Assume that the middle gate has a higher voltage V_2 applied to it than do either of the side gates, and that V_2 is greater than the MOS threshold voltage. If electrons have been introduced into the surface region, they will therefore reside in the channel under the middle gate. The application of a more positive voltage V_3 to the right-hand gate will cause the electrons to be transferred to the channel beneath it (Fig. 7.19*b*). The voltage on the middle gate can now be decreased

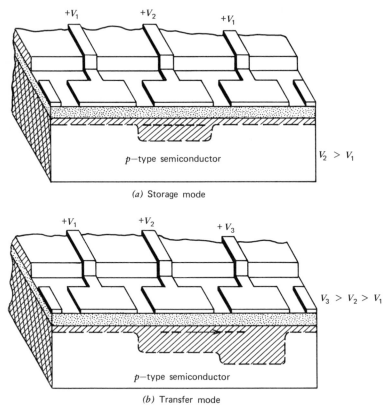

(a) Storage mode

(b) Transfer mode

Figure 7.19 Basic transfer mechanism in a CCD. (a) In the storage mode, charge is held under the center gate, which has a channel beneath it. (b) Application of $V_3 > V_2$ on the right-hand gate causes transfer of charge to the right.

to V_1, and then the voltage at the right-hand gate can be reduced to V_2. The net result is a shift of the channel charge one stage to the right.

In digital applications for CCDs, ones and zeros are usually represented by the presence or absence of channel charge, and long rows of capacitors are laid out side-by-side. An example of nine stages in a CCD is shown in Fig. 7.20 where channel charge exists at stages 1 and 7 in the storage mode of Fig. 7.20a while no charge is stored on stage 4. To initiate a transfer of "information" to the right, the gate voltages are changed as shown in Fig. 7.20b and the signals are seen in Fig. 7.20c to exist on gates 2, 5, and 8. Thus, signals are clocked through a CCD at a rate set by the timing of the gate voltages. The CCD structure is useful as a memory for digital information which is to be stored for a precise time; that is, the time it takes to traverse all stages in a CCD array. Analog (i.e., continuously variable)

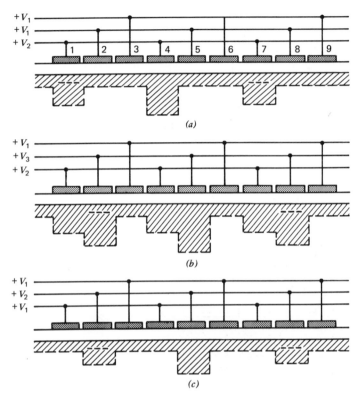

Figure 7.20   Nine gates of a "three-phase" CCD. (*a*) "Ones" are stored on gates 1 and 7; a "zero" on gate 4. (*b*) The binary information is transferred. (*c*) The information has been shifted one stage along the CCD.

signals can also be stored in CCD arrays because the amount of charge in the channel can be varied continuously. Thus, CCDs are useful for analog delay lines. By accessing the signal after differing delay periods, various useful signal processing tricks such as convolution, which involves multiplication of a time-varying signal by a delayed value of itself, can be implemented easily with CCDs.

Another analog use for a CCD array is as an imaging device. For this application, a whole field of CCD stages is biased into deep depletion and exposed to a focused image for a time interval. The generation of carriers under the CCD gate is enhanced during the exposure time according to the brightness of the image. Each channel will therefore become charged to a level that represents the brightness at its location. The analog information can then be clocked out to sense amplifiers that are built on the edges of the CCD image array. This scheme has been successfully demonstrated in a handheld, portable television camera.

A key factor in the production of a CCD is the spacing of adjacent gates. These must be close enough to one another to permit the fringing fields to cause the charge to be transferred when it is desired. Although systems have been built utilizing only metal gates, there are advantages to employing silicon gates singly or in two or three levels, especially to implement more elaborate clocking schemes than we have described in this brief introduction to CCDs. Figure 7.21 shows the electrode structure used to produce a CCD with both silicon and metal gates. A microphotograph of a 40-stage CCD that has been fabricated with two-level silicon and aluminum gates is shown in Fig. 7.22. This device has been designed for use either as a band-pass filter or for an analog delay line. The gates in the structure are roughly 25 μm in width.

Overall, the CCD is a remarkably versatile device having applications to digital memories, signal-processing circuits, and as an imaging device. All of these uses have evolved from a detailed understanding of the technology and electronics of the MOS system.

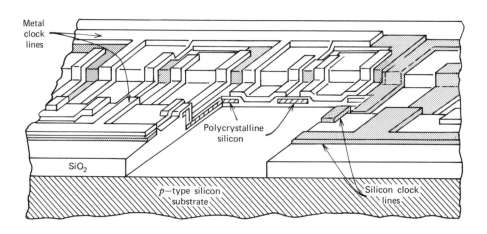

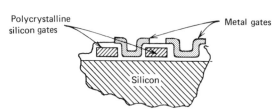

Figure 7.21   The simultaneous use of polycrystalline silicon gates and metal gates allows for close spacing in this "four-phase" CCD structure.

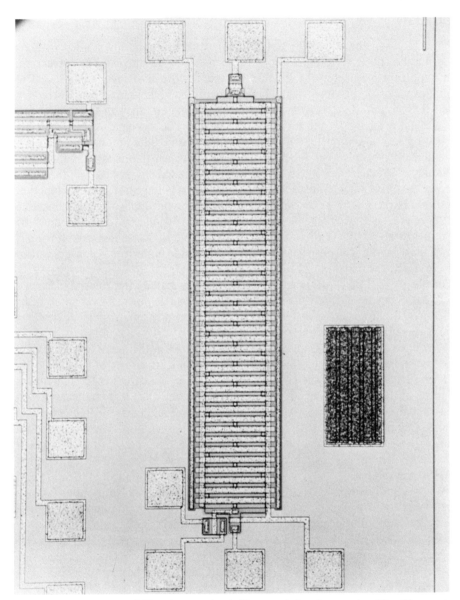

Figure 7.22 A 40-stage CCD using two-level silicon and aluminum gates. The silicon gates have gaps that make it possible to use this device as a bandpass filter as well as an analog delay line. (Courtesy Texas Instruments, Inc.)

SUMMARY

The electronic behavior of the oxide-silicon system is gainfully studied by considering the effects of voltages applied to a metal-oxide-silicon (MOS) structure. An important parameter of the MOS structure is the *flat-band voltage* V_{FB}. The significance of V_{FB} becomes clear when an energy-band diagram is constructed for the MOS system. If V_{FB} is applied between the metal and the substrate, there will be no field or charge at the surface of the silicon. The flat-band voltage is determined by several parameters. For ideal (charge-free) oxides and interfaces, it is a function only of the work-function difference between the metal and the semiconductor. Practical structures, however, contain fixed charges both in the oxide and at the oxide-silicon interface, and the flat-band voltage is a sensitive function of these charges. If the oxide charges are affected by the applied voltage, unstable MOS characteristics result. Generally, variable oxide-charge densities must be kept below about 10^{10} cm^{-2} to produce acceptable MOS devices. Alkali metals, particularly sodium, are especially troublesome oxide impurities because they can be moved by the field applied to an MOS oxide.

In many cases there are practical advantages to making the "metal" electrode in an MOS device of doped silicon. The silicon is usually formed over amorphous silicon oxide in an epitaxial reactor. It is composed of many small crystallites and is therefore known as *polycrystalline silicon*.

A voltage applied across an MOS capacitor between the metal or silicon *gate* and the substrate can accumulate or deplete the silicon surface of bulk majority carriers. It can also bias the silicon to inversion, in which case a *pn* junction is induced at the silicon surface. The charge that is controlled from the gate is then distributed, partly as fixed ions in the depletion layer and partly as free carriers in the inversion layer. The three conditions, *accumulation*, *depletion*, and *inversion* can be sensed by measuring the MOS capacitance. The measured MOS capacitance is frequency dependent because of the time required to establish an inversion layer at the surface.

When a surface is inverted, the inversion layer is often called a *channel*. If the channel exists adjacent to a *pn* junction, it is possible to bias the induced *pn* junction by means of the diffused *pn* junction. This method of biasing can alter the distribution of gate-induced charge between the depletion layer, where the charge is fixed, and the inversion layer, where it is free.

The space-charge layer at the silicon surface can affect the characteristics of *pn* junctions. The activity of surface states depends upon the surface potential, as can be demonstrated through experiments with *gated diodes*. An important direct application of the MOS system is for the production of *IC* capacitors. Arrays of closely spaced MOS capacitors make up *charge-coupled devices* (CCDs). Important applications of CCDs have been made to memory circuits, signal-processing circuits, and for imaging.

REFERENCES

1. R. H. Kingston and S. F. Neustadter, *J. Appl. Phys.*, 26, 718 (1955).
2. S. M. Sze, *Physics of Semiconductor Devices*, Wiley-Interscience, Wiley, New York, 1969, p. 435.
3. C. E. Young, *J. Appl. Phys.*, 32, 329 (1961).
4. W. Shockley, *Bell Syst. Tec. J.*, 28, 435 (1949).
5. J. S. Blakemore, *Semiconductor Statistics*, Pergamon Press, New York, 1962, p. 184.
6. A. S. Grove and D. J. Fitzgerald, *Solid-State Electron.*, 9, 783 (1966).
7. J. L. McCreary and P. R. Gray, *IEEE J. Solid-State Circuits*, SC-10, 371 (1975). Reprinted by permission.
8. W. S. Boyle and G. E. Smith, *Bell Syst. Tec. J.*, 49, 587 (1970).
9. C. N. Berglund, *IEEE Trans. Electron Devices*, ED-13, 701 (1966).
10. C. Jund and R. Poirer, *Solid-State Electron.*, 9, 315 (1966).

PROBLEMS

7.1 Sketch the energy-band diagrams (i) at thermal equilibrium and (ii) at flat band for ideal MOS systems made with aluminum gates (a) to 1 Ω-cm n-type silicon and (b) to 1 Ω-cm p-type silicon.

7.2 Repeat the sketches required in Problem 7.1 for an MOS system with a polycrystalline silicon gate. Assume that the silicon gate has a band structure similar to single-crystal silicon but that (i) the gate over n-type silicon is doped with acceptors until it is just at the edge of degeneracy and (ii) the gate over p-type silicon is doped with donors until it is just at the edge of degeneracy. (These conditions correspond to usual silicon-gate technology for reasons to be described in Chapter 8.)

7.3 Prove that the small-signal capacitance of an MOS capacitor C' that is biased in the depletion mode is given by Eq. (7.2.3). That is, show that C' is equal to the capacitance of a series connection of two capacitors: (1) a capacitor made with one plate in the bulk of the silicon and the other plate at the oxide-silicon interface, and (2) a capacitor that has its plates separated by the oxide. (*Hint. Use Gauss' law to express the charge* $\Delta Q' = \epsilon_{ox} \Delta \mathscr{E}_{ox}$. *Then, show that the voltage across the capacitor is* $\Delta V = \Delta \mathscr{E}_{ox} x_{ox} + \Delta \mathscr{E}_{ox} \epsilon_{ox} x_d / \epsilon_s$ *and evaluate* $C' = \Delta Q' / \Delta V$.)

7.4 Take $V_{FB} = -0.5$ V and use Eq. (7.2.3) to show the behavior of the overall capacitance C' for an MOS system in the depletion region. Sketch a plot of C'/C'_{ox} versus V_G. Consider that the silicon oxide is 1000 Å thick and the silicon is p-type and 1 Ω-cm in resistivity. Locate the flat-band capacitance C'_{FB} using Eq. (7.2.2).

7.5[†] The value of ϕ_s (the surface potential) is frequently needed for experimental studies of MOS systems.
   (a) By using the results of Problem 7.3, show that when the gate voltage V_G is changed on an MOS capacitor biased in the depletion region, it is possible to find the corresponding change in ϕ_s by using the measured capacitance of the MOS system. The change in ϕ_s can be calculated from the relationship

$$\phi_s(V_{G2}) - \phi_s(V_{G1}) = \int_{V_{G1}}^{V_{G2}} \left(1 - \frac{C'}{C'_{ox}}\right) dV_G$$

This technique is known as *Berglund's method* after its originator.[9] It can be used conveniently if V_{G1} is taken to be V_{FB} at which point C' is given by Eq. (7.2.2).

(b) If V_{G1} is taken as V_{FB}, sketch a low-frequency MOS capacitance curve for p-type silicon (normalized to C'_{ox}) and indicate (by shading) an area on the curve that is equal to $\Delta \phi_s$.

7.6† Consider that an MOS system on p-type silicon is biased to deep depletion by the sudden deposition of a total charge Q_G on the gate at $t = 0$. Carrier generation in the space-charge zone at the silicon surface results in a charging current for the channel charge Q'_n as described in the discussion of Eq. (4.3.26). This allows one to write

$$\frac{dQ'_n}{dt} = -\frac{qn_i(x_d - x_{df})}{2\tau_0}$$

where x_d is the (time dependent) depletion-region-width at the surface and τ_0 is the electron lifetime as given in Eq. (4.2.14). The quantity x_{df} is the space-charge region width at thermal equilibrium; that is, when $x_d = x_{df}$, channel charging by generation goes to zero.

(a) Show that a differential equation for Q'_n is

$$Q'_n + \left(\frac{2\tau_0 N_a}{n_i}\right)\left(\frac{dQ'_n}{dt}\right) = -[Q'_G - qN_a x_{df}]$$

(b) Solve this equation subject to $[Q'_n(t = 0) = 0]$ and thus show that the characteristic time to form the surface inversion layer is of the order of $2N_a\tau_0/n_i$. (Reference 10).

7.7 Sketch capacitance-voltage curves of the MOS structures shown in Figs. P7.7a, P7.7b, and P7.7c. The capacitance is the small-signal value normalized to that of the oxide and measured at 100 kHz. In all cases, the gate dc bias is varied slowly. Show (by using dotted curves) what effect an increase in positive Q_{ss} would have on the C-V_G curves. Label each region on the curves (accumulation, depletion, and inversion). Assume that the substrate resistivity is of the order of 10 Ω-cm in each case and make your sketches qualitatively correct.

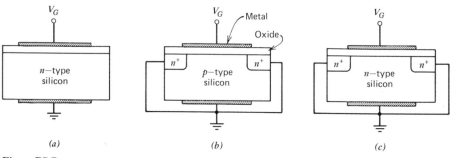

(a)   (b)   (c)

Figure P7.7

7.8 Sketch the curves as described in (a) through (d) below for an MOS capacitor on an n-type substrate that has been biased to inversion. Consider that $V_{FB} = -2$ V mainly because of the presence of fixed oxide charge Q_{ss}. The sketches should show (a) the band diagram, (b) all charge in the system, (c) the electric field, and (d) the potential. (Use the silicon bulk as the reference for potential.)

7.9 Construct a table similar to Table 7.1 to represent the surface-charge conditions for n-type silicon.

7.10 Using the formulas in Section 7.3, prove that for $n_s = 10\,N_a$, ϕ_s is only 58 mV greater than $-\phi_p$.

7.11[†] Consider the dependence on $(V_C - V_B)$ of the expressions for Q_n' [Eq. (7.3.16)] and V_T [Eq. (7.3.18)] in order to sketch a qualitative family of (low frequency) curves for C'/C_{ox}' versus V_G as $(V_C - V_B)$ is varied. This dependence was studied by Grove and Fitzgerald.[6]

7.12 Find the threshold voltage (a) in 1 Ω-cm p-type silicon and (b) in 1 Ω-cm n-type silicon. The MOS systems for each case are characterized by: (i) aluminum gate for which $q\Phi_M = 4.1$ eV, (ii) 1000 Å silicon oxide, (iii) the oxide is free of charge except for a surface density $(Q_{ss}'/q) = 5 \times 10^{10}$ cm^{-2}. The channel is not biased except from the gate ($V_C = V_B = 0$).

7.13[†] Consider the effects of oxide charge on the surfaces of n and p regions as described in Section 7.4. Apply these results to the high resistivity collector region in a double-diffused bipolar transistor. In particular, use sketches and develop arguments that show why these effects make it harder to manufacture reproducible and stable double-diffused pnp bipolar transistors than to produce npn bipolar transistors.

7.14[†] In practical MOS systems, measurements of capacitance versus voltage sometimes show hysteresis effects; that is, the C-V_G curves look like the sketch in Fig. P7.14.* The sketch refers to measurements made when V_G is swept with a very low frequency triangular wave (~ 1 Hz) and the ac measurement frequency is of the order of 1 kHz or higher. The sense of the hysteresis on such a curve can be observed experimentally to be either counter-clockwise, as shown in the sketch, or else clockwise. The hysteresis sense allows one to differentiate between the two most common causes of nonideal behavior. (a) Show this,

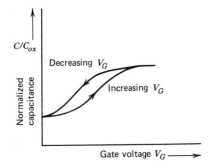

Figure P7.14

---

* A laboratory test for oxide "quality" is to cycle the C-V_G measurements at an elevated temperature and check the total amount of voltage hysteresis present—a few tens of milli-volts at 125°C is frequently the limit allowed.

by considering the following nonideal effects: (i) field-aided movement of positive ions in the insulator and (ii) trapping of free carriers from the channel in traps at the oxide-silicon interface. (b) Using qualitative reasoning, prepare a table with sketches of the expected C-V_G plots for n- and p-type substrates; on each sketch indicate the sense of the hysteresis (i.e., clockwise or counterclockwise).

7.15   Compare the maximum capacitance that can be achieved in an area of $100 \times 100\ \mu m$ by using either an MOS capacitance or a reverse-biased pn-junction diode. Assume an oxide breakdown strength of 8×10^6 V cm^{-1}, a 5 V operating voltage, and a safety factor of two (i.e., design the MOS oxide for 10 V). The pn junction is built by diffusing boron into an n-type semiconductor doped to 10^{16} cm^{-3}.

7.16[†]   Calculate the area density of surface states that would lead the surface generation rate I_S [Eq. (7.5.3)] of a fully depleted surface to equal twice the generation rate in the surface depletion region I_F [Eq. (7.5.2)]. Consider the states to be characterized by a capture cross section of 10^{-15} cm^2 and the thermal velocity to be 10^7 cm s^{-1}. Assume that the surface depletion region is 1 μm in width and that the time constant τ_0 is 1 μs.

CHAPTER 8

THE INSULATED-GATE
FIELD-EFFECT TRANSISTOR

The electronic properties of the metal-oxide-silicon system make possible a different type of transistor from the bipolar junction device that was discussed in Chapters 5 and 6. This new type of transistor is often called a MOSFET or *metal-oxide-silicon, field-effect transistor*. Several other names are, however, in common use including the designation that we shall usually make: *IGFET* or *insulated-gate field-effect transistor.** The simplicity of the IGFET and the high component density possible when the device is employed in integrated circuits have made IGFETs of great commerical importance, especially for applications to digital circuitry.

The concept of the IGFET was actually developed well before the invention of the bipolar transistor. In the early 1930s, patents were issued for devices that resemble the modern silicon IGFET but that made use of combinations of materials other than silicon.[1] Poor control of the insulator-semiconductor interfaces in use at that time and a lack of a full understanding of the insulator-semiconductor

---

* An exception is reference to MOS integrated circuits and MOS memories because these terms are so widely used.

systems made the use of these inventions impossible. It was not until after the advent of the planar process and the technology to produce a well-behaved oxide-silicon interface that the IGFET became practical.

The basic structure of an n-channel IGFET is shown in Fig. 8.1. It consists of the MOS structure that was discussed extensively in Chapter 7 with the channel region of the IGFET extending between two diffused junctions. These diffused junctions are electrically disconnected unless there is an n-type inversion layer at the surface to provide a conducting channel between them. When the surface is inverted, and a voltage is applied between the junctions, electrons can enter the channel at one junction, which is known as the *source*, and leave at the other, which is therefore designated the *drain*.

It is possible that the influence of Φ_{MS} and of oxide charges leads to an inverted surface or channel between the junctions when no gate voltage is applied. If this is the case, the IGFET is called a *depletion-mode* device because the gate is then most usually employed to reduce the conductance of the built-in channel. In most cases, however, gate voltage must be applied to induce a channel and IGFETs of this type are called *enhancement-mode* devices. The discussion in this chapter is generally directed toward enhancement-mode IGFETs, which are much more frequently employed in *IC* applications than are depletion-mode IGFETs.

The use of the terms source and drain is the same for p-channel IGFETs; holes (i.e., carriers in the inversion region) are supplied at the source and removed at the drain. Hence, conventional current flows from the source to the drain in p-channel IGFETs and in the opposite sense for n-channel IGFETs. This comment is one of the few specific references made to p-channel devices. As in Chapter 7, almost all of the theory developed is based on n-channel transistors. The corresponding behavior in p-channel structures is readily grasped. Equations for both p- and

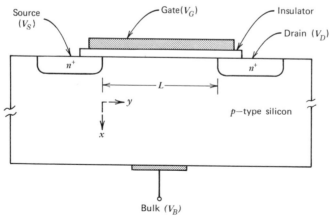

Figure 8.1   Basic elements of an n-channel, insulated-gate field-effect transistor (IGFET).

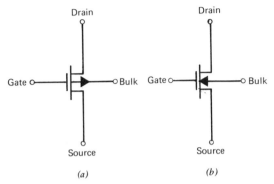

Figure 8.2 Electrical symbols for IGFETs: (*a*) *p*-channel, (*b*) *n*-channel.

n-channel transistors are tabulated at the conclusion of the chapter, and several problems deal with the properties of *p*-channel IGFETs.

If the IGFET in Fig. 8.1 is compared to the device in Fig. 3.16, it can be seen that its structure resembles that of the junction field-effect transistor or JFET. The two transistors both have source and drain contacts through which output currents flow and a gate that controls the magnitude of this current. In the two examples that are sketched, electrons travel from the source to the drain through a channel. The channel, however, is formed differently in the two devices. In the JFET, a channel of *n*-type material is built into the structure. In the IGFET, however, the channel is induced by a field applied from the gate electrode to the silicon through the insulating silicon oxide. Thus, the IGFET utilizes the electronic properties of the MOS system that were described in Chapter 7. In particular, the influence of the gate voltage on Q_n, the free charge in the channel, results in gate control of the conductance between the drain and source electrodes.

The symbols that have been adopted for IGFETs are sketched in Fig. 8.2*a* for a *p*-channel device and in Fig. 8.2*b* for an *n*-channel device. Note that the distinction between the two types of IGFETs is made in terms of an arrow that represents the *pn* junction between the bulk and the induced channel at the surface. Unless an asymmetric geometry is used, the IGFET is a bilateral device at its output terminals, and the source and drain electrodes are drawn identically.

8.1 CURRENT-VOLTAGE CHARACTERISTICS

The *n*-type source and drain diffusions in the IGFET shown in Fig. 8.1 are separated by a lateral distance known as the *channel length*, which is denoted by the symbol *L*. The channel length extends along the *y* axis in the figure, while the direction into the silicon perpendicular to the oxide is the *x* dimension. The channel width in a dimension perpendicular to the figure (in the *z* direction) is given the symbol *W*.

In practical devices, typical channel lengths are in the order of 5 μm or greater*
while channel widths are selected by considering the requirements of the circuit
design. The circuit designer chooses a value for W in order to achieve a desired
conductance at a given bias condition in the IGFET. The thickness of the oxide
that separates the gate from the channel is about 1000 Å although it can range
from roughly one half to twice this value.

The IGFET is a four-terminal device with connections to the source, drain, gate,
and silicon bulk (substrate). The voltages at these terminals with appropriate
subscripts are shown in Fig. 8.1. It is valuable for circuit applications to derive
equations that permit these voltages to vary independently, and the general analysis presented will do this. It is, however, easier to understand IGFET electronics
under the condition that V_S and V_B (the source and substrate voltages) are the
common circuit ground. This condition is encountered frequently and, hence, our
first physical description of IGFET operation is under this assumption. Later,
equations are derived for more general conditions.

Charge-Control Analysis

As in the analysis of the JFET, we consider the IGFET first at low drain-source
voltages (V_{DS} positive but small) so that the channel charge does not vary strongly
with position between the source and drain electrodes. Under this condition the
surface space-charge region is uniform along the channel (Fig. 8.3a).

If both V_S and V_B are at zero volts and V_G is increased from zero, the source and
drain are initially isolated by a back-biased pn junction. When V_G becomes larger
than V_{FB}, a depletion region forms along the channel. The depletion region then
widens to accommodate the increasing density of electrical flux lines that emanate
from the gate. As V_G continually increases, the mid-gap energy E_i at the surface will
eventually be pulled below the Fermi level, causing the surface potential ϕ_s to
change from negative to positive values (Eq. 7.3.2) and surface inversion will
commence. When this occurs, electrons in the surface inversion layer can flow
and carry current by drifting between the source and the drain. For surface poten-
tials less than the threshold value $|\phi_p|$, however, only small, so-called *subthreshold
currents* will be able to flow between the drain and the source. These subthreshold
currents are not large enough to be useful for most applications, but they are
sufficiently large to cause troublesome leakage, which can complicate the design
of some circuits. Subthreshold currents will be considered in greater detail in
Section 8.4.

At higher gate voltages, the surface potential of the semiconductor reaches its
strong inversion value $|\phi_p|$.† Once strong inversion occurs, higher gate voltages

---

* Five microns is not a fundamental limit; technological advances in microfabrication are
  already shrinking the channel length toward 1 μm in experimental devices.
† Referenced to the bulk, the surface potential at the onset of strong inversion is $\phi_s = \phi_p - (-\phi_p) = 2|\phi_p|$.

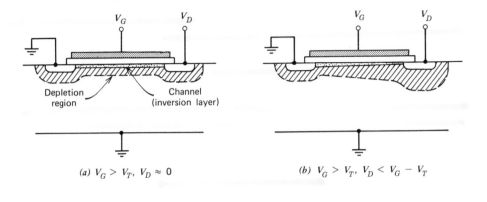

(a) $V_G > V_T$, $V_D \approx 0$

(b) $V_G > V_T$, $V_D < V_G - V_T$

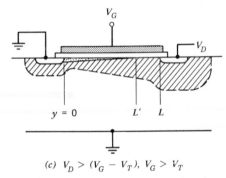

(c) $V_D > (V_G - V_T)$, $V_G > V_T$

Figure 8.3 IGFET cross sections showing the effects of various bias conditions on the space-charge regions. (a) The drain voltage is very small and the space-charge region is nearly uniform along the channel. (b) The drain voltage is large enough to cause significant variation in the thickness of the space-charge region. (c) The drain voltage exceeds the saturation value and the channel extends only to $L' < L$.

cause little change in Q_d, the depletion-layer charge, and additional gate voltage primarily increases the electron concentration in the inversion layer Q_n. The electron concentration in the inversion layer then varies approximately linearly with applied gate voltage in excess of the threshold voltage, as does I_D the current flowing from drain to source. The current also increases linearly with drain-source voltage as long as this voltage difference remains small.

It is straightforward to derive equations to represent IGFET behavior under these conditions by employing charge control, as was done for the bipolar transistor in Chapter 6. The drain current I_D is related to the channel charge Q_n by the transit time T_{tr} through the equation

$$I_D = -\frac{Q_n}{T_{tr}} \tag{8.1.1}$$

Because current flow in the channel is by drift, T_{tr} is just the channel length L divided by the drift velocity $v_d = -\mu_n \mathscr{E} = \mu_n V_D/L$ or

$$T_{tr} = \frac{L^2}{\mu_n V_D} \tag{8.1.2}$$

The mobility μ_n in Eq. (8.1.2) is typically about 50% of the bulk value because of differing free-carrier scattering mechanisms at the surface (Section 8.4). The channel charge Q_n in Eq. (8.1.1) can be most simply written by using Eq. (7.3.19) together with the channel dimensions

$$Q_n = -C'_{ox}(V_G - V_T)WL \tag{8.1.3}$$

where the threshold voltage V_T is given by Eq. (7.3.18) with $V_C = V_B = 0$ for the case considered here. Hence, the current for very low drain bias is approximately given by

$$I_D = \mu_n \frac{W}{L} C'_{ox}(V_G - V_T)V_D \tag{8.1.4}$$

Equation (8.1.4) agrees with our intuitive analysis; for very low drain bias, the drain current is linear in V_D with a conductance value $[\mu_n W C'_{ox}(V_G - V_T)/L]$ that is proportional to $(V_G - V_T)$.

If V_D is now increased in value until it is not negligible compared to V_G, the analysis just given loses accuracy. This is because the drain voltage V_D affects the channel bias near the drain and acts to reduce Q_n as was described in Section 7.3. The space-charge zone and channel charge for this situation are sketched in Fig. 8.3b. A first order, approximate analysis incorporating this effect can be carried out by considering the average voltage (above threshold) that is applied between the gate and the channel to be $(V_G - V_D/2)$. This bias leads to a channel charge Q_n that differs from the value given in Eq. (8.1.3). The new value is

$$Q_n = -C'_{ox}\left(V_G - V_T - \frac{V_D}{2}\right)WL \tag{8.1.5}$$

Therefore, instead of Eq. (8.1.4), the drain-current equation becomes

$$I_D = \mu_n \frac{W}{L} C'_{ox}\left(V_G - V_T - \frac{V_D}{2}\right)V_D \tag{8.1.6}$$

Curves of I_D versus V_D based on Eq. (8.1.6) are sketched in Fig. 8.4a for varying values of $(V_G > V_T)$. The curves have maximum slope in the vicinity of the origin where the conductance is greatest, and Eq. (8.1.4) applies. For higher V_D, the conductance diminishes and the curves consist of a series of downward-facing parabolas having maxima at $V_D = (V_G - V_T)$. In Fig. 8.4a, the curves are drawn as

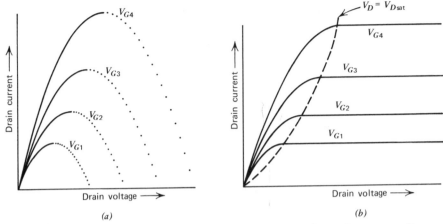

Figure 8.4 (a) Drain current as a function of drain voltage for various gate voltages as predicted by Eq. (8.1.6). The dotted portions of the curve are unreasonable on physical grounds. (b) Overall $I_D - V_D$ curves as predicted by Eqs. (8.1.6) and (8.1.8). The dashed curve represents values of $V_{D\,sat}$ from Eq. (8.1.7). The gate voltage increases from bottom to top in both curve families.

dotted lines for $V_D > (V_G - V_T)$ because they do not apply in this range of drain bias. The negative incremental conductance $(\partial I_D/\partial V_D)$, indicated in this region, is an intuitive clue to their being physically unreasonable. The solutions do not apply in this range because there is no free-charge density near the drain when V_D exceeds $(V_G - V_T)$ [Eq. (7.3.19)]. Hence, the analysis we have carried out, which is based on ohmic flow all the way along the channel from the source to the drain, is not valid.

Although the theory breaks down for $V_D > (V_G - V_T)$, it is quite possible to deduce the behavior of I_D versus V_D in this bias range. To do this, we recognize that electrons in the channel "see" no barrier as they approach the depletion region near the drain. On the contrary, they approach a high-field region in which they are accelerated (usually reaching limiting drift velocities) when $V_D > (V_G - V_T)$. The drain current I_D is thus determined by the rate at which these electrons arrive at the edge of the depletion region. In the first order analysis, this rate is insensitive to V_D, and the drain current thus becomes constant or *saturates* for $V_D > (V_G - V_T)$. Hence, the maxima in the curves of Fig. 8.4a occur at a voltage known as the *drain saturation voltage* $V_{D\,sat}$ where

$$V_{D\,sat} = V_G - V_T$$

$$= V_G - V_{FB} - 2|\phi_p| - \frac{1}{C'_{ox}}\sqrt{4\epsilon_s q N_a |\phi_p|} \qquad (8.1.7)$$

The constant current for $V_D > V_{D\,\text{sat}}$ is sketched in Fig. 8.4b. In this range of bias, the end of the channel is no longer at $y = L$, but rather occurs at the point $y = L'$ where the channel voltage $V_C(y) = V_{D\,\text{sat}}$. The current when $V_D > V_{D\,\text{sat}}$ (denoted $I_{D\,\text{sat}}$) can therefore be calculated by using Eq. (8.1.7) in Eq. (8.1.6):

$$I_{D\,\text{sat}} = \frac{\mu_n W C'_{ox}}{2L}(V_G - V_T)^2 \tag{8.1.8}$$

Since both Eqs. (8.1.8) and (8.1.6) are valid (within the assumptions made) at $V_D = V_{D\,\text{sat}}$, a plot of the locations of the $V_{D\,\text{sat}}$ points on the $I_D - V_D$ curves is an upward-facing parabola (dashed curve in Fig. 8.4b). The overall $I_D - V_D$ curves (excluding breakdown effects) are thus plotted by using Eq. (8.1.6) for $V_D < V_{D\,\text{sat}}$ and Eq. (8.1.8) for $V_D > V_{D\,\text{sat}}$ as shown in Fig. 8.4b.

The channel charge and depletion regions for $V_D > V_{D\,\text{sat}}$ are shown in Fig. 8.3c. This figure emphasizes that L' is reduced as V_D increases. This causes $I_{D\,\text{sat}}$ to increase slightly with V_D. The dependence on V_D owing to this effect could be incorporated by rewriting Eq. (8.1.8) with L' in place of L and then expressing L' as a function of V_D. This modification will be considered at a later point in the discussion.

The analysis given here has been especially simple because it has only considered variations in conditions along the channel in terms of an average channel potential, and has calculated the current [Eq. (8.1.6)] in terms of the corresponding charge given in Eq. (8.1.5). Actually, both the threshold voltage V_T and the channel voltage V_C are functions of the channel position y. Nonetheless, the equations derived are sufficiently accurate in many cases to be adequate for circuit analysis, and they are widely used. They do, however, represent approximations that are not always allowable. Furthermore, the equations have not been cast in a sufficiently general form because they have been derived on the assumption that the source and substrate electrodes are tied to a common ground point. These restrictive assumptions will be removed in the derivation carried out in the next section.

Distributed Analysis

The charge-control analysis in the previous section has provided a basic understanding of IGFET operation. To represent the device more exactly, it is necessary to account for the variation of channel conditions by writing differential equations. The differential equations can be derived by following the same procedure as was used in Chapter 3 for the analysis of the JFET. The incremental voltage drop along the channel is represented as a function of the channel current. Integration of this incremental voltage along a path extending from the source to the drain will lead to an equation for I_D in terms of the applied voltages.

The *gradual-channel* approximation is assumed to be valid as was the case in the JFET analysis in Chapter 3. This approximation states that the fields in the direction of current flow are much smaller than are fields in the direction perpendicular to the silicon surface, mathematically that $|\partial\phi/\partial y| \ll |\partial\phi/\partial x|$. This assumption validates the use of a one-dimensional MOS analysis (as carried out in Chapter 7) to find the carrier concentrations and the dimensions of the depletion region under the channel.

For the present analysis, we also assume that the channel is appreciably longer than the depletion region at the drain (i.e., $L'/L \approx 1$). In very short channel devices, this may not be true and the channel charge can be influenced strongly by the drain voltage as well as by the gate voltage. This short-channel effect is considered further in Section 8.4. Finally, we consider that since drain current consists of electron flow between n-regions, it is carried exclusively by the drift process.

The equations are written in terms of the terminal voltages indicated on Fig. 8.1 with an arbitrary voltage reference. Keeping the reference arbitrary permits the equations to be applied to such practical cases as the series-connected IGFET pair, which is frequently encountered in integrated circuits. In the series pair, one device may have no source bias with respect to the bulk while the other has its source at the drain voltage of the first. Hence, equations to describe both IGFETs must be written in the most general terms.

Figure 8.5 shows an n-channel IGFET that is biased below saturation. An incremental length dy along the channel sustains a voltage drop dV_C that can be expressed (Problem 8.2) as the product of the drain current I_D and the incremental resistance dR:

$$dV_C = I_D\, dR = -\frac{I_D\, dy}{W\mu_n Q_n'(y)} \tag{8.1.9}$$

where $Q_n'(<0)$ is a function of the channel voltage $V_C(y)$ as well as the gate voltage

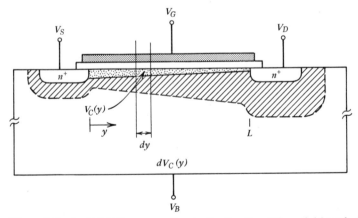

Figure 8.5 An IGFET cross section indicating the differential length dy along the channel. An ohmic voltage drop $dV_C = I_D\, dR$ is sustained across dy. The channel is W units wide.

V_G. From Eq. (7.3.16)

$$Q'_n = -C'_{ox}(V_G - V_{FB} - 2|\phi_p| - V_C) + \sqrt{2\epsilon_s q N_a(2|\phi_p| + V_C - V_B)} \quad (8.1.10)$$

Equation (8.1.10) can be used in Eq. (8.1.9) to express Q'_n as a function of V_C. Then Eq. (8.1.9) can be separated and the resultant equation integrated from the source to the drain

$$I_D \int_0^L dy = I_D L = -\mu_n W \int_{V_S}^{V_D} Q'_n(V_C) dV_C \quad (8.1.11)$$

The integration has many terms, but is easily carried out (Problem 8.2). The result can be written

$$I_D = \mu_n \frac{W}{L} \left\{ C'_{ox} \left(V_G - V_{FB} - 2|\phi_p| - \frac{1}{2}V_D - \frac{1}{2}V_S \right)(V_D - V_S) \right.$$
$$\left. - \frac{2}{3}\sqrt{2\epsilon_s q N_a} \left[(2|\phi_p| + V_D - V_B)^{3/2} - (2|\phi_p| + V_S - V_B)^{3/2} \right] \right\} \quad (8.1.12)$$

As noted in the last section, the analysis leading to Eq. (8.1.12) is valid as long as inversion charge exists along the entire length of the channel. When V_D is increased sufficiently to cause $Q'_n(L)$ to be reduced to zero, the drain current saturates. To find the corresponding drain voltage $V_{D\,sat}$, therefore, we let $V_C = V_D$ in Eq. (8.1.10) and solve for $V_{D\,sat}$ by letting $Q'_n(L)$ equal zero.

$$Q'_n(L) = 0 = -C'_{ox}(V_G - V_{FB} - 2|\phi_p| - V_{D\,sat})$$
$$+ \sqrt{2\epsilon_s q N_a(2|\phi_p| + V_{D\,sat} - V_B)} \quad (8.1.13)$$

From Eq. (8.1.13), the saturation drain voltage is

$$V_{D\,sat} = V_G - V_{FB} - 2|\phi_p| - \frac{\epsilon_s q N_a}{C'^2_{ox}} \left[\sqrt{1 + \frac{2C'^2_{ox}}{\epsilon_s q N_a}(V_G - V_{FB} - V_B)} - 1 \right] \quad (8.1.14)$$

Note that the saturation drain voltage is independent of the source voltage.

To compare these results from the distributed analysis with the equations derived earlier by approximate analysis, we must take corresponding voltages $V_S = V_B = 0$. Under this condition, Eq. (8.1.12) reduces to

$$I_D = \frac{\mu_n W}{L} \left\{ C'_{ox} \left(V_G - V_{FB} - 2|\phi_p| - \frac{1}{2}V_D \right) V_D - \frac{2}{3}\sqrt{2\epsilon_s q N_a} \right.$$
$$\left. \times \left[(2|\phi_p| + V_D)^{3/2} - (2|\phi_p|)^{3/2} \right] \right\} \quad (8.1.15)$$

which can be compared to Eq. (8.1.6) when V_T is explicitly included in the formulation. If that is done, Eq. (8.1.6) takes the form

$$I_D = \mu_n \frac{W}{L} \left[C'_{ox} \left(V_G - V_{FB} - 2|\phi_p| - \frac{1}{2}V_D \right) V_D - 2V_D\sqrt{\epsilon_s q N_a |\phi_p|} \right] \quad (8.1.16)$$

Equations (8.1.15) and (8.1.16) have been used to plot the curves showing I_D versus V_D that are given in Fig. 8.6. The calculations were made for an IGFET described by the following (typical) parameter values: $\mu_n W/L = 12000$ cm^2 V^{-1} s^{-1}, $C'_{ox} = 3.98 \times 10^{-8}$ F cm^{-2} (870 Å oxide), $N_a = 2 \times 10^{16}$ cm^{-3}, and $V_{FB} = -0.5$ V. The results of the charge-control analysis have been calculated from Eq. (8.1.16) and are plotted as the heavier-weight curves. The currents predicted by the distributed analysis [Eq. (8.1.15)] are plotted as lighter-weight curves. The predictions from this theory have been shown to agree well with measured characteristics.[2] It is clear from inspection of Fig. 8.6, however, that the simpler charge-control equations [Eqs. (8.1.16) and (8.1.7)] overestimate both the currents in the saturation region and the magnitude of $V_{D\,sat}$ at every value of gate voltage. The general behavior of the two sets of curves is, however, similar, and the currents predicted by both theories become the same as the drain voltage approaches zero. This agreement is expected because the basic assumption of the charge-control analysis, a constant field in the channel, is met at very small values of V_D (Problem 8.5). The two theories are observed to begin to deviate strongly when I_D becomes roughly 20% of $I_{D\,sat}$ [with $I_{D\,sat}$ predicted by Eq. (8.1.8)] for the IGFET analyzed in Fig. 8.6.

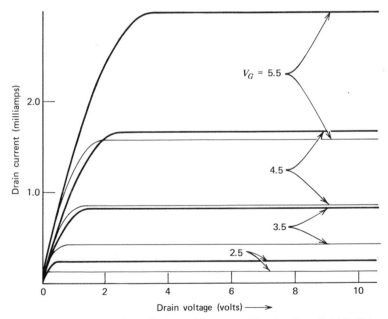

Figure 8.6 Theoretical predictions of I_D versus V_D using Eqs. (8.1.15) (lighter-weight curves) and (8.1.16) (heavier-weight curves) for an IGFET having the parameters given in the text. (I_D is taken to be equal to $I_{D\,sat}$ for $V_D > V_{D\,sat}$.)

A closer correspondence can be obtained between the predictions of the two theories if the prefactor $(\mu_n W/L)$ is adjusted to a lower value in Eq. (8.1.16) than is used in Eq. (8.1.15). There is some justification for latitude in specifying this prefactor because the exact value of mobility (μ_n) is not a precisely controlled *IC* parameter. (For a given MOS process, however, μ_n is nearly constant and the relative values of $(\mu_n W/L)$ between different IGFETs are accurately specified by the W/L ratios.) In Fig. 8.7, the drain currents predicted by the distributed analysis [Eq. (8.1.15)] are replotted using the same parameters as were used in Fig. 8.6. Equation (8.1.16), however, has been replotted with μ_n reduced to roughly 50% of its former value. This empirical modification is seen to bring the values of saturation current as predicted by both theories into correspondence. The values of $V_{D\,sat}$, however, are not matched and currents in the ohmic region are underestimated by the charge-control analysis. Despite this disparity, there are many cases where design is possible with the less accurate representation of the charge-control analysis, and it is frequently used. In general, reducing either the oxide thickness or the silicon dopant density (or both simultaneously) will cause the two formulations to correspond more closely. This is illustrated in Fig. 8.8, which is a plot similar to that of Fig. 8.6 except that the oxide thickness has been reduced by 50% and the dopant density has been reduced by an order of magnitude. Note that these changes increase the output currents as well as the values of $V_{D\,sat}$.

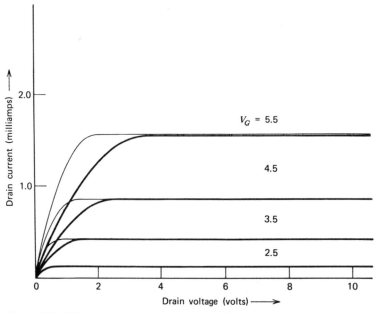

Figure 8.7 Theoretical $I_D - V_D$ curves similar to those of Fig. 8.6, except that μ_n in Eq. (8.1.16) has been adjusted so that $I_{D\,sat}$ at the highest value of V_G coincides for the two models.

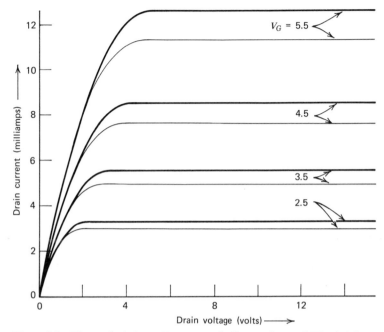

Figure 8.8   Theoretical $I_D - V_D$ curves similar to those of Fig. 8.6 for an IGFET having $N_a = 2 \times 10^{15}$ and $x_{ox} = 435$ Å. (This oxide is thinner than typically used.)

Summary. Because of the length of this section, it is worthwhile to review briefly the results obtained and to give them perspective. First, IGFETs are characterized by a gate threshold voltage that must be reached to induce a conducting surface channel. After the channel has been formed, curves of drain current versus drain voltage exhibit two distinct regions: (1) an ohmic region in which the channel extends all the way from the source to the drain, and (2) a saturated region for $V_D > V_{D\,\text{sat}}$ in which the drain current flows across a high-field space-charge region near the drain. This general behavior is predicted by both the charge-control and the more exact, distributed theory. There are very important circuit uses for the IGFET as a controlled resistor when it is biased in the ohmic region of operation. For use as an amplifier or as a closed switch, it is more frequently biased into the saturated region of operation. Finally, it was shown that by adjusting the value of μ_n the simpler equation for I_D in the charge-control theory can be made to correspond closely to the equation from the more exact theory in the prediction of $I_{D\,\text{sat}}$. Thus, a reasonable expression for $I_{D\,\text{sat}}$ for many applications is

$$I_{D\,\text{sat}} = k' \frac{W}{L} (V_G - V_S - V_T)^2 \qquad (8.1.17)$$

where k' is proportional to $\mu_n C'_{ox}$ but, in general, is smaller than this product. In the next section, considerable use will be made of Eq. (8.1.17) because it so greatly simplifies the representation of the IGFET under saturated conditions.

8.2 IGFET PARAMETERS

The most intensive circuit application of IGFETs is to digital circuits where they are usually designed either to switch as rapidly as possible between cut-off and saturation or are used as resistive elements. IGFETs are, however, also used in analog circuits, particularly for applications that demand an amplifier with an extremely high input impedance. Because the gate of an IGFET is a very high quality capacitor, the device is ideal for use as an electrometer amplifier (i.e., an amplifier that senses charge directly). For both digital and analog applications, it is necessary to develop understanding of the speed limitations of the transistor as well as to characterize its dc behavior. For analog applications, it is convenient to develop the small-signal gain parameters of the IGFET.

In this section, we consider some practical aspects concerning the dc parameters of the transistor first. Then we develop expressions for the small-signal ac gain and investigate the limits on the switching speed of the device. After an IGFET model to be used in circuit design and analysis is presented, we consider some aspects of IGFET technologies.

Determination of dc Parameters

A method frequently employed to obtain the dc parameters that describe an IGFET is to tie together the gate and the drain of the transistor, and then to measure drain current as a function of the applied drain voltage. The experimental arrangement is shown in Fig. 8.9 (inset). Because V_D is set equal to V_G in this circuit, we can see from Eq. (8.1.7) that the transistor is in the saturated region of operation. Hence, Eq. (8.1.17) can be applied, and it becomes apparent that a plot of $\sqrt{I_D}$ versus V_G should be linear. Typical measured data obtained in this way are shown in Fig. 8.9. The utility of this analysis technique is apparent: the threshold voltage V_T is readily obtained from the intercept with the voltage axis and the slope of the plotted data can be used to determine the value of $k'W/L$. The dotted section of the curve in Fig. 8.9 results from currents that flow below the conventional threshold. We discuss these currents further in Section 8.4.

A circuit similar to that shown in the inset of Fig. 8.9, except with a reverse bias applied between the source and the substrate (or *bulk*), can be used to measure the variation in the threshold voltage resulting from the source-substrate bias (often called the *body effect*). This circuit is shown in the inset of Fig. 8.10. As seen from the plots of $\sqrt{I_D}$ versus $(V_G - V_S)$ or $(V_D - V_S)$ in Fig. 8.10, source-to-substrate bias increases the threshold voltage and the curves of $\sqrt{I_D}$ are accordingly translated along the voltage axis. The change in threshold voltage ΔV_T for a case of constant

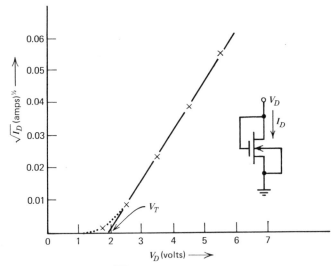

Figure 8.9 Plot of $\sqrt{I_D}$ versus V_D for an *n*-channel IGFET in saturation. Inset shows the circuit arrangement. The threshold voltage is indicated by the intercept of the straight line with the voltage axis.

substrate doping can be calculated from Eq. (7.3.18):

$$\Delta V_T = \frac{\sqrt{2\epsilon_s q N_a}}{C'_{ox}} \left(\sqrt{2|\phi_p| + |V_S - V_B|} - \sqrt{2|\phi_p|} \right) \qquad (8.2.1)$$

In practice, the dependence predicted by Eq. (8.2.1) is observed as long as the channel length of the IGFET is considerably larger than the depletion width of the reverse-biased, source, substrate *np* junction. If this is not the case, the one-dimensional analysis used to solve Poisson's equation [which led to Eq. (7.3.18)] becomes seriously in error. A two-dimensional theory for the space-charge con-figuration would then be needed to obtain an accurate theoretical expression for ΔV_T. The curves in Fig. 8.10 apply to an IGFET with the characteristics of Fig. 8.6 (8 μm channel) and the agreement between Eq. (8.2.1) and the observed ΔV_T is very good. Note in Fig. 8.10 that the variation in V_T is roughly proportional to $\sqrt{|V_S - V_B|}$. Although not exact, this dependence is a reasonable approximation for typical IGFET parameters, and it is therefore the basis for a rough "rule of thumb" in MOS circuit design: that $\Delta V_T \simeq K\sqrt{|V_S - V_B|}$ where K is an empirical constant.

A "threshold voltage" can be obtained by observing the gate voltage at which a significant drain current flows, and IGFETs are often described by giving the "threshold voltage" for 1 μA of drain current. This threshold is, however, sensitive to the W/L ratio of the structure, and the determination of V_T by the intercept

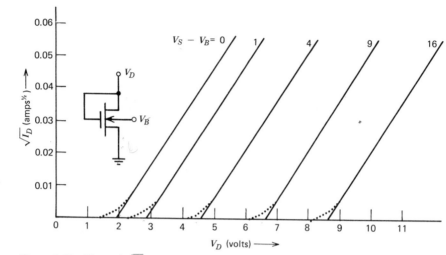

Figure 8.10 Plots of $\sqrt{I_D}$ versus V_D showing the effects of a bias $V_S - V_B$ between the source and the bulk. The observed shift in threshold is predicted by Eq. (8.2.1).

method, as illustrated in Fig. 8.9, is preferable for modeling the device. The I_D versus V_D characteristics can, however, be used to determine $[\mu_n C'_{ox}(W/L)]$. To do this most conveniently, we note from Eq. (8.1.6) that the slope at the origin ($\partial I_D / \partial V_D$ or the zero-bias conductance) is given by

$$\frac{\partial I_D}{\partial V_D}\bigg|_{V_D \to 0} = \mu_n C'_{ox} \frac{W}{L}(V_G - V_T) \qquad (8.2.2)$$

Note that it is not proper to use the slope of the $\sqrt{I_D}$ versus V_G curve in saturation in order to find $\mu_n C'_{ox}(W/L)$ as is sometimes erroneously done. That slope gives the value of $k'(W/L)$ from Eq. (8.1.17) which, as discussed in Section 8.1, is not the same as $\mu_n C'_{ox}(W/L)$ because of the inaccuracy of the charge-control model.

Gain and Speed of Response

The small-signal amplification of an IGFET is most usefully characterized by its transconductance because the output (drain) current typically varies in response to a changing input (gate) voltage. From Eq. (8.1.12), we find

$$g_m \equiv \frac{\partial I_D}{\partial V_G} = \mu_n C'_{ox} \frac{W}{L}(V_D - V_S) \qquad (V_D < V_{D\,sat}) \qquad (8.2.3)$$

which increases linearly with drain voltage but is independent of gate voltage.

When $V_D > V_{D\,sat}$, the transconductance (as obtained from the distributed analysis) is given by Eq. (8.2.3) with $V_D = V_{D\,sat}$ [from Eq. (8.1.14)]

$$g_{m\,sat} = \mu_n C'_{ox} \frac{W}{L} \left\{ V_G - V_{FB} - 2|\phi_p| - V_S \right.$$
$$\left. - \frac{\epsilon_s q N_a}{C'^2_{ox}} \left[\sqrt{1 + \frac{2C'^2_{ox}}{\epsilon_s q N_a}(V_G - V_{FB} - V_B)} - 1 \right] \right\} \qquad (8.2.4)$$

Thus, $g_{m\,sat}$ is independent of V_D, but nearly linearly dependent on V_G. If we use the more convenient representation for $I_{D\,sat}$ that is given in Eq. (8.1.17), we obtain the simpler expression

$$g_{m\,sat} = 2k' \frac{W}{L}(V_G - V_S - V_T) \qquad (8.2.5)$$

which clearly exhibits the proportionality between transconductance and the effective gate bias $(V_G - V_S - V_T)$. The transconductance is also proportional to the ratio of channel width to length and to the product of mobility and oxide capacitance per unit area because an increase in any of these terms increases the output current per unit change in gate-to-source voltage.

Speed of Response. There are two intrinsic limits on the speed of response of an IGFET. First, as in all current amplifiers, there is a basic limit set by the time for charge transport along the channel: that is, the transit-time limitation (described for the bipolar transistor in Section 6.3). Second, a limit is imposed by the charging of capacitances that are inherent in the device structure. In practical applications, a third limit on speed can be set by parasitic capacitances that are unavoidable in a given integrated circuit, but not inherent in the device itself. Limitations of this third type are properly considered by adding components external to the transient model for the intrinsic device, and we shall comment on them briefly later.

As the analysis carried out thus far has shown, there are really two conditions under which the speed limitations can be considered: either with the IGFET under ohmic bias or else when it is biased into saturation. We shall consider it sufficient to analyze the speed of response of the device in the saturated region of bias. This restriction leads to considerable simplification, and is really all that is warranted for most applications.

When the IGFET is operated in the current-saturated region, an approximate solution for the field along the channel $\mathscr{E}_y(y)$ can be derived fairly easily. This is accomplished by writing the differential equation for channel voltage [Eq. (8.1.9)] with the use of Eq. (7.3.19) to express $Q'_n(y)$. The equation then takes the form

$$dV_C(y) \simeq \frac{I_D \, dy}{\mu_n C'_{ox} W[V_G - V_T - V_C(y)]} \qquad (8.2.6)$$

which is only approximately true because V_T is really also a function of y. If, however, V_T is treated as a constant, a solution for $\mathscr{E}_y(y) = -(dV_C/dy)$, from Eq. (8.2.6), is

$$\mathscr{E}_y(y) = -\frac{(V_G - V_T)}{2L} \frac{1}{\sqrt{1 - (y/L)}} \tag{8.2.7}$$

The transit time along the channel is then found directly by using Eq. (8.2.7) in the expression

$$T_{tr} = \int_0^L \frac{1}{v_y} \, dy = -\int_0^L \frac{1}{\mu_n \mathscr{E}_y} \, dy \tag{8.2.8}$$

which leads to

$$T_{tr} = \frac{4}{3} \frac{L^2}{\mu_n(V_G - V_T)} \tag{8.2.9}$$

If we consider an n-channel IGFET having $L = 10 \, \mu m$, $\mu_n = 600 \, cm^2 \, V^{-1} \, s^{-1}$, and $(V_G - V_T) = 5$ V, Eq. (8.2.9) predicts a transit time of 4.4×10^{-10} s. This time is at least an order of magnitude shorter than the fastest switching times that are obtained in IGFETs of these dimensions (usually measured in ring-counter structures). Accordingly, we can conclude that the speed of response of the IGFET is governed not by the channel transit time but by the time needed to charge the capacitances associated with the device. Based upon the understanding of IGFET behavior that has been developed, the construction of a circuit model for the device is straightforward. The model consists of elements that represent the equations derived in Section 8.1 together with elements that represent the inherent capacitances and resistances within the structure of the transistor.

IGFET Model

The capacitances in an IGFET are shown schematically in Fig. 8.11. Of the six capacitors indicated in the figure, only one is essential to the operation of the transistor. This is the capacitance between the gate and the channel (sketched in bold lines) that controls the conductance between the drain and the source. The speed limitations associated with charging this capacitance are fundamentally related to the transit time of charge in the channel (Problem 8.8). The other capacitances result from the IGFET structure. Reducing their sizes will improve the switching speed of the IGFET without any corresponding degradation in the gain of the device. The two capacitors C_{RS} and C_{RD}, linking the gate and the source and the gate and the drain, are residual capacitances that result from misalignment and overlap of the gate with respect to the source and drain diffusions. The three capacitors connected between the bulk and the source, channel, and drain, respectively (C_{TS}, C_{TC}, and C_{TD}), are depletion-region capacitances that are present

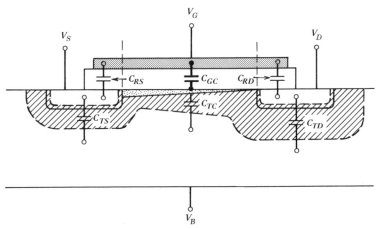

Figure 8.11 Schematic illustration of the capacitances in an IGFET. The gate-to-channel capacitance C_{GC} is essential for operation of the device and is shown in bold lines. The performance of the transistor would be improved by reducing the sizes of all other capacitors.

because of the reverse-biased *pn* junctions in these regions. For an accurate model, it is also generally necessary to add series resistors at the source and drain (typically a few tens of ohms each) to account for ohmic voltage drops in these regions. The resistance between the substrate contact and the edge of the space-charge zones may be accounted for by resistors which typically have values of the order of 100 Ω.

An overall equivalent circuit that incorporates these elements is shown in Fig. 8.12. In the figure, the gate to channel capacitance is divided between the source and drain terminals (C_{GS} and C_{GD}). When the IGFET is saturated, C_{GS} becomes $\frac{2}{3} C_{ox}$ and C_{GD} becomes nearly zero (Problem 8.8), representing the fact that few electrical flux lines link the gate to the drain. The current generator I_D can be specified by the equations developed in Section 8.1 for large-signal operation or through Eq. (8.2.5) for small-signal applications. The analysis of circuits with a model such as that shown in Fig. 8.12 is easily implemented on a computer, and predictions from it have been shown[3] to agree extremely well with measurements of large-signal transient performance. One refinement to the model presented in Fig. 8.12 was added for the analysis carried out in reference 3. This is the representation of the finite slope of the saturated current $I_{D\,sat}$ for $V_D > V_{D\,sat}$ resulting from modulation of the channel length. The slope was accounted for adequately by using a linear representation, that is, by writing

$$I_{D\,sat} = k' \frac{W}{L} (V_G - V_S - V_T)^2 (1 + l|V_D - V_S|) \tag{8.2.10}$$

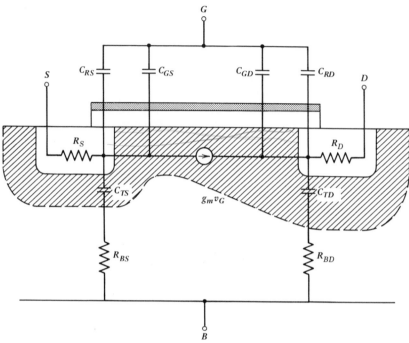

Figure 8.12 Equivalent circuit for the IGFET.

where l is an empirical constant taken to be 0.1 in reference 3. In Section 8.4, we consider channel-length modulation in more detail.

For most amplifying applications, many of the elements in Fig. 8.12 are superfluous. The source and substrate are usually common ac ground points so that C_{TS} is unimportant. The resistances R_S and R_D are likely to be much smaller than the load resistance and can be neglected. The current generator can then be represented by $g_{m\,sat}\,v_G$, which leads to a simple equivalent circuit for the device.

IGFET Technologies

Although the theory presented has emphasized n-channel IGFETs, the first large-scale development of integrated circuits employing IGFETs was for p-channel devices made with aluminum gates. Semiconductor manufacturers chose to make p-channel IGFETs first because the oxide-charge densities (which are generally positive) were sizable and quite variable in early IGFET technologies that had evolved directly from bipolar IC processes. Since positive charge tends to accumulate an n-type surface but to deplete a p-type surface, typical oxide-charge densities above a p-type substrate often result either in an inverted surface at zero bias (depletion-mode IGFET) or else in an enhancement-mode IGFET with an uncontrollably small threshold voltage.* The silicon wafers used were $\langle 111 \rangle$-oriented and typical p-channel, aluminum-gate, threshold voltages were approximately -4 V.

A major innovation in processing was the successful utilization of polycrystalline silicon for the gate material. When the gate is constructed of silicon, it can be deposited prior to the source and drain diffusions, and the gate itself can serve as a mask for these diffusions. With this technology the gate is nearly perfectly aligned over the channel. The only overlap at the source and drain is due to lateral diffusion of the dopant atoms; this *self alignment* reduces the parasitic overlap capacitances C_{RD} and C_{RS} in Fig. 8.11, and thus improves transistor performance. Additionally, fabrication with self-aligned silicon gates results in a Φ_{MS} term that aids inversion of an n-type silicon surface (see Problem 7.2). This feature enabled the production of p-channel silicon-gate IGFETs on $\langle 111 \rangle$ silicon with thresholds of roughly -2 V. A threshold near this value is necessary to make IGFET circuits that are compatible with bipolar transistor-transistor-logic (T^2L) circuits—a necessity for the design of many systems. From the point of view of integrated circuits, there is another advantage to silicon-gate technology. The refractory nature of the gate material permits the complete encapsulation of the IGFET in a glassy layer. Not only does this afford excellent protection and stability to the sensitive IGFET channel region, but it also allows the polycrystalline silicon to be used in areas other than IGFET gates. This is important because the polycrystalline silicon can then provide an additional layer of interconnections that can be crossed by the standard metal interconnection or by yet another layer of polycrystalline silicon. Figure 8.13 illustrates a cross section of the glass-over-coated, p-channel, silicon-gate IGFET. The stability of the structure and the connection of the metal to a polycrystalline-silicon interconnection line are apparent in the figure.

After it was discovered that surface-state charge is reduced by roughly a factor of 3 when $\langle 100 \rangle$-oriented material is used, silicon with this orientation became widely used for IGFETs in place of $\langle 111 \rangle$-oriented silicon, especially for n-channel devices.

A further major innovation in the processing of IGFETs was the introduction of ion implantation for threshold adjustment. When dopant atoms are implanted in the channel regions of an IGFET, it is possible to adjust the depletion-layer charge Q_d [see Eq. (7.3.7)] to a precisely determined value. By this means, the threshold voltage can be set very accurately through an ion-implantation step after the channel oxide has been formed. The use of ion implantation has made possible the reliable production of n-channel IGFETs. It has also permitted the

---

* A more quantitative discussion of threshold control in p- and n-channel IGFETs appears in Section 8.5.

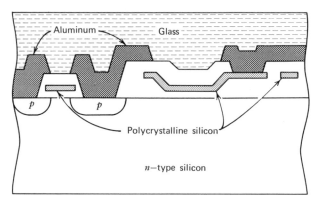

Figure 8.13 Cross section showing the use of oxide-passivated polycrystalline silicon both for silicon-gate IGFETs and for electrical crossovers of interconnections. (Courtesy Intel Corporation.)

use of lightly doped silicon without suffering inadvertent surface inversions in regions where they would be undesirable. The lower is the silicon doping, the smaller will be the useless capacitance between the active regions of the IGFET and the silicon bulk (C_{TS}, C_{TC}, and C_{TD} in Fig. 8.11). Likewise, surface mobility increases as doping concentrations are reduced. Presently, n-channel IGFETs are typically made on $\langle 100 \rangle$ silicon in which $N_a \approx 10^{15}$ cm^{-3}. Surface mobilities as high as 700 cm^2 V^{-1} s^{-1} are attained. The ion-implantation technology that makes this possible is described more fully in Section 8.5.

8.3 MOS INTEGRATED CIRCUITS

A first practical consideration about IGFETs for integrated-circuit applications might be: why use them in place of bipolar transistors? Indeed, all of the information needed to answer this question is still not available. Making lasting comparisons about rapidly evolving processes and designs has (not surprisingly) proven difficult at best. Two basic device advantages of IGFETs over bipolar transistors in integrated circuits have, however, been apparent from the earliest development of stable IGFETs. These are (1) smaller size per element, and (2) simpler processing. As the complexity of integrated circuits has grown, both of these advantages have increased in importance.

Figure 8.14 is from an early study comparing IGFETs to bipolar transistors.[4] It shows the cross sections of a conventional diffused npn bipolar transistor that is isolated by p-type diffusions extending to the substrate. Below the bipolar transistor, and to the same scale, is a p-channel IGFET. Marked on the sketch by Δ is the feature with the smallest dimension on the silicon surface, the emitter-stripe width for the BJT and the gate width for the IGFET. The device layouts

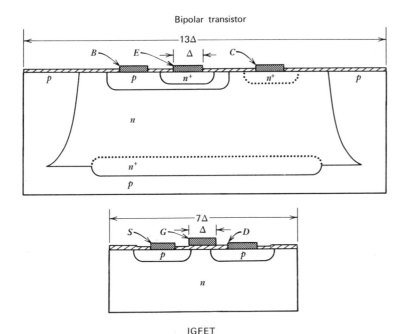

Figure 8.14 Comparison of cross sections of a junction-isolated bipolar transistor and a metal-gate IGFET.[4]

conform to the design rules for the planar process which originate from a consideration of masking tolerances and photolithographic constraints. The IGFET is seen to require only about half the surface dimension needed for the bipolar structure. This size reduction occurs mainly because the IGFET is self-isolating. A gate is needed to cause a surface channel between nearby diffused regions, whereas adjacent bipolar structures will interact unless minority carriers are prohibited from traveling between them.

Another characteristic of IGFETs bears on the relative size of bipolar and IGFET *IC*s. This is the use of IGFETs as load elements (or resistors), in addition to employing them as switches or as amplifying elements. To achieve typical values in the kilohm range, an IGFET load resistor can be built in appreciably less space than is required by a diffused resistor as is conventionally used in bipolar integrated circuits. Minimizing size is a first order priority in designing integrated circuits because it is the key to increasing circuit function, yield, and ultimately to maximizing profitability. Thus, the inherent size advantage of IGFETs is a very major attribute.

Because IGFETs can perform the dual functions of load resistors and switching elements, very large integrated circuits are built containing only IGFETs with varying W/L ratios. Design of such circuits is simpler than if several elements were needed, as is true for most bipolar circuits. An IGFET IC process is also simpler than is a bipolar process. Basically, only a single diffusion (for the source and drain) is required.* Epitaxy is not necessary; in fact, typical IGFET processing requires only about half as many steps as are needed for the production of bipolar ICs. Other advantages and a detailed discussion of processing for IGFET ICs are given in reference 5.

Arrayed against these advantages is an inherently greater gain and speed in bipolar transistors. When operated at the same current levels, bipolar transistors typically exhibit values of g_m that are a factor of 10 or more greater than are those obtained in IGFETs (Problem 8.9). Because of their higher transconductance, bipolar transistors can be made in which transit time across the base determines the speed of response of the device, whereas capacitance charging causes the speed limitation in IGFETs. The bipolar transistor is also more effective as a current switch; collector current varies exponentially with base-emitter voltage in the BJT while drain current in an IGFET increases at most as the square of the gate bias.

Blurring these comparisons, however, is the influence of continuing developments in technologies and in circuit design. In digital bipolar integrated circuits, for example, the new high-speed family of logic circuits known as *integrated injection logic* (I^2L) that was described in Section 6.7 has challenged the cell sizes of equivalent IGFET circuits by merging *pnp* and *npn* transistors and by doing away with the need for many isolating junctions. Similarly, major improvements in speed and packing densities have been achieved by using localized oxidation for BJTs (Fig. 4.19) instead of junction isolation. An exhaustive discussion of all of these topics is clearly outside the realm of our present interests. The major conclusion that can be drawn from these brief remarks is that advantages and disadvantages exist for both BJTs and IGFETs as IC components. There will continue to be applications for both types of devices in the forseeable future. For any particular system design, the choice between circuits using one or another of these devices will involve careful consideration of the tradeoffs between them.

MOS Memory[†]

The application of IGFETs to semiconductor memories has been extremely intensive and productive. Memory circuits consist mainly of a regular, repeated array of an information storage cell together with a considerably smaller number of circuits for coding and decoding of memory location addresses and for reading and writing information into the cells. The suitability of MOS devices to this task can be measured by the fact that storage of information on MOS memory chips

---

* However, most production processes also employ a diffusion or implantation step to stop unwanted channels under the thick *field oxides* between transistors.

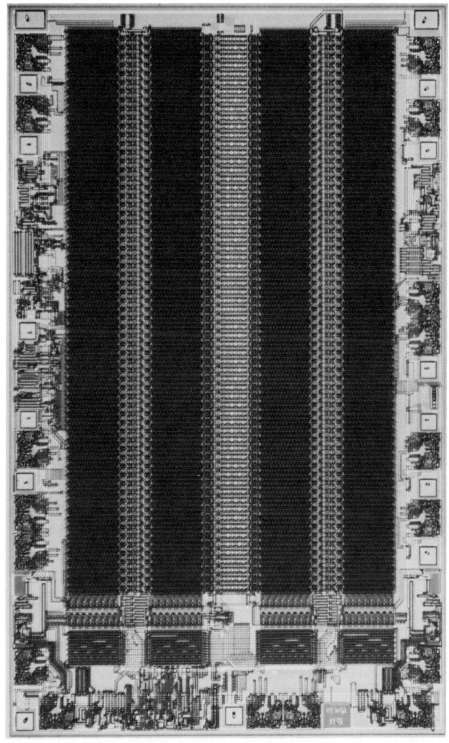

Figure 8.15 A large-scale MOS integrated circuit; this MOS random access memory stores 16,384 bits of information and uses about 35,000 MOS transistors for storage, addressing, reading and writing of information. (Photograph courtesy of Intel Corporation.)

has grown from 256 (2^8) bits to 16,384 (2^{14}) bits within just eight years. The chip area has increased by only a factor of 3 (from 2.7 by 2.7 mm to 3.68 by 5.95 mm) to accommodate this factor of 64 increase in memory capacity. Pictured in Fig. 8.15 is a 16,384-bit n-channel MOS memory. It makes use of two separate layers of buried polycrystalline silicon; one layer forms a silicon-gate IGFET and the other is used to form an MOS capacitor for information storage. The storage cells measure only 13 by 35 μm.

The subject of MOS memory ICs is very much wider in scope than is appropriately covered here. Further discussion is available in sources such as reference 6. However, before leaving the topic, we turn from a focus on LSI or *large-scale integration* to a description of a single specialized MOS memory element.

The invention of a device with important applications as a memory element was made possible by the mastery of a totally encapsulated silicon-gate technology, such as that illustrated in Fig. 8.13. The novel memory device[7] consists simply of an IGFET with a "floating" or electrically disconnected gate, as sketched in cross section in Fig. 8.16. The only way for charge to enter the isolated gate in the device is by the injection of energetic charge carriers through the oxide. The energetic electrons can be provided by causing avalanche selectively at the desired drain-substrate pn junction in an array of these devices. The memory element is called a *floating-gate, avalanche-injection MOS* transistor or FAMOS. Once the FAMOS gate has been charged, it will induce a conducting channel between the source and the drain, exactly as if a gate voltage had been applied. Thus, binary information is stored by the FAMOS according to the absence or presence of a conducting channel in a given device. The buried silicon layer in a FAMOS retains the electrons for any practical time period unless some source of external energy is able to excite the electrons that are held on the buried gate. If they are given sufficient energy to pass over the potential barrier at the interface between the buried silicon layer and the oxide, they can escape to the substrate and thus return the FAMOS to the off-state. Complete erasure of a FAMOS memory can therefore be accomplished by irradiating the surface with ultraviolet light.

Complementary MOS[†]

Logic circuits with nearly zero standby power consumption can be built using paired n- and p-channel IGFETs. Circuits of this type are called *complementary MOS* transistor circuits or simply CMOS circuits. A basic CMOS inverter cell is sketched in Fig. 8.17a with a plot of the output voltage as a function of input voltage shown in Fig. 8.17b. To understand the operation of this inverter, assume that the input voltage is lower than the n-channel threshold, but still sufficiently negative with respect to the bulk of the p-channel IGFET to turn it on. Then, the p-channel IGFET will be conducting and the n-channel device will be turned off. The output terminal is typically tied to gates associated with a succeeding stage and hence will draw no current. Therefore, the output voltage will rise to the supply

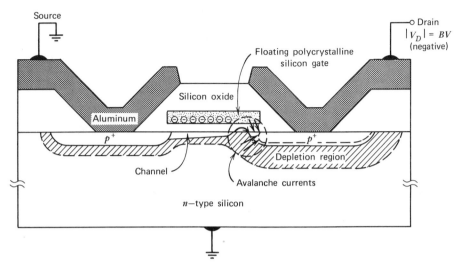

Figure 8.16 Cross section showing the mechanism of charge injection into the gate by avalanche in a FAMOS memory element.[7]

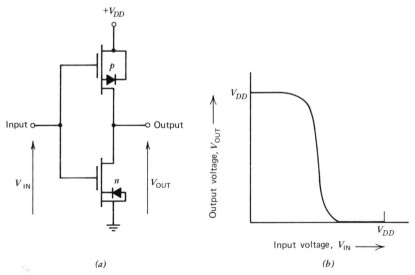

Figure 8.17 (*a*) A basic CMOS inverter circuit. (*b*) The transfer characteristic (voltage out versus voltage in) of a CMOS inverter circuit.

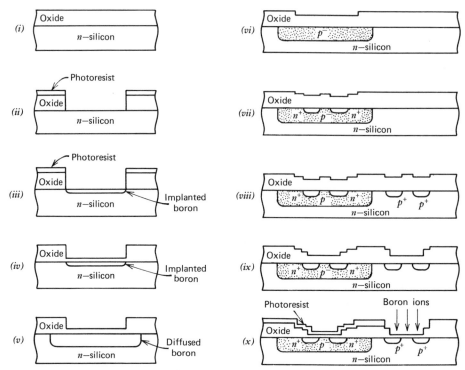

Figure 8.18   Processing sequence for an 0.8 V threshold CMOS process.[8]

(*i*) *n*-type starting silicon (3Ω-cm ⟨100⟩) is oxidized.

(*ii*) Oxide is removed by photoengraving where *p*-wells are required; the photoresist is left in place.

(*iii*) Boron is implanted in the *p*-well.

(*iv*) The resist is removed and an oxide 750 Å thick is grown to protect the silicon during the diffusion of the implanted boron.

(*v*) The implanted boron deposit is diffused to the appropriate depth (approximately 10μm) in a dry N$_2$ ambient.

(*vi*) The complementary substrate is oxidized to a suitable field oxide thickness.

(*vii*) The silicon in the *n*-channel source-drain and interconnect regions is exposed. Phosphorus is deposited, and the silicon reoxidized.

(*viii*) The silicon in the *p*-channel source-drain and interconnect regions is exposed and boron is deposited, diffused, and reoxidized. (The phosphorus in the *n*-channel regions is simultaneously diffused.)

(*ix*) Oxide is removed from the desired *n*-and *p*-channel regions and the gate oxide is grown and annealed.

(*x*) Photoresist is applied and masked to expose the *p*-channel device regions, and boron ions are implanted in the *p*-channels to reduce the *p*-channel threshold voltage.

(*xi*) The resist is removed, contact holes cut, aluminum is evaporated and etched to the interconnection pattern, contacts are sintered and the overglass passivation is applied. (Step (*xi*) is not shown in the sketch.)

voltage value V_{DD}. As the input voltage increases to the threshold value of the n-channel device, that IGFET will turn on and force the output voltage toward ground. The p-channel device will then be turned off. Thus, quiescently, there is no path except for junction leakages to carry current from the supply. Almost all of the power dissipation in CMOS circuits takes place during the switching transients.

For optimal CMOS circuit performance, the threshold voltages of the two types of IGFETs should be complements of one another (i.e., $V_{Tp} = -V_{Tn}$). This has proven possible in a manufactured IC with ion implantation. As a practical example, the processing sequence for a low threshold CMOS watch circuit that dissipates only a few microwatts of power is shown in Fig. 8.18.[8] Note that the process demands the creation of a p-well for the n-channel IGFET. For this circuit, the IGFET threshold voltages have been held to within ± 0.3 V of 0.8 and -0.8 V for n- and p-channel IGFETs, respectively. Integrated circuits making use of CMOS are inherently less dense and more complicated to make than are MOS integrated circuits that make use of only one transistor type. The extra cost to obtain them, however, is frequently justified.

Many other topics might be included in a discussion of MOS integrated circuits. An example is MNOS or metal-nitride-oxide-silicon memory elements in which the threshold voltage of the transistor can be intentionally controlled by storing or expelling charge from the nitride-oxide interface. One might also discuss SOS or silicon-on-sapphire technology, in which switching speeds are greatly improved and the body effect is eliminated by doing away with the depletion-layer capacitances between the junctions and the substrate of the IGFETs. Further additions to the list could expand this discussion of MOS integrated circuits to book length. The intent of this section is served if an appreciation has been gained of the wide impact of MOS devices on integrated circuits.

8.4 SECOND-ORDER CONSIDERATIONS[†]

It is of value to consider some aspects of the electronics of MOS devices in slightly greater depth, and to introduce several other topics that might reasonably be omitted from a first-order analysis. Accordingly, in this section consideration is given to the following topics: oxide breakdown under high-field stress, surface mobility, channel-length modulation, and subthreshold current. Some modifications of the basic IGFET theory that are imposed by the construction of very short channels are considered in a concluding subsection.

Oxide Breakdown. Breakdown of an insulating oxide at high fields causes its physical destruction and, therefore, must be avoided.* For carefully grown thermal silicon oxide, the breakdown field is approximately 7×10^6 V cm^{-1}. Hence, a

---

* By contrast, pn-junction breakdown is not destructive unless high power concentrations occur.

typical gate oxide thickness of 1000 Å breaks down when about 70 V are applied across it. Because the oxide thickness can vary somewhat and also because there are typically "weak spots" in a practical oxide, the gate voltage is usually restricted to less than half of this value. When the gate lead of an IGFET is exposed (as in discrete devices or for inputs to integrated circuits), it is possible to cause oxide breakdown by the buildup of static electricity on the gate. Because of the small size and capacitance of the gate, this can occur with distressing ease (Problem 8.10). For this reason, IGFETs that do have exposed gates are often designed with voltage-regulating diodes connected in parallel with the gate. The diode will conduct by avalanche or by tunneling when the reverse voltage applied to it exceeds its breakdown value. The breakdown voltage of the regulator diode is designed to be low enough to protect the IGFET gate oxide. These gate-protection diodes are not used unless they are necessary because they introduce parasitic capacitance and leakage that degrades IGFET performance.

Surface Mobility. The performance of an IGFET is clearly improved by increasing the value of surface mobility. A mobility increase results in higher gain and higher drain current for given device dimensions. This is one reason that n- and not p-channel MOS is the major focus for the design of integrated circuits. As was mentioned earlier, surface mobility in IGFETs is limited to only about half of the bulk value. Furthermore, mobility decreases as V_G increases above the threshold voltage. Despite a great deal of theoretical work, a complete theory that adequately predicts surface mobility as a function of MOS parameters is not yet available. Generally, however, it is understood that the reduced mobility results from extra loss mechanisms associated with collisions at the oxide-silicon interface. The scattering and concomitant mobility lowering is intensified at higher gate biases because the carriers are attracted more strongly to the surface. A reasonable empirical fit to the observed behavior of electron mobility is given by

$$\mu_{\text{eff}} = \mu_{so}\left(1 + \frac{\mathscr{E}_s}{\mathscr{E}_{so}}\right)^{-r} \tag{8.4.1}$$

where μ_{so} is the low-field surface mobility, $\mathscr{E}_{so}$ is typically about 6×10^4 V cm^{-1}, and r is between 0.1 and 0.2.[9] The surface field in the semiconductor $\mathscr{E}_s$ can be found from Eq. (7.3.13) with $\phi_s = -\phi_p + V_C - V_B$.

Although the effects of scattering at the oxide-silicon interface decrease as the gate voltage decreases toward the threshold value, it is found experimentally that mobility does not continue to improve as bias is lowered. Instead, there is a maximum of mobility in the neighborhood of the threshold voltage and then a sharp decrease as the surface enters the region of very weak inversion (V_G slightly below threshold). In this subthreshold region, it appears that the free-carrier densities are so low that Q_{ss} and other randomly distributed oxide charges cause the carriers to become unevenly distributed in the channel, leading to a mobility reduction.[10]

Conduction effects in the weak inversion region do not have much relevance to IGFET operation although transport mechanisms are important in this region, particularly for CCD applications. For most IGFET applications, it is sufficient to consider that mobility continually decreases from its value near threshold.

Channel-Length Modulation. In the discussion of IGFET current-voltage characteristics in Section 8.1, it was pointed out that the drain current is only approximately independent of drain voltage in the saturation region ($V_D > V_{D\,sat}$). In fact, the drain current varies weakly with V_D in this range because the end of the ohmic channel moves away from the drain junction when the drain bias exceeds $V_{D\,sat}$. This, in turn, causes the distance L' from the source to the end of the conducting channel to decrease. As can be seen from Eq. (8.1.12), there will then be a corresponding increase in I_D. An accurate quantitative theory for channel-length modulation is quite complicated because of the two-dimensional nature of the problem. The channel length is influenced by the gate electrode as well as by the reverse-biased drain junction.[11]

A simple physical model has been presented[12] to treat this problem approximately. The model has been shown to predict the channel conductance in the saturated region for both *n*- and *p*-channel IGFETs having a wide range of substrate doping and oxide thicknesses. It is found that for V_D close to $V_{D\,sat}$, the channel length L' can be approximated by considering that a one-dimensional, space-charge solution corresponding to that in an abrupt diode applies. From the theory developed in Chapter 3, we can therefore write

$$L - L' = \sqrt{\frac{2\epsilon_s}{qN_a}(V_D - V_{D\,sat})} \qquad (8.4.2)$$

Equation (8.4.2) is accurate only if the oxide thickness is much less than the width of the space-charge region $x_{ox} \ll (L - L')$. If this inequality does not apply, an accurate prediction of channel-length modulation requires consideration of the influence of the gate electrode.[12] In many cases, this sort of accuracy is not needed and even simpler equations than Eq. (8.4.2) may prove adequate for a particular analysis. An example was noted in the discussion of IGFET modeling, where a linear representation of channel-length modulation [Eq. (8.2.10)] was successfully employed.

Subthreshold Current. The conventional IGFET theory presented in Section 8.1 approximates the channel conductance by assuming that the channel does not contain any free carriers unless the gate voltage exceeds the threshold value. The theory therefore fails to predict so-called *subthreshold currents* arising from the inversion charge that exists at gate voltages below the conventional threshold. An approximate theory to indicate the behavior of currents in this region can be developed by considering the case of a low drain-source bias and taking ϕ_s to be close to $-\phi_p$. Under this condition, Poisson's equation can be written with terms expressing the inversion charge as well as the fixed-acceptor charge. Its form is

$$\frac{d^2\phi}{dx^2} = -\frac{\rho}{\epsilon_s} \approx \frac{q}{\epsilon_s}(N_a + n) = \frac{qN_a}{\epsilon_s}\left[1 + \exp\left(\frac{q(\phi - |\phi_p|)}{kT}\right)\right] \quad (8.4.3)$$

Multiplying Eq. (8.4.3) by $d\phi$ and noting that $\mathscr{E}(d\mathscr{E}/d\phi) = d^2\phi/dx^2$, we obtain an expression for the field in the silicon near the oxide-silicon surface of the form

$$\mathscr{E}_s = \left\{\frac{2qN_a}{\epsilon_s}\left[\phi_s + |\phi_p| + \frac{kT}{q}\left(\exp\frac{q(\phi_s - |\phi_p|)}{kT} - \exp\frac{-2q|\phi_p|}{kT}\right)\right]\right\}^{1/2} \quad (8.4.4)$$

As in Section 7.3, Eq. (8.4.4) can be used to relate the induced semiconductor charge Q_s' to the surface field through Gauss' law. The free-charge density Q_n' is then obtained by subtracting the fixed depletion-layer charge $Q_d' = -\sqrt{2q\epsilon_s N_a(\phi_s + |\phi_p|)}$ from Q_s'. When this procedure is carried out, and approximations suitable to the range of surface potential under consideration are made, an expression for Q_n' is seen to have the form

$$Q_n' \approx -\frac{kT}{2q}\sqrt{\frac{2q\epsilon_s N_a}{\phi_s + |\phi_p|}}\exp\left[\frac{q(\phi_s - |\phi_p|)}{kT}\right] \quad (8.4.5)$$

which indicates that just below threshold Q_n' varies nearly exponentially with the surface potential. The procedures carried out in Section 7.3 can again be employed to express Q_n' as a function of gate and threshold voltage. The result is

$$Q_n' \approx -\frac{kT}{2q}\sqrt{\frac{q\epsilon_s N_a}{|\phi_p|}}\exp\left[\frac{q(V_G - V_T)}{\eta kT}\right] \quad (8.4.6)$$

where $\eta = [1 + (1/2C_{ox}')(q\epsilon_s N_a/|\phi_p|)^{1/2}]$

Thus, both the free-electron density and the current increase exponentially as V_G approaches V_T. At the onset of strong inversion ($V_G = V_T$), Q_n' becomes linearly dependent on V_G. Measurements that show this behavior for a typical IGFET are plotted in Fig. 8.19. Since the current decreases exponentially below V_T, subthreshold currents are made negligible in practical circuits by biasing an "off" transistor about a half volt below V_T.

Short-Channel Effect on Threshold Voltage. A basic assumption that helped to simplify the derivation of the IGFET equations was that the charge in the surface space-charge layer was a function only of the voltage between the gate and the substrate. This assumption is in error near the source and drain depletion regions where the *pn* junctions, as well as the gate, affect the surface space charge. When the channel length is appreciably greater than the thickness of the junction depletion layers, this assumption should cause negligible error. As the channel length is

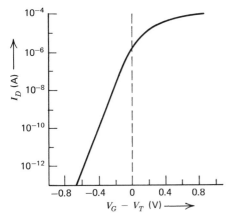

Figure 8.19 Measured values of drain current near and below conventional threshold for an *n*-channel IGFET. The behavior is exponential for $V_G < V_T$ as predicted by Eq. (8.4.6).

reduced, however, one can expect that overlooking the effect of the source and drain depletion layers on the total space charge will lead to considerable inaccuracy. In particular, error should arise in predictions of threshold voltage because a significantly smaller amount of the charge in the space-charge region will be linked to the gate at threshold than is predicted by simple theory. The remainder is coupled to the *pn*-junction depletion regions.

A straightforward geometrical technique for analyzing this problem has led to a theory that agrees well with experiment.[13] This technique considers the charge induced by the voltage on the gate to be approximately contained in a volume whose cross section is the trapezoid of width $x_{d\,max}$ and length varying from L at the surface to L_1 at the substrate side of the depletion region (Fig. 8.20). The cross-sectional area is shown dotted on the figure. If this charge is called Q_{d1}, then

$$Q_{d1} = qx_{d\,max}WN_a \frac{L + L_1}{2} \qquad (8.4.7)$$

where W is the width of the channel and N_a is the dopant density. The charge Q_{d1} in Eq. (8.4.7) is taken to be the depletion-layer charge that must be induced by the gate at threshold. If the channel length is large so that the space-charge zones at the source and drain are very much smaller than L, then L_1 in Fig. 8.20 will approach L. In that case by Eq. (8.4.7) Q_{d1} would equal $Q_d = qx_{d\,max}N_aWL$ as was assumed in the first-order theory of Section 8.1. For shorter channels, L_1 becomes appreciably less than L, and Q_{d1} is therefore less than Q_d, as expected from our qualitative arguments.

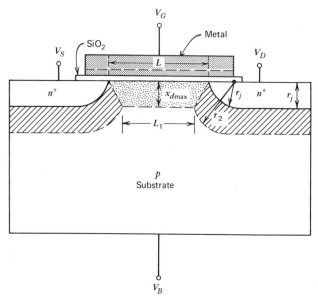

Figure 8.20   Cross section of a short-channel IGFET illustrating a simple geometrical model to account for threshold reduction.[13] The depletion-region charge induced by the gate is estimated to be that contained in the dotted region.

For a useful theory, L_1 must be related to the geometry of the IGFET. From Fig. 8.20, $r_2 = r_j + x_{d\,max}$ where r_j is the junction radius and r_2 is the boundary of the space-charge zone. A geometrical exercise leads to

$$f \equiv \frac{Q_{d1}}{Q_d} = 1 - \frac{r_j}{L}\left(\sqrt{1 + \frac{2x_{d\,max}}{r_j}} - 1\right) \tag{8.4.8}$$

The parameter f is therefore a unique function of the IGFET geometry. The expression for the threshold voltage is written directly from Eq. (7.3.18).

$$V_T = V_{FB} + 2|\phi_p| + V_S - \frac{fQ'_d}{C'_{ox}}$$

$$= V_{FB} + 2|\phi_p| + V_S + \frac{f}{C'_{ox}}\sqrt{2\epsilon_s q N_a(2|\phi_p| + V_S - V_B)} \tag{8.4.9}$$

Equation (8.4.9) has accurately predicted V_T in a series of experimental devices having various channel lengths.[13] As an indication of the size of this effect, Fig. 8.21 shows the variation in threshold voltage predicted by Eq. (8.4.9) for p-channel IGFETs having various values of substrate doping, $V_B - V_S = 5$ V, $x_{ox} = 500$ Å,

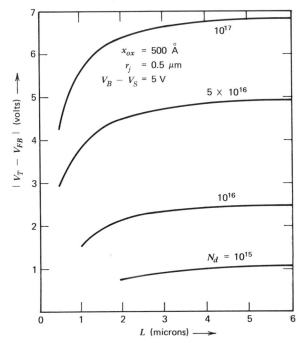

Figure 8.21 Theoretical variation in threshold voltage as a function of channel length for *p*-channel IGFETs having various substrate dopant concentrations.[13] The predictions were made using Eq. (8.4.9). (Reprinted by permission.)

and $r_j = 0.5$ μm. The figure makes it evident that threshold-voltage reduction resulting from the short-channel effect analyzed here becomes significant when the channel length is less than about 3 μm.

Other effects that arise as channel lengths are shortened, such as punch-through between the source and drain junctions and high-field phenomena, are beyond the scope of this discussion.

8.5 DEVICE: THE ION-IMPLANTED IGFET

The control of the threshold voltage of IGFETs is a major concern in the production of MOS integrated circuits. Not only are the threshold voltages of the IGFETs themselves of importance, but so also are the threshold voltages of the metal or polycrystalline-silicon patterns that may induce undesired inversion layers when used as conducting interconnections. An example is the region of the silicon surface under the polycrystalline-silicon interconnections seen in the cutaway drawing of Fig. 8.13. Because typical MOS *IC*s are so densely packed with metal lines and junctions, it is very likely that the *IC* will malfunction if the voltages applied to interconnections could cause inversion of the silicon surface regions that are below them. Hence, the magnitude of the threshold voltages below the interconnection lines must generally be higher than the highest voltages that will be applied to the lines. To obtain high thresholds, the oxides under the lines (usually called *field oxides*) are made appreciably thicker than the gate oxides in IGFETs. Ratios of 10 or more to one between the oxide thicknesses and, therefore, between the capacitance values associated with the gate and field oxides, are common.

Threshold Control. For the successful design of any MOS *IC*, the first level of threshold control is to be able to produce either enhancement or depletion-mode IGFETs reliably, according to circuit requirements. For most applications, enhancement-mode devices are desired. Hence, for *p*-channel IGFETs V_T must be less than zero, while for *n*-channel devices V_T must be greater than zero. The requirement for *p*-channel devices is easily met, but that for *n*-channel IGFETs can cause difficulty. Consideration of the equations for the threshold voltage makes clear why this is true. Under the condition that V_S is zero, the threshold equations with the substrate taken as a reference can be written (Table 7.1)

$$V_T = V_{FB} + 2|\phi_p| + \frac{|Q_d'|}{C_{ox}'}$$
(8.5.1)

for the *n*-channel IGFET and

$$V_T = V_{FB} - 2|\phi_n| - \frac{|Q_d'|}{C_{ox}'}$$
(8.5.2)

for the *p*-channel IGFET. In both equations Q_d' represents the depletion-charge density that exists when the surface becomes inverted.

The flat-band voltage will typically be negative both for *p*- and *n*-channel IGFETs (Problem 8.1). Since the remaining terms in Eq. (8.5.2) are also negative, V_T will assuredly be negative and therefore the production of enhancement-mode, *p*-channel IGFETs poses no difficulties. For the *n*-channel device, however, the two added terms in Eq. (8.5.1) must sum to a larger value than V_{FB} to produce enhancement-mode IGFETs. This requirement sets a lower limit on the allowable dopant concentration for *n*-channel devices. For typical 1000 Å oxides and densities of Q_{ss}'/q near 10^{11} cm^{-2}, the level of doping for marginal operation is in the low 10^{15} cm^{-3} range. To design with a safety factor (e.g., to hold threshold voltages to nearly a volt) the dopant density should be of the order of 10^{16} cm^{-3}. Such a high dopant density is undesirable because it leads to high substrate capacitance and relatively low junction breakdowns. Compounding this problem is the fact that boron, the commonly used p-type dopant, tends to segregate into the oxide as the oxide grows, thus reducing the doping N_a at the surface below that in the substrate. An alternative means of changing the threshold in the desired direction is to decrease C'_{ox} by increasing the gate oxide thickness. This is undesirable, however, because it reduces the gain factor [k' in Eq. (8.1.17)] of the IGFET.

Another way to alter the threshold voltage is to apply a substrate bias that will shift V_T by the effect that was illustrated in Fig. 8.10. This technique (sometimes called *body biasing*) is employed in a number of commercial n-channel MOS circuits. Since it requires the provision of a separate voltage supply for the *IC*, body bias is not as attractive as "building in" the desired threshold voltage. A process that permits adjusting the threshold voltage by changing the doping selectively in the channel region of the IGFET is clearly very desirable. This is possible with the use of ion implantation.

Ion Implantation.[†] A brief description of ion implantation was given in Chapter 1. The technique employs a very pure beam of ionized dopant atoms that are accelerated in a strong electric field to energies of the order of 100 keV and allowed to strike the *IC* surface. The ions penetrate into the silicon surface (or the oxide or photoresist film) to a depth that is usually a small fraction of a micron. The ions that land in the oxide are generally electrically inert. The ions implanted in the silicon become active donors or acceptors when the wafer is annealed at a temperature that is low relative to that used for diffusion steps. Hence, the junction profiles obtained in diffusion steps are not significantly modified by the ion-implantation process.

When ion implantation is used to adjust the threshold voltage, the process step typically consists of exposing the gate region to an ion beam and firing the ions right through the gate oxide as shown in step (x) of the CMOS processing sequence outlined in Fig. 8.18. The implant can, of course, be used either to increase or to decrease the net dopant concentration in the channel region. The CMOS processing sequence of Fig. 8.18 is an example of the latter. In that case, p-type dopant atoms (boron) are implanted in the next-to-the-last step of the process in order to reduce the magnitude of the threshold voltage of the p-channel IGFET to slightly less than one volt.

If the implantation dose is sufficiently high, the threshold voltage can be shifted to the opposite polarity (a pn junction would then be created at the surface) so that the IGFET becomes a depletion-mode device. This is, in fact, an important use of ion implantation because depletion-mode IGFETs are very desirable when used as load devices in MOS logic circuits.[6]

As a specific example of the use of ion implantation, we consider in more detail the particular application that introduced this topic: namely, the fabrication of n-channel IGFETs.[14] Ion implantation for this application usually consists of

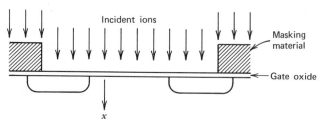

Figure 8.22 Threshold control by ion implantation. Dopant ions are implanted through the gate oxide into the channel region of the IGFET.

increasing the acceptor density in the channel region so that substrate material with background acceptor concentrations of the order of 10^{15} cm^{-3} or lower can be used without needing body bias. As in the CMOS example cited previously, photoresist is usually used to define the implanted regions. A slight overlap of the implant over the source and drain regions is commonly employed to insure total coverage of the channel. The implant step is indicated in the sketch of Fig. 8.22.

Figure 8.23 shows the distribution of dopant atoms in a typical implant as a function of depth into the wafer. After the implant (solid line) the atoms are distributed with a Gaussian distribution. Their concentration $N_i(x)$ is

$$N_i(x) = \frac{N'}{\sqrt{2\pi}\,\Delta R_p} \exp\left[-\frac{(x - R_p)^2}{2(\Delta R_p)^2}\right] \tag{8.5.3}$$

where N' is the number of implanted atoms per unit area (the *dose*), R_p is the average distance an implanted atom penetrates into the semiconductor (the *range*), and ΔR_p is the RMS or standard deviation that describes the "width" of the distribution. Both the range and the width of the implant depend on the energy of the implanting beam as well as the implanted species. Extensive tabulations of R_p and ΔR_p for all ions of interest at implantation energies that are commonly employed have been published.[15]

After the activating anneal, the implanted distribution becomes slightly broader, as shown by the dashed curve in Fig. 8.23. The calculation of the effect of this implant on the threshold voltage of the IGFET is much simplified by approximating the actual distribution by the "box" distribution shown by the dotted lines in Fig. 8.23. In this approximation, the implanted dopant is assumed to have a constant density N_{ai} from the surface to a depth x_i. The total dose N' per unit area is related to N_{ai} and x_i by the equation: $N' = N_{ai}x_i$. The dopant concentration in the region nearest the surface is $(N_{ai} + N_a)$ where N_a is the background acceptor concentration in the substrate. For $x > x_i$, the effective acceptor concentration is just N_a.

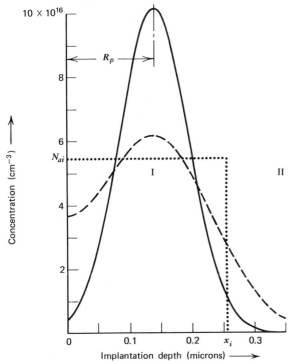

Figure 8.23 Distribution of the implanted dopant (*a*) immediately after the implant (solid line), (*b*) after activating anneal and diffusions (dashed curve), (*c*) approximate model used for threshold calculations (dotted curve).

With respect to shifts in the threshold voltage, there are two cases to consider. In the first case, the effective depth of the implant is greater than the width of the depletion region at threshold. Thus, the threshold voltage is simply given by Eq. (7.3.18) with N_a replaced by $N_{ai} + N_a$. It is advisable to avoid this type of an impurity profile, however, because the increased doping near the substrate edge of the junction leads to higher substrate capacitance and lower breakdown voltage in the channel region as well as to greater sensitivity of the threshold to source-substrate bias (*i.e.*, a more troublesome *body effect*).

The second case, in which the implanted ions are all contained within the surface space-charge region, is therefore preferable. Calculations for the threshold voltage in the second case are more complicated, but they follow the same basic pattern that was used previously. The first step is to find a solution for Poisson's equation when the space charge changes abruptly as in the approximate "box" distribution.

When the solution is found, and the depletion approximation is used, an expression for the depletion-layer width x_d can be written in terms of the surface potential ϕ_s.

$$x_d = \left[\frac{2\epsilon_s}{qN_a} (\phi_s + |\phi_p|) - x_i^2 \frac{N_{ai}}{N_a} \right]^{1/2} \tag{8.5.4}$$

Real solutions for Eq. (8.5.4) are only possible if the second term in the brackets is smaller than the first because we have assumed that the highly doped, implanted region is contained within the depletion zone. Imaginary solutions occur for cases that violate this assumption. To find the threshold, ϕ_s in Eq. (8.5.4) must be set to the inversion value appropriate to the heavier doped surface region: $\phi_{ps} = (kT/q) \ln [(N_{ai} + N_a)/n_i]$. This leads to an expression for $x_{d\,max}$ that can be used to find the depletion-charge density at inversion Q'_d.

$$\begin{aligned} Q'_d &= -qN_{ai}x_i - qN_a x_{d\,max} \\ &= -qN_{ai}x_i - [2qN_a\epsilon_s(|\phi_{ps}| + |\phi_p| + V_S - V_B) - q^2 x_i^2 N_a N_{ai}]^{1/2} \end{aligned} \tag{8.5.5}$$

Equation (8.5.5) can easily be expressed in terms of the implantation dose by making use of the expression $N_{ai} = N'/x_i$.

As in Section 7.3, the gate voltage is then related to Q'_d, and the threshold voltage can be written as

$$V_T = V_{FB} + V_S + |\phi_p| + |\phi_{ps}| + \frac{qN'}{C'_{ox}}$$

$$+ \frac{1}{C'_{ox}} [2qN_a\epsilon_s(|\phi_{ps}| + |\phi_p| + V_S - V_B) - q^2 x_i N_a N']^{1/2} \tag{8.5.6}$$

As expressed in Eq. (8.5.6), the threshold voltage has an arbitrary reference point. The zero for potential can therefore be taken at any electrode.

By comparing Eq. (8.5.6) to Eq. (7.3.18), it can be seen that the implantation affects the threshold V_T in three ways. (1) Instead of the voltage drop across the surface depletion region at inversion being $2|\phi_p| + (V_S - V_B)$, it is $|\phi_{ps}| + |\phi_p| + (V_S - V_B)$. (2) There is a linear dependence on dose through the term qN'/C'_{ox}. (3) The square-root expression for the depletion-charge term is altered principally by the added term $-q^2 x_i N_a N'$. Of these three changes, the first depends on the dose only logarithmically and is therefore not of much consequence. The second change, which depends linearly on the dose N', is the dominant effect. It enters the expression for V_T as a direct addition to the term representing fixed surface charge Q'_{ss}, as is evident from the expression for the flat-band voltage (without distributed oxide charge) $V_{FB} = \Phi_{MS} - Q'_{ss}/C'_{ox}$. The third change in the threshold-voltage expression is the only one that depends on x_i, the implantation depth. The dependence is not strong, as can be seen from inspection of Fig. 8.24 which represents the threshold voltage as a function of implantation depth. The values of V_T plotted in Fig. 8.24 have been calculated from Eq. (8.5.6) for a dose N' of 3.5×10^{11} cm^{-2} in a substrate having a background doping of 2×10^{15} cm^{-3}. The IGFET analyzed has a gate oxide that is 867 Å thick, and a fixed charge density $Q'_{ss}/q = 10^{11}$ cm^{-2}. Because the dependence on x_i is not strong, a reasonable estimate of V_T can be obtained by disregarding the effect of the implant on the square-root term, and this is often done in practice. The threshold voltage without implantation would nominally be -0.19 V and too close to zero for the production of reliable n-channel IGFETs. Hence, these calculated results emphasize the importance of the implant in fabricating a useful n-channel device.

The effectiveness both of ion implantation and of body bias in adjusting the threshold voltage of n-channel IGFETs is illustrated in Fig. 8.25. The data points on the figure represent the threshold voltages measured as the body bias ($V_S - V_B$) was varied for a series of n-channel IGFETs that were implanted with varying doses of boron. The curves on the figure were calculated using Eq. (8.5.6) with V_S taken as the reference. The x-axis in Fig. 8.25 corresponds to the magnitude of the square root in Eq. (8.5.6) and the curves are therefore linear. The substrate material was $\langle 100 \rangle$ silicon that was doped with 1.2×10^{16} boron atoms cm^{-3}. The IGFETs had an oxide thickness of 1000 Å, an oxide charge density $Q'_{ss}/q = 8 \times 10^{10}$ cm^{-2}, and $\Phi_{MS} = -0.92$ V.

In order to match the experimental data to Eq. (8.5.6), increasing values of body bias ($V_S - V_B$) are seen from the figure to be necessary as the dose is increased. The

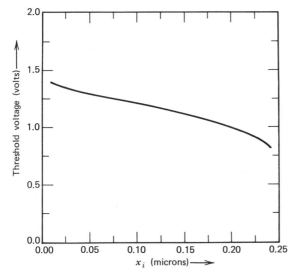

Figure 8.24 Dependence of the threshold voltage on the implantation depth x_i as predicted by Eq. (8.5.6) for an IGFET with an 867 Å-thick gate oxide, $Q'_{ss}/q = 10^{11}$ cm^{-2}, $N' = 3.5 \times 10^{11}$ cm^{-2}, and $N_a = 2 \times 10^{15}$ cm^{-3}. Both V_S and V_B are taken to be zero.

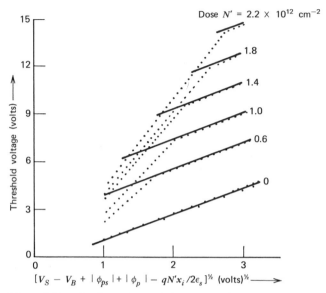

Figure 8.25 The dependence of the threshold voltage (referenced to the source electrode) for ion-implanted, *n*-channel IGFETS. The dots represent experimental values measured as $(V_S - V_B)$ was varied. The solid lines are plotted from Eq. (8.5.6). The properties of the IGFETs are given in the text. (Courtesy Hewlett-Packard Corporation.)

higher body bias is needed to cause the depletion layer to penetrate completely through the implanted silicon: a condition required to validate the analysis presented. In the region that the analysis is valid, the correspondence between theory and experiment is good, and Eq. (8.5.6) is therefore a useful expression for the threshold voltage in an ion-implanted IGFET.

SUMMARY

The insulated-gate field-effect transistor is an active device in which the conductance of the silicon surface can be controlled by the voltage applied to the metal electrode of an MOS system. In the IGFET, the conductance in the *channel* between *source* and *drain* electrodes is modulated by a voltage applied to the *gate*. An IGFET in which the channel can carry current between the source and drain when the gate-source bias equals zero is a *depletion-mode* transistor. If a finite value of gate voltage is necessary to induce a conducting channel, the IGFET is called an *enhancement-mode* transistor.

The dependence of drain current on drain voltage in an IGFET can be divided into two regions: a region at low drain biases where ohmic flow over the entire channel characterizes the device, and a current-saturation region at higher drain biases. In the saturation region, the conducting channel is detached at the drain end and the current is transported there across a high-field region. A simple derivation of the equations for IGFET currents can be carried out by using charge-control analysis and by approximating the threshold voltage in the channel region as being constant. The equations derived by this method are simple to use and widely applied, although they are only accurate at low drain bias voltages.

For an accurate representation of the current-voltage dependences in the IGFET, it is necessary to carry out a distributed analysis in which the variation of depletion-layer charge along the channel is taken into account. The results of the distributed analysis and those obtained from the approximate derivation can be brought into correspondence in the current-saturation region by reducing the value of the effective mobility used in the charge-control model.

Several of the parameters that describe the IGFET are conveniently obtained by measuring drain current when the gate and drain are both tied to a voltage that is varied. This technique is useful in measuring the *body effect*, which is the variation in threshold voltage that occurs when the bias between the source and the substrate is changed. The body effect increases as the substrate doping increases but it varies inversely with the oxide capacitance. The transconductance of an IGFET increases linearly with drain voltage until saturation occurs, but is independent of gate voltage below saturation. Once saturation is reached, g_m becomes a linear function of gate voltage, but independent of drain voltage. The speed of response in IGFETs, as they are presently made, is determined not by the channel transit time but by the charging and discharging times of capacitances which are present in the device. A circuit model for the IGFET consists of elements representing the dc circuit equations for the device together with associated capacitances and resistances.

Although the planar process, as evolved for the production of bipolar transistor circuits, is the basis for producing MOS circuits, a number of important modifications have been introduced to improve the performance of IGFETs. These include the use of $\langle 100 \rangle$-oriented silicon wafers to reduce oxide-charge densities, the fabrication of polycrystalline silicon on oxide surfaces for IGFET gates and interconnection patterns, and the adjustment of threshold voltages by ion implantation. It is a complex problem to make comparisons between bipolar and MOS transistors for integrated circuits, but smaller size and simpler processing are two basic advantages of MOS devices. MOS memory circuits and low-power CMOS circuits are two important areas of IGFET applications to integrated circuits.

In an IGFET, the drain current does not actually become constant at high drain voltages because the channel length, and hence the drain current, is modulated by the changing voltage between the drain and the end of the channel. This effect becomes increasingly important as the built-in channel length is reduced. Although it is complicated to analyze exactly, channel-length modulation can often be accounted for in circuit models by approximating techniques. The *subthreshold currents* that flow below conventional threshold voltages can be analyzed by modifying the expression for Poisson's equation to include an inversion-charge term. The resulting theory predicts an exponential decrease in current below the threshold voltage in agreement with observed behavior. Control of the threshold voltage is one of the major requirements of any MOS process, and the difficulties are greater with *n*-channel than with *p*-channel IGFETs. Ion implantation is an extremely useful technique for adjusting the threshold voltage. With ion implantation, it is possible to produce depletion and enhancement-mode IGFETs in the same integrated circuit as well as to produce reliable *n*-channel MOS circuits in large-scale integrations (*LSI*).

REFERENCES

1. For patent enumeration, see J. T. Wallmark and H. Johnson *Field-Effect Transistors*, Prentice-Hall., Englewood Cliffs, New Jersey, 1966.
2. A. S. Grove, *Physics and Technology of Semiconductor Devices*, Wiley, New York, 1967, p. 324.
3. H. Schichman and D. A. Hodges, *IEEE J. Solid-State Circuits*, SC-3, 285 (1968).
4. R. M. Warner, *IEEE Spectrum*, *4*, 50 (June 1967). Reprinted by permission of the publisher.
5. W. M. Penney and L. Lau (Editors), *MOS Integrated Circuits*, Van Nostrand-Reinhold, New York, 1972, Chapter 1.
6. Ibid, Chapter 7.
7. D. Frohman-Bentchkowsky, *IEEE J. Solid-State Circuits*, SC-6, 301 (1971). Reprinted by permission of the publisher.
8. P. J. Coppen, K. G. Aubuchon, L. O. Bauer, and N. E. Moyer, *Solid-State Electron.* *15*, 165 (1972).
9. J. E. Meyer, *RCA Review*, *32*, 42 (March 1971).
10. J. T. C. Chen and R. S. Muller, *J. Appl. Phys.*, *45*, 828 (1974).
11. J. E. Schroeder and R. S. Muller, *IEEE Trans. Electron Devices*, ED-15, 954 (1968).
12. D. Frohman-Bentchkowsky and A. S. Grove, *IEEE Trans. Electron Devices*, ED-16, 108 (1969).
13. L. D. Yau, *Solid-State Electron.* *17*, 1059 (1974).
14. R. H. Dennard et al., *IEEE J. Solid-State Circuits*, SC-9, 256 (1974).
15. J. F. Gibbons, W. S. Johnson, and S. W. Mylroie, *Projected Range Statistics*, Dowden, Hutchinson and Ross, Inc., Stroudsberg, PA, distributed by Halsted Press, Wiley, 1975.

PROBLEMS

8.1 Construct a table showing the threshold voltage V_T (with $V_S = V_B = 0$) as a function of dopant concentration for both *n*- and *p*-channel IGFETs. Take values of N_a and N_d of 10^{15}, 10^{16} and 10^{17} cm^{-3} and assume that there is a surface density of fixed positive charge $Q'_{ss}/q = 10^{11}$ cm^{-2} at the oxide-silicon interface in all cases. The oxides are 1000 Å thick ($C'_{ox} = 3.45 \times 10^{-8}$ F cm^{-2}) in all cases and all gates are made of aluminum so that $\Phi_M - X = (4.1 - 4.15) = -0.05$ V. Do any of the dopant concentrations result in depletion-mode IGFETs? (The equation summary in Table 8.1 should be helpful for this problem.)

8.2 Derive Eqs. (8.1.9) and (8.1.12).

8.3 An IGFET with $W/L = 5$, a SiO$_2$ thickness of 800 Å, and $\mu_n = 600$ cm^2 V^{-1} s^{-1} is to be used as a controlled resistor.

(a) What electron charge density Q'_n/q is required for the device to present a resistance of 2.5 kΩ between its terminals at low values of V_{DS}?

(b) What gate voltage in excess of the threshold voltage is needed to produce the desired resistance?

8.4   Use Eq. (8.2.6) to derive expressions for $V_C(y)$ and $\mathscr{E}_y(y)$. Note that I_D is a constant and take $V_S = V_B = 0$. From the expressions derived, obtain equations for $V_C(y)$ and $\mathscr{E}_y(y)$ that are valid in saturation.

8.5   Using the results of Problem 8.4, sketch $\mathscr{E}_y$ versus y over the range $y = 0$ to L for $V_G > V_T$ and $V_D = 0.25, 0.50, 0.75$, and 1 times $V_{D\,\text{sat}}$. Note that the field approaches a constant value when V_D is very small.

8.6[†] Considering the plots sketched in Problem 8.5, comment on the effect that a saturating velocity for electrons (as noted in Chapter 1, Fig. 1.13) has on the validity of the theory of Section 8.1.

8.7[†] (a) Show that the use of Eq. (8.2.7) in Eq. (8.2.8) leads to the expression for the transit time given in Eq. (8.2.9).

(b) Obtain the transit time for the IGFET in saturation by applying the expression $T_{tr} = -Q_n/I_D$ where Q_n is the total channel charge. Show that the result corresponds to Eq. (8.2.9). [*Hint*: Q_n should be calculated from the expression $Q_n = W \int_0^L Q'_n(y)\,dy$]

8.8   (a) From the expression obtained for Q_n in Problem 8.7, show that the small-signal capacitance between the gate and the source under saturation conditions is given by

$$C_{GS} \equiv \left|\frac{\partial Q_n}{\partial V_{GS}}\right| = \frac{2}{3} C_{ox} = \frac{2}{3}\frac{\epsilon_{ox}}{x_{ox}} WL$$

(b) Use Eq. (8.1.8) together with this result to prove that $T_{tr} = 2C_{GS}/g_{m\,\text{sat}}$. [Note that except for the factor 2, a similar result was found between τ_F and the diffusion capacitance in the bipolar transistor [Eq. (6.5.4)]. The functional dependence between input capacitance and τ_F is a basic charge-control result that applies generally to triode devices.]

8.9   Consider that a bipolar transistor and an IGFET are being evaluated for use in a linear amplifier. The quiescent current in the amplifier is to be 1 mA. What would the ratio between the transconductances of the two devices be if the IGFET were to be biased with $(V_G - V_T) = 1$ V? [Eq. (8.1.8) may be assumed to be valid for the IGFET.] (The superior g_m is obtained in the bipolar transistor at any output current.)

8.10 (a) Show that the number of electrons on an MOS gate that can cause an oxide to break down is a function only of the gate area and the permittivity of the oxide.

(b) If an IGFET having a channel length of 6 μm and a W/L ratio of 5 is charged by static electricity, how many electrons could be transferred to the gate before it would sustain the oxide breakdown field of 7×10^6 V cm^{-1}?

(c) How long would it take to deliver these electrons to the gate if an average current of 1 pA flowed because of the transfer of static electricity?

8.11[†] Complete the analysis of subthreshold currents that is begun in the text by deriving Eqs. (8.4.5) and (8.4.6).

8.12[†] Carry out the steps necessary to derive Eq. (8.4.8).

8.13   Consider symmetrical n-channel IGFETs biased in the three configurations shown in Fig. P8.13. Use the charge-control theory to estimate I_D as a function of the variable voltage V_S in each case and sketch a graph of the three relations on the same axes. Assume that the threshold voltage of the enhancement-mode IGFETs in sketches (a) and (b) is 2 V above the source voltage and that the threshold voltage of the depletion-mode IGFET in sketch (c) is 2 V below the source voltage. Take $V_{GG} = 14$ V, $V_{DD} = 10$ V, and $\mu_n C'_{ox}(W/L) = 3 \times 10^{-4} \ \Omega^{-1} \text{V}^{-1}$ for all IGFETs.

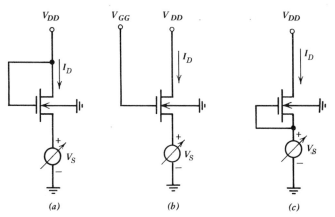

(a)   (b)   (c)

Figure P8.13

8.14   In a typical p-channel, silicon-gate, MOS integrated circuit there are sometimes three different MOS structures present as shown in Fig. P8.14. Assume a substrate resistivity of 5 Ω-cm and $\langle 111 \rangle$-oriented silicon, $2|\phi_n| = 0.58$ V, $Q'_{ss}/q = 2 \times 10^{11} \text{cm}^{-2}$ and oxide thicknesses of 0.12 μm, 1.0 μm, and 1.5 μm. Find the threshold voltages of the useful transistor (0.12 μm oxide) and that of the two parasitic structures. Assume that the energy-band structure of polycrystalline silicon is the same as that of similarly doped single-crystal silicon and that the p-type, polycrystalline-silicon gates are heavily doped so that $E_f \simeq E_v$.

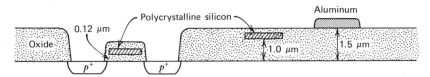

5 Ω–cm n–type silicon

Figure P8.14

8.15   A p-channel IGFET with a heavily doped p-type polycrystalline-silicon gate has a threshold voltage of -1.5 V with the source and substrate grounded. When a 5 V reverse bias is applied to the substrate, the threshold voltage changes to -2.3 V.

(a)   What is the dopant concentration in the substrate if the oxide thickness is 1000 Å?

(b) What is the threshold voltage when 25 V reverse bias is applied to the substrate? (*Hint: Note that ϕ_n changes slowly with N_d.*)

8.16   An *n*-channel, silicon-gate IGFET has an oxide thickness of 1200 Å and Q'_{ss}/q of 5×10^{10} cm^{-2}. The substrate is *p*-type silicon of 1 Ω-cm resistivity, and the gate is heavily doped with phosphorus. The source and substrate are grounded.

(a) Calculate the flat-band voltage V_{FB} and the threshold voltage V_T.

(b) For $W/L = 1$ and $V_G = 5$ V, I_D was measured to be 18.7 μA at $V_D = 0.4$ V. If a circuit requires a current of 1.6 mA at $V_G = 5$ V and $V_D = 0.4$ V, what is the minimum width W of the device if we assume that the polycrystalline-silicon gate length is 10 μm and the n^+ source and drain each diffuse sideways 0.75 μm under the gate.

8.17[†] A silicon IGFET has a gate that is itself a nichrome resistor having contacts above both ends of the channel (Fig. P8.17). The voltages at the two ends of the gate are V_{G0} and V_{G1}, respectively. Assume that the flat-band voltage is constant and take the oxide to be x_{ox} units thick. Using the depletion approximation, set up the differential equation having a solution that relates the drain current to the drain voltage below saturation. *Do not attempt to solve the equation.*

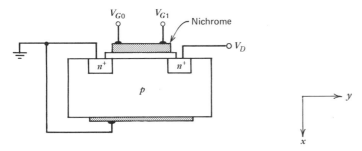

Figure P8.17

8.18[†] We can derive a voltage for an IGFET that is analogous to the "Early voltage" V_A used with bipolar transistors (Section 6.1). To obtain an effective Early voltage for an IGFET, consider the variation in the effective channel length with V_D when V_D is greater than $V_{D\,sat}$. Assume that we can write $\partial I_D/\partial V_D = I_D/V_A$ and that Eqs. (8.1.8) and (8.4.2) apply. Show that under these conditions, V_A varies in proportion to $\sqrt{V_D - V_{D\,sat}}$. Why is this result in error when $V_D \sim V_{D\,sat}$?

8.19[†] Derive Eq. (8.5.4) for the depletion-layer width for the "approximate" dopant profile from the ion implantation process discussed in Section 8.5.

8.20   Carry through the steps indicated in the text to obtain Eqs. (8.5.5) and (8.5.6) for the depletion-charge density at inversion and the threshold voltage, respectively, in an ion-implanted IGFET.

Table 8.1 Formulas for IGFETs

	n-channel	p-channel
Depletion charge density at threshold	$Q'_d = -qN_a x_{d\,max} = -\sqrt{2\epsilon_s q N_a(2|\phi_p| + |V_S - V_B|)}$	$Q'_d = +qN_d x_{d\,max} = +\sqrt{2\epsilon_s q N_d(2|\phi_n| + |V_S - V_B|)}$
Flatband voltage	$V_{FB} = \Phi_{MS} - \dfrac{Q'_{ss}}{C'_{ox}} - \dfrac{1}{C'_{ox}}\displaystyle\int_0^{x_{ox}} \dfrac{x\rho(x)\,dx}{x_{ox}}$	
Threshold voltage	$V_T = V_{FB} + |V_S| + 2|\phi_p| + \dfrac{|Q'_d|}{C'_{ox}}$	$V_T = V_{FB} - |V_S| - 2|\phi_n| - \dfrac{|Q'_d|}{C'_{ox}}$
Charge-control (simplified) theory ($V_S = V_B = 0$) **Current equations below saturation**	$I_D = \mu_n \dfrac{W}{L} C'_{ox}\left[V_G - V_T - \left(\dfrac{V_D}{2}\right)\right]V_D$	$I_D = -\mu_p \dfrac{W}{L} C'_{ox}\left[V_G - V_T - \left(\dfrac{V_D}{2}\right)\right]V_D$
Saturation voltage	$V_{D\,sat} = (V_G - V_T)$	
Current equations above saturation	$I_D = k'_n(V_G - V_T)^2$	$I_D = -k'_p(V_G - V_T)^2$

$[k'$ is proportional to $\mu C'_{ox}(W/L)]$

Current equations: distributed theory below saturation

$$I_D = \mu_n \frac{W}{L}\left\{ C'_{ox}\left[V_G - V_{FB} - 2|\phi_p| - \left(\frac{V_D}{2}\right) - \left(\frac{V_S}{2}\right)\right](V_D - V_S)\right.$$

$$-\frac{2}{3}\sqrt{2\epsilon_s q N_a}\left[(2|\phi_p| + |V_D - V_B|)^{3/2} - (2|\phi_p|\right.$$

$$\left.\left. + |V_S - V_B|)^{3/2}\right]\right\}$$

$$I_D = -\mu_p \frac{W}{L}\left\{ C'_{ox}\left[V_G - V_{FB} + 2|\phi_n| - \left(\frac{V_D}{2}\right) - \left(\frac{V_S}{2}\right)\right](V_D - V_S)\right.$$

$$-\frac{2}{3}\sqrt{2\epsilon_s q N_d}\left[(2|\phi_n| + |V_D - V_B|)^{3/2} - (2|\phi_n|\right.$$

$$\left.\left. + (|V_S - V_B|)^{3/2}\right]\right\}$$

Saturation Voltage

$$V_{D\,sat} = V_G - V_{FB} - 2|\phi_p|$$
$$-\frac{\epsilon_s q N_a}{C'^2_{ox}}\left[\sqrt{1 + \frac{2C'^2_{ox}}{\epsilon_s q N_a}(V_G - V_{FB} - V_B)} - 1\right]$$

$$V_{D\,sat} = V_G - V_{FB} + 2|\phi_n|$$
$$+\frac{\epsilon_s q N_d}{C'^2_{ox}}\left[\sqrt{1 + \frac{2C'^2_{ox}(V_G - V_{FB} - V_B)}{\epsilon_s q N_d}} - 1\right]$$

Transconductance ($|V_D| \ll |V_{D\,sat}|$)

$$g_m = \mu_n C'_{ox} \frac{W}{L}(V_D - V_S)$$

$$g_m = -\mu_p C'_{ox} \frac{W}{L}(V_D - V_S)$$

Other MOS equations are given in Table 7.1 on page 324.

Index